INDUSTRIAL ENVIRONMENTAL CONTROL

PULP AND PAPER INDUSTRY

INDUSTRIAL ENVIRONMENTAL CONTROL

PULP AND PAPER INDUSTRY

ALLAN M. SPRINGER
Department of Paper Science and Engineering
Miami University
Oxford, Ohio

A Wiley-Interscience Publication
JOHN WILEY & SONS
New York · Chichester · Brisbane · Toronto · Singapore

Library of Congress Cataloging in Publication Data:

Springer, Allan M.
Industrial environmental control.

Includes bibliographies and index.
1. Paper industry—Environmental aspects—Handbooks, manuals, etc. 2. Wood-pulp industry—Environmental aspects—Handbooks, manuals, etc. 3. Water—Pollution—Handbooks, manuals, etc. 4. Air—Pollution—Handbooks, manuals, etc. I. Title.
TD195.P37S67 1985 676.042 85-6506
ISBN 0-471-80756-7

Printed in the United States of America

10 9 8 7 6 5 4 3 2 1

**To my wife Sandy,
without whose considerable help this
could not have been accomplished**

Preface

This is the first comprehensive reference from a major publisher on pollution control in the pulp and paper industry that contains material on both air and water pollution abatement and covers both in-plant methods and external treatment. It grew out of teaching courses in environmental control in the Paper Science and Engineering program at Miami University and industrial courses presented through the Technical Association of the Pulp and Paper Industry. No good text was available for the purpose, so extensive notes and literature had to be amassed to cover the subject. The most continuous series of literature on environmental control for the paper industry exists in the Technical Bulletins of the National Council of the Paper Industry for Air and Stream Improvement, an organization by which I was employed for 5 years. These bulletins contain useful information but represent a series of specialized projects rather than a unified presentation of basic information. They do not contain much commonplace, needed information. Much other research has also been reported, but the literature is so scattered that current workers often do not bother to make the effort required to find it; or, when found, the material is not always presented in a fashion useful to the practicing scientist or engineer. This book has been created to fill these needs.

This book is designed to be a useful reference tool for the practicing engineer and contains, where available, quantitative information as well as qualitative descriptions of the treatment processes. Environmental engineers outside the paper industry will also find this book useful, for it contains descriptions of general treatment process technology while also highlighting the experience of a particular industry with that technology. The information it presents should save most environmental process engineers 6 months to a year in gaining background information for new projects or areas of work. It is my hope that this book will promote further implementation of cost-effective environmental control by industry so that the world will become a better place for future generations to live in.

ALLAN M. SPRINGER

Oxford, Ohio
March 1985

Contents

1
Ecology Perspective

When walking by a clear, pollution-free mountain stream, many people are struck by the natural beauty of the situation and believe that the stream should be preserved in its natural state for future generations to enjoy. Yet what few people realize is that if this stream were visited 500 to 1,000 years in the future, it would be unrecognizable as the same area. The stream may have cut a deep gully by then, and the surrounding vegetation would have evolved into a vastly different mix of plant life. Natural systems are not static but are slowly evolving systems that engage in a set of interactions that are often lumped together under the heading "balance of nature." The study of the earth's environmental system is usually termed the "study of ecology." The word "ecology" comes from the Greek word *oikos*, which means house; thus, ecology is the study of the patterns of relations between organisms and their "house", or environment.

People often think of ecology as something relevant only to the ardent conservationist or the avid environmentalist, or as something of concern only to the proponents of the back-to-nature movement. This is an incorrect impression, because the study of ecology should provide a rational basis for the control of all pollution. It is impossible to prevent change, but at least we can be aware of what changes are likely to occur in the environment, and we can try to make these changes beneficial rather than harmful. Those who have children are especially sensitive to the concern of making the world a better place to live in. We have a moral obligation at least to leave the environment in a livable state for the future. Even the industrial environmentalist should have some understanding of the rudiments of ecology.

In order to read the ecology literature, one needs to be familiar with the meanings of a few commonly used terms. "Population," in the ecological sense, means a group of one kind of individual—for instance, a group of people, or of owls, or of bees. Each of these groups is a different population. All of the different populations living in a given area make up a community. Thus a "community" is a group of populations living in a certain area. The community and the nonliving environment or area together make up an "ecosystem." Thus an ecosystem contains everything, living and nonliving, in a finite area. Some examples of ecosystems are a pond, stream, meadow, or forest. All of the ecosystems together make up the "ecosphere" or "biosphere." Therefore the biosphere encompasses all of the earth's living organisms and the physical environment in which they interact. The advent of space travel, especially the first human landing on the moon, made it clearer to all of us that we are in a closed finite system traveling through space.

FINITE PLANET CONCEPT

Early in human evolution, the prespective was very different. In terms of the geological age of the earth, humanity is a relatively recent phenomenon, and its effect on the biosphere was relatively small until the invention of agriculture some 8,000 years ago. Early in their evolution, humans were victims of predators, and wild beasts had a chance of causing their extinction. Now the situation is reversed; humans have eliminated many plants and animals, and may do the same to many more.

The traditional view of the ecosphere was that nature was awesome and immense and humans puny and insignificant (1). This view is manifest in landscape paintings of the romantic style. In this period in which humans were few and their technology was simple, they had little impact on nature, and "the solution to pollution was dilution." In this period, wise men preached adaptation to and harmonization with nature; this is still a good philosophy, but for an entirely different set of reasons. Aldo Leopold (2), who spent his life learning to understand the ecosphere, viewed the world in terms of Paul Bunyan's Round River. In his travels, this folk hero of logging lore discovered a river that had no source and no mouth, but just flowed continuously into itself—a round river. The oceans of the world are like the Round River, and the land masses contained in the oceans are like islands in the river. All human activities can be thought of in terms of this concept. Geology is the study of the origins and changes with time of the Round River, biology of its plants and fauna, engineering of how best to use its resources. Ecology is the study of the entire system.

INTEGRATED MANAGEMENT CONCEPT

The round river concept makes us realize that we are in a closed system that can be damaged beyond repair if we are not careful. A total management concept is needed in which all factors and all fields of endeavor are considered in making decisions. This is probably an idealistic and impractical endeavor, but at the very least, those involved in environmental protection should take an interdisciplinary approach to the solution of problems and be cognizant of the interactive, cross-disciplinary nature of environmental problems.

Human activities can certainly have adverse environmental consequences, but so can natural disasters. It is rare that humans cause, by any single activity, even with malevolent intent, as much disruption of nature as is caused by the eruption of Mount St. Helens or by a Guatemalan earthquake. The ash fallout from the eruption of Mount St. Helens put much more particulate matter into the streams in its area than any human activities. These natural disasters are not usually termed pollution because there is no known means of controlling them, nor are they caused by humans. People have the responsibility to control their actions so as to mintain themselves with minimum harm to the environment and in such a manner that nature can adjust to their interactions. Meeting this requirement can pose many ecological dilemmas (3).

One way of meeting this goals that is often discussed is a requirement of zero discharge. The goal of zero discharge is part of Public Law (PL) 92–500 and is normally interpreted to mean zero discharge of pollutants rather than of water or air. This is a somewhat more reasonable approach but requires that "pollutants" be defined. This issue will be covered in the next chapter.

Zero discharge of pollutants may be theoretically possible, but the costs need to be weighed against the benefits gained. Resources and energy are required to remove all of the pollutants from an air or water stream, and obtaining these resources and energy involves environmental consequences. It may be possible to recycle the resources, but additional energy input would be required. Thermodynamics tells us that the creation of order in any system requires an input of energy (4). Even the earth's ecosystem cannot function without the energy from the sun that is necessary to drive the system. Thus the benefits of zero disharge must ultimately be weighed against the energy required to achieve it and the consequences of obtaining and using that energy. "We must face the cold hard fact that our activity has a net damaging effect on the environement regardless of our good intentions. However, we are obliged to determine how our actions can be the least damaging" (4).

EXTENT OF THE WATER POLLUTION PROBLEM

The human impact on water pollution has been far-reaching. In 1976 the U.S. Environmental Protection Agency (EPA) estimated that one third of all stream miles in the United States were polluted and that less than 10% of the watersheds could be characterized as unpolluted or only moderately polluted. The main pollution problem appears to be organic waste, the largest source of which is industrial waste.

The pulp and paper industry ranks third in terms of freshwater withdrawal after primary metals and the chemical industry. It is believed that by the year 2000 the paper industry will become the largest manufacturing user of water (5).

The pulp and paper industry ranks fifth among the major industries in its contribution to the water pollution problem. The industry has had significant impact on the nation's streams, but the situation is improving rather than growing worse. Since 1976 many mills have built secondary biological waste treatment plants, so that by 1985 it is estimated that 99% of the pulp and paper industry in this country will have secondary treatment or its equivalent. The fact that the majority of the industry has wastewater treatment has had a visibly beneficial effect on water quality.

A case in point is the lower Fox River in Wisconsin, which flows from Lake

Winnebago into Green Bay, an arm of Lake Michigan. "Winnebago" is an Indian word meaning "stinking water," so it can be inferred that this lake was never pure. The conditions in this shallow, nutrient-rich lake encouraged algal blooms. Nevertheless, the Fox River and the lake formerly had abundant fish life. By 1968, a 25-mile stretch of the lower Fox River had 18 pulp and paper mills located along its banks and was devoid of all fish life. When the alewife problem occurred in the Great Lakes, an eminent aquatic biologist was asked if it would spread to Lake Winnebago. He remarked that if the alewife could survive the lower Fox River, nothing would be able to stop it. The alewife was a small sardine-sized fish that infested the Great Lakes. Needless to say, the alewife did not make it to the lake. However, in 1983 fish life returned to the river, and good fishing is now found along its banks.

This is just one example of a general trend that is occurring in water quality. The shade are running again in the Hudson River, and commercial fishing is now practical in Lake Erie, which was once thought to be almost dead. Pollution control efforts have allowed nature to recover from the adverse effected of the previous era of poor management.

The water resources available to humans are fixed and of sufficient quantity to meet their needs far into the future. The problem will be to maintain the quality of this resource well enough to meet the myriad human needs for water—agriculture, process water, drinking, and recreation.

REFERENCES

1 J. Wei, "Technologists and the environment," *Chemtech*, 8:340–343 (June 1978).

2 A. Leopold, "The land ethic," *A Sand County Almanac*, Oxford University Press, Oxford, England, 1966.

3 A. Wolman, "Ecologic dilemmas," *Science*, 193:740–742 (1976).

4 P. MacBerthouex and D. F. Rudd, "Fundamental concepts," *Strategy of Pollution Control*, Wiley, New York, 1977.

5 S. G. Yulke, J. T. O'Rouke, and T. Asano, "Water reuse in the pulp and paper industry in California," *Proceedings of the Water Reuse Symposium II*, AIChE, New York, 1981.

2

Overview of Water Pollutants and Their Impact: The Pulp and Paper Industry

MEANING OF POLLUTION

The aim of all pollution control efforts is to leave the natural environment unaltered. "Water pollution" might be defined as the introduction into a body of water of substances whose character and quantity so alter its natural quality as to impair its usefulness or to render it offensive to the sense of sight, taste, or smell. The term "usefulness' implies that the water has an intrinsic value to humans. Some individuals might feel that any alteration of natural water quality is inherently bad, but this viewpoint appears unsupportable when one realizes that even in a system not influenced by humans, water quality changes with time. An example is the natural eutrophication of a lake or stream. Humans are just one additional factor in the changing dynamic system often termed the "balance of nature." Humans cannot prevent change, but they can try to ensure that their actions promote beneficial change compatible with some common definition of utility.

DEFINITIONS OF POLLUTANTS

Defining a "pollutant" as a substance that can alter the natural environment makes all known materials potential pollutants, which they are, but a narrower classification is helpful in practice. The EPA broadly classifies pollutants into eight types: oxygen-demanding substances, disease-causing agents, synthetic organic compounds, plant nutrients, inorganic chemicals and mineral substances, sediments, radioactive substances, and thermal discharges. Each of these classes is discussed separately below.

Oxygen-Demanding Substances

The largest group of materials that demand oxygen are organic in character, and tests that have been developed usually reflect the organic content of the

Table 2-1 Theoretical Oxygen Demand (ThOD)

$$C_5H_{12}O_6 + 6O_2 \rightarrow 6CO_2 + 6H_2O$$

$$200\,\text{mg/l} \times \frac{1\,\text{mole}}{180{,}000\,\text{mg}} \times \frac{6\,\text{moles O}_2}{\text{mole glucose}} \times \frac{32{,}000\,\text{mg}}{\text{mole O}_2} = \frac{213\,\text{mg O}_2}{1}$$

waste. Most existing laboratory tests do not measure the quantity of organic matter directly, but rather relate it to the amount of oxygen required to oxidize it. If the exact composition of the organic matter is known, so that a chemical formula for it can be written, the theoretical oxygen demand can be determined by writing a reaction in which the organic matter is oxidized to carbon dioxide and water. For example, let's consider the chemical reaction for glucose shown in Table 2-1. In it 6 moles of oxygen are consumed to oxidize 1 mole of glucose. For a 200 mg/1 solution of glucose, the theoretical oxygen demand could be calculated on the basis of this reaction, as shown in the table. The theoretical oxygen demand, or ThOD, is what most laboratory tests attempt to measure. The laboratory methods commonly used are biochemical oxygen demand (BOD), chemical oxygen demand (COD), total organic carbon (TOC), and total oxygen demand (TOD).

BOD. The most widely used measure of organic matter is BOD. The reason for its widespread acceptance and use is that it reflects somewhat the fate of organic matter in a biological treatment system. BOD is a measure of the oxygen required to oxidize the organic matter in a sample through the action of microorganisms contained in the sample. The classic dilution method has been to determine the dissolved oxygen both before and after a period of incubation at 20°C. The usual time period in the past has been 5 days. Other time intervals and temperatures can be used provided the temperature is held constant. Thus, BOD_5 is used to designate an incubation period of 5 days, with the temperature understood to be 20°C unless otherwise stated.

This test was originally developed to simulate the fate of a pollutant introduced into English rivers. The 5-day period was chosen because no stream in England was over 5 days' flow from the sea.

In the dilution method, if the oxygen demand of the sample is greater than the available dissolved oxygen, a dilution is made. The amount of dilution depends on the original amount of dissolved oxygen. For wastes with an unknown demand, it is necessary to make up a number of dilutions in order to be sure that one will meet the requirement.

There are a number of factors that influence the rate of oxidation of organic matter by bacteria and, therefore, the 5-day oxygen demand. The type of dilution water, the pH value, and the bacteria content are the most important. The dilution water is "seeded" with a bacterial culture that has acclimated, if necessary, to the organic matter present in the water. The seed culture that is used to prepare the dilution water for the BOD test is a mixed culture. Such cultures contain large numbers of saprophytic (decay) bacteria and other organisms that oxidize the organic matter. In addition, they contain certain autotrophic bacteria that oxidize noncarbonaceous matter, for example, ammonia to nitrate. Autotrophic bacteria use CO_2 as a carbon source.

After incubation, the dissolved oxygen of the sample is measured and the BOD is calculated by the formulas given in Table 2-2. Biochemical oxidation is a slow process and theoretically takes an infinite amount of time to be completed. Within a 20-day period, oxidation is about 95–99% complete, and in the 5-day period used for the BOD test, oxidation is usually 60–70% complete.

Table 2-2 Equations for Calculating BOD from Test Results

For percent mixtures:

$$\text{BOD (mg/l)} = \left[(DO_\text{b} - DO_\text{i}) \frac{100}{\%} \right] - (DO_\text{b} - DO_\text{s})$$

For direct pipetting:

$$\text{BOD (mg/l)} = (DO_\text{b} - DO_\text{i}) \frac{\text{vol. of bottle}}{\text{ml of sample}} - (DO_\text{b} - DO_\text{s})$$

where DO_b and DO_i are the dissolved oxygen values found in the blank and dilution of the sample, respectively, at the end of the incubation period and DO_s is the dissolved oxygen originally present in the undiluted sample

The kinetics of the BOD reaction are usually formulated in accordance with first-order reaction kinetics, although some believed that a second-order approach might be more applicable (1–5). Table 2-3 summarizes the first-order kinetics. L represents the BOD present at time $t = 0$, that is, the total or ultimate BOD initially present. The amount of BOD that has been expended at any time t is y, which is equal to $L - L_t$. If $t = 5$, y is the amount expended in 5 days. The values of K' vary significantly; the range may be from 0.02 to 0.15 per day or more. The reaction rate constant is temperature dependent, and an approximate relationship derived from the van't Hoff–Arrhenius relationship—the last equation in Table 2-3—may be used. The value of Θ in this equation is 1.056 in the temperature range 20–30°C and 1.135 in the range 4–20°C (6).

The total time–dependent BOD is usually broken down into two stages. The first-stage BOD is due to carbonaceous organic matter and the second stage to noncarbonaceous matter such as ammonia. Figure 2-1 illustrates this phenomenon, and the second hump in the figure is the sum of carbonaceous and nitrogenous demand. Ammonia is produced during the hydrolysis of proteins. Some autotrophic bacteria are capable of using oxygen to oxidize the ammonia to nitrites and nitrates. It normally takes 6–10 days for these bacteria to reach significant numbers and to exert a measurable oxygen demand. The lag time for nitrification leads to its being termed a second stage. The nitrogeneous

Table 2-3 Kinetics of the BOD Reaction

$$\frac{dL_t}{dt} = -K'L_t$$

which integrates to

$$\ln L_t|_0^t = -K't$$

rearranging by taking the exponential of each side

$$\frac{L_t}{L} = e^{-K't}$$

$$Y_t = L - L_t = L(1 - e^{-K't})$$

at $t = 5$ days

$$Y_5 = L(1 - e^{-K'5})$$

To adjust K' for different temperatures, use

$$K'_T = K'_{20}\Theta^{(T-20°\text{C})}$$

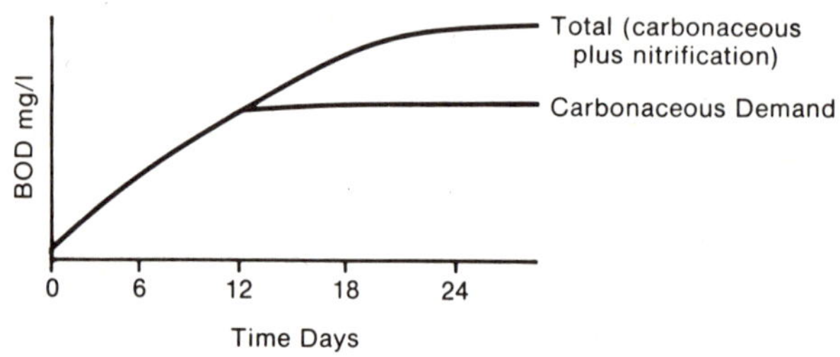

Figure 2-1 BOD demand curve

demand can be eliminated by adding inhibitory agents such as methylene blue, thiourea, allylthiourea or 2-chloro-6-(trichloromethyl)pyridine (7).

The value of K' is needed if BOD_5 is to be used to obtain the ultimate or 20-day BOD. The usual procedure followed when these values are unknown is to determine K' and L from a series of BOD measurements by using one of the following methods:

1 Least squares method (8, 9)
2 Method of moments (10)
3 Daily difference method (11)
4 Rapid ratio method (12)
5 Thomas method (13)

If you need to do this, consult the appropriate reference listed with the method.

For precise details on how to make a BOD determination, the book *Standard Methods for the Examination of Water and Wastewater* (14) should be consulted. This is a basic reference that should be used for procedures whenever applicable.

The BOD test, although very advantageous in that it relates to natural processes, does have several limitations:

1. It requires a high concentration of active acclimated seed bacteria.
2. Special care must be taken to eliminate all materials from the sample that destroy microorganisms, such as heavy metals, alkali, acid, and toxicants.
3. It requires an arbitrary and long period of time to obtain results.
4. Only the biodegradable organics are measured.
5. The test does not have stoichiometric validity after the soluble organic matter present has been used.

These limitations, especially the need for a more rapid test, have led to the development of other methods to measure organic matter.

COD. In order to get a more rapid measure of oxygen demanding organics, wet chemical techniques have been developed that measure the amount of oxidant consumed by a waste sample using strong oxidizing agents such as potassium permanganate or dichromate. Potassium dichromate has been found to be excellent for this purpose and for the COD test that uses it. The test is performed at an elevated temperature in the presence of a catalyst (silver sulfate). The COD test can be used to measure organic matter in waste that is toxic to biological life. The COD of a waste is in general higher than the BOD because more compounds can be chemically oxidized than can be biologically oxidized. For many types of wastes, it is possible to correlate COD with BOD. There is, however, no inherent constant relationship between COD and BOD. If a correlation can be established for a given waste, the COD measurements can be used to good advantage for treatment plant control and operation. The COD can be determined in 3 hours compared with the 5 days required for BOD.

TOC. Instrumental methods have further reduced the time required for analysis of the quantity of organic matter present. The TOC test is performed by injecting a known quantity of sample into a high-temperature furnace, where the organic carbon is oxidized to carbon dioxide in the presence of a catalyst. The carbon dioxide that is produced is quantitatively measured by means of an infrared analyzer. Acidification and aeration of the sample before analysis eliminate errors due to the presence of inorganic carbon. The test can be performed very rapidly and is becoming more popular. Certain resistant organic compounds may not be oxidized, however, and the measured TOC value will be slightly less than the actual amount present in the sample. The technique was developed and patented by the Dow Chemical Company and exclusively

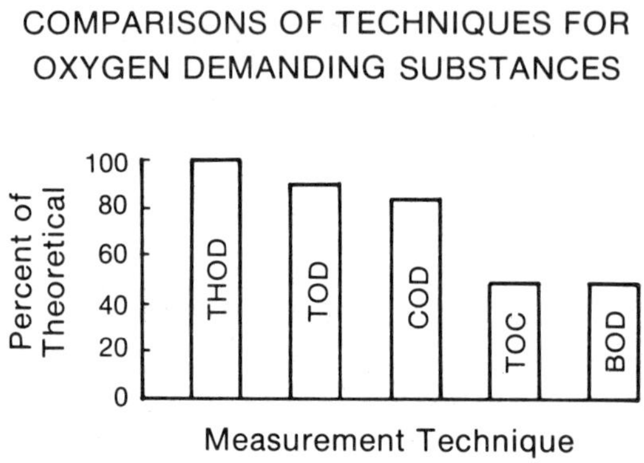

Figure 2-2 Alternative oxygen demand measurements

licensed to Beckman Instruments, Inc., which has marketed the process as the carbonaceous analyzer.

This method has been further improved by using persulfate as a chemical oxidizing agent in the presence of ultraviolet (UV) light to perform the oxidation at a low temperature. This approach to TOC is marketed by Dohrmann and Astro. The low-temperature oxidation seems to produce a more dependable, trouble-free instrument.

TOD. Another instrumental method that can be used to measure the organic content of wastewater is the TOD test (15). In this test, organic and, to a minor extent, inorganic substances are converted to stable end products in a platinum-catalyzed combustion chamber. The TOD is determined by monitoring the oxygen content present in the nitrogen carrier gas. This test can be carried out rapidly, and results have been correlated with the COD.

The relationships among the various techniques for monitoring oxygen-demanding substances can be graphically represented as shown in Figure 2-2. Organic matter is a major problem in pulp and paper mill effluents.

Disease-Causing Agents

The main concern is with pathogenic microorganism—bacteria, fungi, or viruses—that cause diseases in humans. If the pulp and paper mill has a segregated sewer system in which sanitary waste is handled separately, there should be no reason for pathogenic organisms to be presented in the mill effluent.

Because the numbers of pathogenic organisms present in wastes and polluted waters are few and difficult to isolate, the coliform organism, which is more numerous and more easily tested for, is used as an indicator organism. The presence of coliform organisms is taken as an indication that pathogenic organisms may also be present and their absence as an indication that water is free from such organisms. The coliform bacteria include the genera *Escherichia* and *Aerobacter*. The use of coliforms as indicator organisms is complicated by the fact that *Aerobacter* and certain *Escherichia* (e.g. *E. coli*) are not entirely of fecal origin, a high-incubation-temperature test at 44.5°C has been developed to attempt to differentiate this group. The differentiation is not complete, and it would be better to term the test the "elevated-temperature coliform test" rather than the "fecal coliform test," as is common practice. In particular, *Klebsiella pneumoniae* has been shown to respond positively to the test and to be present in surface waters and mill effluents (16). Thus a positive fecal coliform test does not always indicate contamination with fecal matter.

There are two accepted methods for determining the number of coliform organisms present in a given volume of water, the most probable number (MPN) technique and the membrane filter technique. The MPN technique is based on statistical analysis of the number of positive and negative results obtained when testing multiple portions of equal volume and in proportions constituting a geometric series for the presence of coliform. MPN gives not the absolute concentration of organisms that are present but only a statistical estimate of that concentration.

In the membrane filter technique, a known volume of water sample is passed through a filter of very small pore size. The bacteria are retained on the filter because they are larger than the pores. The bacteria are then contacted with agar that contains nutrients necessary for their growth. After incubation, the coliform colonies can be counted and the concentration in the original water sample determined. The membrane filter technique has the advantages of being faster than the MPN procedure and of giving a direct count of the number of coliforms. Detailed procedures for both methods are given in *Standard Methods* (14).

Synthetic Organic Compounds

In the class of pollutants known as "synthetic organic compounds," four subclasses can be defined—detergents, pesticides, oils and greases, and industrial chemicals. An element that all of these materials have in common is their potential toxicity. Bioassay tests are usually used to assess the toxicity of a material. The specific objectives of bioassay tests are:

1. To determine the concentration of a given waste that will kill 50% of the test organisms in a specified time period
2. To determine the maximum concentration causing no apparent effect on the test organism in 96 hours

These objectives are achieved by introducing fish or other appropriate test animals into test aquaria containing various concentrations of the waste in question and observing their survival with time. Although fish are usually used as the test organism, there is interest in seeing Daphnia come into common usage (17) because a test using Daphnia would be less costly to conduct.

Observations are usually made after 24, 48, and 96 hours. The test can be conducted under static conditions in which the waste remains fixed in the test vessel or under dynamic conditions in which the waste is flowing through the test vessel. For details of the procedure, consult *Standard Methods* (14) or ASTM Procedure D1345.

The prescribed measure of acute toxicity is the median tolerance limit (TL_m), defined as the concentration of toxicant in which 50% of the test animals are able to survive for a specified period of exposure. The median tolerance limit is sometimes abbreviated LD_{50}. As a rule of thumb, levels of 0.05–0.1 of the 96-hour LD_{50} concentration commonly produce no stress in the aquatic environment.

Detergents. Detergents are potentially toxic materials and, in addition, may cause foam problems. Detergents are used in the industry in certain cleaning applications such as felt washing. A wide variety of surface-active materials are used in the industry as dispersing agents to keep the system free of deposits and as additives in boil-out solutions used to thoroughly clean out the system. The determination of surfactants is usually accomplished by measuring the color change in a standard solution of methylene blue dye. There are natural surfactants produced and freed in the pulping process that will react positively with this test. If the specific surfactant of interest is known, it is best to develop a gas chromatographic procedure to detect its presence specifically.

Pesticides. Pesticides are not used in the paper industry unless their normal definition is broadened to include slimicides. Slimicides are used in the industry to keep the microbial population in the water system within limits in order to prevent slime accretion. Among the biocides used are chlorine, chlorinated phenols, some amines, quaternary ammonium compounds, organo-sulfur compounds, and silver compounds. Among the present criteria for the selection of a biocide are its susceptibility to biological treatment and its persistence in the environment. Industry suppliers have been able to provide biocides that adequately meet these requirements. Slimicides are normally detected by specific gas chromatographic procedures.

In the past, the industry used polychlorinated biphenyls (PCBs) in the manufacture of carbonless papers. PCBs are no longer so used, but there is still some contamination in the waste paper supply that will gradually disappear with time.

Oils and Greases. Oil and grease discharges, in addition to their potential toxicity, cause difficulties by reducing stream reaeration and light penetration; they also represent a potential fire hazard. The paper industry has potential problems in this area due to the discharge of spent lubricating oils and possible spills. Spills are particularly a problem if fuel oil is used as a boiler fuel. Proper spill prevention and management procedures will usually keep the problem under control.

The normal procedure for determining the presence of oils and grease in wastewater is filtration of a sample followed by hexane extraction of the filter pad for 4 hours to remove the oil and grease. The hexane is then evaporated and the weight of oil and grease measured (14).

Industrial Chemicals. The term "industrial chemicals" is used to encompass the remainder of the chemical entities produced in the pulping and bleaching processes or extracted from the wood during pulping. Tables 2-4 through 2-8 give the major nonvolatile toxic constituents from the debarking, Kraft pulping, sulfite pulping, mechanical pulping, and bleaching processes. Volatile reduced sulfur compounds such as hydrogen sulfide, methyl mercaptan, and dimethyl-

Table 2-4 Principal Debarking Toxic Chemicals

Resin acids: abietic, dehydroabietic, isopimaric, palustric, pimaric, sandaracopimaric, neoabietic
Unsaturated fatty acids: oleic, linoleic, palmitoleic
Diterpene alcohols: pimarol, isopimarol, abienol, 12E-abienol, 13-epimanool

Source: Reference 18.

Table 2-5 Principal Toxic Chemicals of Kraft Pulping

Resin acids: abietic, dehydroabietic, isopimaric, palustric, pimaric, sandaracopimaric, neoabietic
Unsaturated fatty acids: oleic, linoleic, linolenic, palmitoleic

Source: Reference 19.

Table 2-6 Principal Toxic Chemicals of Sulfite Pulping

Resin acids: abietic, dehydroabietic, isopimaric, palustric, pimaric, sandaracopimaric, neoabietic
Unsaturated fatty acids: oleic, linoleic, linolenic, palmitoleic
Juvabiones: juvabione, juvabiol, $\Delta^{1'}$-dehydrojuvabione, $\Delta^{1'}$-dehydrojuvabiol
Lignin degradation products: eugenol, isoeugenol, 3, 3′-dimethoxy-4, 4′-dihydroxystilbene

Table 2-7 Principal Toxic Chemicals of Mechanical Pulping

Resin acids: abietic, dehydroabietic, isopimaric, palustric, pimaric, sandaracopimaric, neoabietic
Unsaturated fatty acids: oleic, linoleic, linolenic, palmitoleic
Diterpine alcohols: pimarol, isopimarol, abienol, 12E-abienol, 13-epimanool
Juvabiones: juvabione, juvabiol, $\Delta^{1'}$-dehydrojuvabione, $\Delta^{1'}$-dehydrojuvabiol

Source: Reference 20.

Table 2-8 Principal Toxic Chemicals from Bleaching Plant Caustic Extraction–Stage Effluent

Chlorinated resin acids: monochloro- and dichlorodehydroabietic
Unsaturated fatty acid derivatives: epoxystearic acid, dichlorostearic acid, 3, 4, 5-trichloroguaiacol, 3, 4, 5, 6-tetrachloroguaiacol

Source: Reference 21.

sulfide also contribute to the toxicity of Kraft mill effluents (22). A good review of toxicity data and research in the industry has been published (23).

In addition to their toxicity, these chemicals can cause difficulties by imparting unpleasant tastes and odors to the receiving water. From Kraft mill effluents, sulfides, organic sulfur compounds, resin acids, and terpines have been identified as causing difficulties (24). The taste- and odor-producing potential of chlorinated phenolic bodies present in pulping wastes has also been discussed (25).

Color. Color bodies also need to be considered under the classification of industrial chemicals. "Color" is defined as either "true" or "apparent" color. In *Standard Methods* (14), the true color of water is defined as the color of water from which the turbidity has been removed. Apparent color includes not only the color due to substances in solution, but also due to suspended matter. A simplified technique for the determination of true color has been developed (26).

In the various chemical pulping processes, lignin and lignin derivatives are solubilized and removed from the wood during the cooking process. The spent cooking liquors containing these highly colored compounds are removed from the pulp in a subsequent washing sequence. The wash water is highly colored, and large amounts of color are ultimately discharged to the receiving stream despite some recovery operations.

Colored effluents may have the following detrimental effects upon receiving waters:

1. Color retards sunlight transmission and may interfere with photosynthesis, thereby reducing the productivity of the equatic community.
2. Natural stream color is altered, thus detracting from the visual appeal and recreational value of the receiving waters.
3. Color has effects upon downstream municipal and industrial water users, such as higher water treatment costs, difficulties with water treatment, and a multitude of industrial process operating problems.
4. Color bodies complex with metal ions, such as iron or copper, forming tar-like residues that remove the metals from the stock available to stream organisms for normal metabolism. These complexes also can have direct inhibitory effects on some of the organisms low in the food chain in the aquatic community and thereby reduce the productivity of the receiving water.

5 Color, derived from lignin, is an indicator of the presence of potentially inhibitory compounds and, in addition, may have direct inhibitory effects on some of the lower organisms in the food chain.

6 Color in receiving waters may affect fish movements and fish productivity.

7 Color bodies exert a long-term BOD (20–100 days) that is not measured by the BOD_5 test.

The physical and chemical nature of color bodies has been studied (27).

Plant Nutrients

The principal concern with plant nutrients is the amount of nitrogen, phosphorous, and trace metal ions found in the waste. Pulp and paper industry effluents are traditionally low in nitrogen and phosphorus, and must be supplemented with these elements before biological treatment. Thus there should be little difficulty here unless excess nutrients are supplied in the waste treatment process. The one exception to this generalization may occur when ammonia is used as the base for the pulping chemicals in the ammonium base sulfite process. However, ammonium base sulfite pulp is only a minor fraction of all of the pulp produced in the United States.

Inorganic Chemicals and Mineral Substances

Three classes of inorganic chemicals and mineral substances are of concern—acids, alkalis, and heavy metals.

The pH—acidity or alkalinity—of a mill effluent will depend on the type of pulping and bleaching used and on the type of product manufactured. It can range from highly acid for acid sulfite pulping to highly alkaline for Kraft pulping. In all cases, the pH will need to be adjusted to the range of 6–8 before biological treatment and/or discharge. This pH range is commonly believed to sustain the survival of most aquatic life forms.

The main difficulty with heavy metals is their toxicity, and the concepts of TL_m and LD_{50} apply, as previously discussed. The principal heavy metal ions of concern to the pulp and paper industry are aluminum (Al), chromium (Cr), copper (Cu), nickel (Ni), titanium (Ti), iron (Fe), mercury (Hg), and zinc (Zn). On the whole, heavy metals that might be present in pulp and paper mill effluents originate from one or more of three sources: chemicals used in the pulping process, additives used in papermaking, or products of equipment corrosion.

Pulping processing chemicals, mainly sulfur, salt cake, limestone, sodium hydrosulfite, chlorine and chlorine compounds, and caustic soda, are normally low in heavy metals, and are so used that those present would likely be precipitated in the process and discharged, either with the product produced or with a solid waste.

Zinc has been identified in the effluent from bleached groundwood operations and is due to the use of zinc hydrosulfite as a bleaching chemical. The quantity used is sufficiently low that the zinc probably does not reach toxic levels in receiving waters (28). However, most mills are switching to sodium salt to avoid this contamination.

In papermaking, some heavy metal oxides such as titanium dioxide are used as pigments, but these are in inert form and are not of major concern as pollutants.

While corrosive chemicals are handled and some corrosion takes place in pulp and papermaking equipment, metal losscs great enough to cause an appreciable concentration of metals such as nickel and chromium to appear in effluent are unlikely. Construction materials now employed throughout the industry have a high resistance to corrosion. Some iron might appear in low concentration due to corrosion from disc refiners, jordans, and similar equipment.

In the past, the paper industry had some difficulty with mercury due to its presence in a popular slimicide—phenylmercuric acetate—but this agent has not been used for several years. There was also the problem of mercury contamination of caustic soda produced by the use of mercury cells in the process. Caustic soda is no longer produced in the United States by this method, so this source of mercury contamination has also been eliminated.

It is most unlikely that toxic heavy metals reach critical levels in pulp and paper mill effluents. The absence of any history of difficulty arising from heavy metals bears out this contention.

Methods for determining the concentration of heavy metals vary in complexity according to the interfering substances that may be present (14). The concentration of many of these metals can be determined by such instrumental methods as polarography and atomic absorption spectroscopy.

Sediments

The class of pollutants known as "sediments" is concerned with the settleable solids component of waste that may lead to the formation of sludge beds on stream bottoms. The following attributes of these sludge beds are detrimental to a stream:

1. Coverage of the stream bottom alters the composition of the existing benthic life.
2. The beds exert a demand on the oxygen resources of the stream.
3. Anaerobic decomposition of the beds produces odor problems.
4. Gases from the anaerobic decomposition process float bed solids to the stream surface.

When solids in waste are considered in a broader sense, they are usually broken down into three classes: total solids, dissolved solids, and suspended solids. Analytically, the total solids content of a wastewater is defined as all of the matter that remains as residue upon evaporation at 103–105°C. The suspended or filterable solids are defined as those solids that are above about 1 μm in diameter

Figure 2-3 Imhoff cone

and will be retained on a filter medium. The material passing through the filter medium is considered as dissolved solids, and contains both colloidal materials and true dissolved solids. Colloidal solids are usually considered to be in the range of $1-10^{-3}$ μm and dissolved solids are below 10^{-3} μm in size.

The suspended solids fraction can be broken down into settleable and non-settleable solids. Settleable solids are those that will settle to the bottom of a cone-shaped container called an "Imhoff cone" in 60 minutes. Figure 2-3 illustrates an Imhoff cone. Settleable solids are an approximate measure of the quantity of sludge that will be removed by sedimentation.

Each category of solids may be further classified on the basis of its volatility at 600°C. The organic fraction will oxidize and be driven off as gas at this temperature, and the inorganic fraction remains behind as ash. At 600°C the decomposition of inorganic salts is restricted to magnesium carbonate, which decomposes into magnesium oxide and carbon dioxide at 350°C. Calcium carbonate is stable up to 825°C.

For details of the various solids procedures consult *Standards Methods* (14) or *Methods for Chemical Analysis of Water and Wastes* (29). The method for determination of suspended solids has received further detailed study (30) due to variations in results observed between mills.

Turbidity of a wastewater is often a useful parameter in characterizing the amount of suspended and colloidal matter present. Instrumental techniques to monitor turbidity have been developed (14, 29).

Radioactive Materials

Radioactive materials are potentially very hazardous pollutants, but are of little concern to the pulp and paper industry because there is no radioactive material involved in any main industry processes. There are some sensors that use small quantities of radioactive material, but this material is handled very cautiously and is properly disposed of when replaced.

Thermal Pollution

The addition of water containing thermal energy to a stream causes the stream temperature to rise. This rise in temperature causes the metabolic rate of the organisms living in the stream to increase, which exerts a greater demand on the oxygen resources of the stream. The increased temperature also causes the oxygen solubility to decrease, thus amplifying the problem. The increased temperature may cause population shifts in the kinds of organisms present. If the turbulence in the water body is not sufficient to produce complete mixing and a uniform temperature, free-swimming organisms such as fish may experience thermal shock upon entering the warm area.

The EPA recommends a maximum average receiving stream water temperature of 90°F, with a maximum permissible rise above the naturally existing temperature of 5°F. For lakes, the average allowable temperature rise may be only 3°F. Pulp and paper mill effluents are generally above ambient temperature, so there is a potential for thermal pollution problems, but such problems normally would occur only in small receiving streams. Generally, EPA standards can be met by the use of a good diffuser system on the discharge line.

SUMMARY OF POLLUTION PARAMETERS AND POLLUTION LOAD

The principal pollution parameters of concern to the paper industry are BOD, pH, and total suspended solids (TSS) in all categories of manufacturing. For groundwood, chemimechanical, and thermochemical operations, the heavy metal ion zinc is of concern. For certain selected categories of pulp manufacture, namely, Kraft, soda, and neutral sulfite semichemical pulping (NSSC), the color of the effluent is of concern. The issue of toxic substances is also of concern, but which operations are most affected is uncertain. These parameters—BOD, TSS, pH, zinc, and color—have all been identified in the effluent limitation guidelines of the EPA. Technologies to control these pollutants in the effluent have been developed for all except toxic substances, and the research on managing the latter is underway. These technologies will be discussed in later chapters.

ORIGIN OF RAW WASTE LOAD

To appreciate the sources of pollution loads in a mill system, we will consider an idealized paper mill using bleached Kraft pulp, as represented in Figure 2-4. This type of mill normally produces 500–1,000 tons of product per day. The process can be divided into five separate areas—wood yard, pulping, recovery, bleaching, and paper manufacturing. For the sake of simplicity, only the effluent volumes, TSS, and BOD are shown.

A study of Figure 2-4 produces several important observations. The largest volume of discharge comes from the paper mill, but almost as much is produced by the bleaching plant. Thus volume reduction programs in the paper industry should be aimed first at these two sources. By a large margin, the largest BOD loads are produced by the bleaching plant, with the pulp mill second and the paper mill a close third. The paper mill produces more than twice as much TSS as any other manufacturing area; the recovery area is a distant second. This simple analysis indicates that the bleaching plant and the paper mill should receive primary attention in pollution abatement activities.

A more in-depth analysis of pollution loads from the industry was conducted by the EPA as part of its guidelines development process. This process will be discussed in Chapter 4. It was not until the 1960s that pollution control entered into the planning of pulp and paper mills, and it did not become a serious consideration until the 1970s. Thus most of the processes presently used in the industry were designed without sufficient regard for minimization of effluent volume and pollutant concentration. The only concern was to manufacture the best product possible in the most simple and cost-effective manner; unfortunately, this decision resulted in the production of significant volumes of effluent.

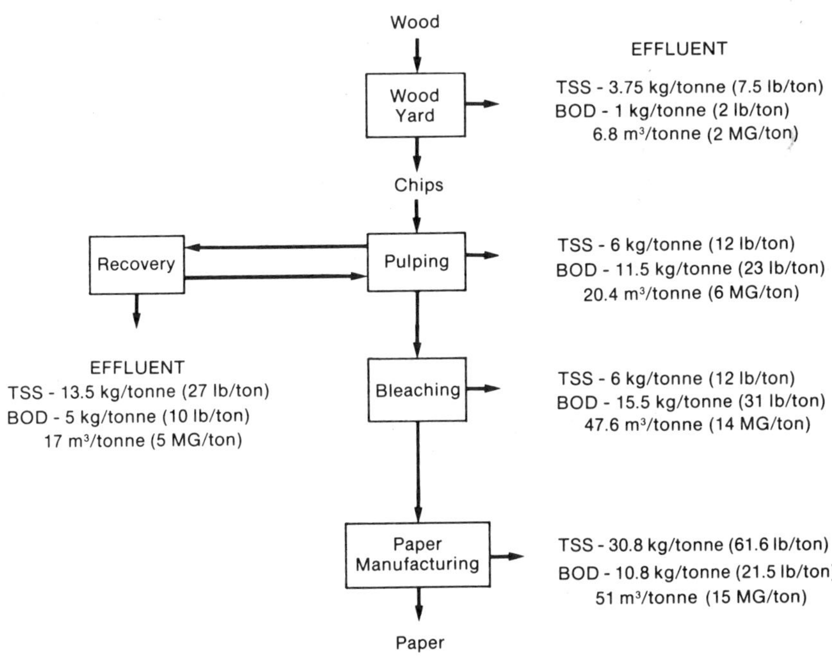

Figure 2-4 Representative bleached Kraft mill loads

Pulping

A wide variety of pulping processes are used in the industry, including Kraft, sulfite, soda, NSSC, thermomechanical, chemimechanical, and groundwood processes. Each process is different, and the nature of the effluent produced also varies. A generalization that seems to hold approximately for the average of each type of effluent is that the amount of BOD_5 discharged is inversely proportional to the pulp yield. Figure 2-5 illustrates this phenomenon. The higher the process yield, the less BOD_5 per ton of production seems to be produced. This conclusion is reasonable for pulping processes, because the major source of BOD is escape of spent liquor, which contains all of the organic matter dissolved during the pulping process. The lower the yield, the more material present in the liquor and the less pulp produced from a given amount of wood.

Thus the least polluting processes, on the basis of kilograms of BOD_5 produced per 1000 kilograms (kkg) of product, are the groundwood and thermomechanical processes, and the most polluting are the chemical processes. The poorest chemical pulping process is the calcium-based sulfite process, in which the yield is low and the economics of the process do not justify liquor recovery; therefore, all of the spent liquor is discharged. The sulfite processes are responsible for the sharp rise in the curve of Figure 2-5 at low yields. These processes are gradually being abandoned because of the high cost of effluent management.

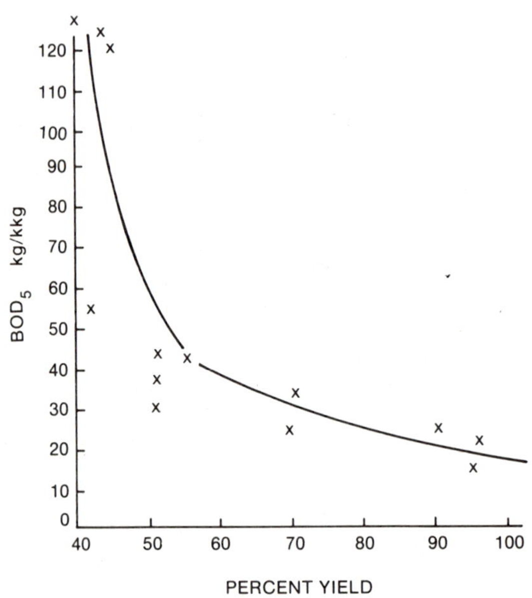

Figure 2-5 BOD_5 load as a function of pulp yield. Copyright © 1981. TAPPI. Reprinted from *Introduction to Environmental Management for the Pulp and Paper Industry,* Study Guide, Volume 6, with permission

They present a classic example of processes that were designed in a period during which environmental considerations were not of concern.

The TSS released by various processes are mostly related to the efficiency of the equipment used to remove solids. All of the processes remove suspended solids, so that theoretically only a minimum amount should appear in the effluent.

Papermaking

In the papermaking process, water is used to transport the stock and additives through the refining, cleaning, and paper-forming stages of the operation. Any materials added to the furnish are for a specific purpose and ideally should remain in the product. If, instead, some fraction is carried by the white water, it is recycled through the several for reuse. Both the paper machine and the saveall are devices that remove suspended solids, so solids should be at a low level in effluent streams.

Three sources of effluents originating the papermaking process are:

1 Cleaner rejects
2 Saveall clarified water
3 Spills

Spills constitute a significant portion of the load, and every means should be used to minimize them.

There is no obvious trend in BOD_5 load per kkg of product for the different types of paper manufacturing. All processes appear to produce about 10–12 kg BOD_5/kkg. There is, however, a trend in the volume of effluent discharged. The lower the quantity specifications for the product and the less sensitive it is to manufacture, the more internal process water reuse is practiced. Thus secondary fiber operations making combination paperboard have the lowest volume of effluent discharged per ton of product, while nonintegrated tissue manufacturing operations have the highest. This result can be seen in the nonintegrated operations portion of Table 2-9.

RAW WASTE DISCHARGED BY CATEGORY

The origin of the information presented in Table 2-9 is the EPA survey for establishing effluent limitation guidelines, referred to above. Table 2-9 summarizes the raw waste loadings for different types of manufacturing (31). The category breakdown is the same as used in the development guidelines and has been questioned by some industry representatives. The data were carefully screened and verified, and represent as good a summary as was available in 1975. They were gathered from seven sources:

Table 2-9 Raw Waste Loads for Different Manufacturing Categories

	Flow		BOD_5			TSS		
	kl/kkg	(kgal/ton)	kg/kkg	(lb/ton)	mg/l	kg/kkg	(lb/ton)	mg/l
Bleached Kraft: dissolving pulp	241	(57.8)	55.0	(110)	228	113	(226)	469
Bleached Kraft: market pulp	171	(41.0)	40.0	(80.0)	234	71.5	(143)	418
Bleached Kraft: paperboard, coarse and tissue	151	(36.2)	37.9	(75.8)	251	70.5	(141)	467
Bleached Kraft: fine papers	133	(31.8)	32.4	(64.7)	244	82.0	(164)	618
Soda	144	(34.5)	43.3	(86.5)	301	142.5	(285)	991
Groundwood: chemimechanical pulp	113	(27.0)	95.5	(191)	846	52.0	(104)	462
Groundwood: thermomechanical pulp	99	(23.8)	28.0	(56.0)	282	48.5	(97)	489
Groundwood: fine papers	91	(21.8)	16.9	(33.8)	186	52.0	(104)	572
Groundwood: coarse papers, molded products, and newspapers	99	(23.8)	17.4	(34.8)	175	70.0	(140)	705
Sulfite: paper	220	(52.8)	126.5	(253)	575	89.5	(179)	406
Sulfite: market pulp	244	(58.5)	123	(246)	504	32.6	(65.2)	134
Sulfite: Low alpha dissolving pulp	251	(60.2)	134	(268)	534	92.5	(185)	368
Sulfite: High alpha dissolving pulp	247	(59.2)	243.5	(487)	986	92.5	(185)	375
Deinking	104	(25.0)	82.5	(165)	791	178.5	(357)	1,712
Nonintegrated fine paper	63	(15.2)	10.8	(21.5)	170	30.8	(61.6)	486
Nonintegrated tissue	96	(22.9)	11.6	(23.1)	121	34.1	(68.1)	357
Nonintegrated tissue from waste paper	94	(22.6)	13.0	(26.0)	138	110.5	(221)	1,173
Unbleached Kraft	53	(12.6)	16.9	(33.8)	—	21.9	(43.8)	—
NSSC-ammonia	35	(8.3)	33.5	(67.0)	—	17	(34)	—
NSSC-sodium	43	(10.3)	25.2	(50.4)	—	12.3	(24.6)	—
Kraft-NSSC	58	(14.0)	19.4	(38.8)	—	20.5	(41)	—
Paperboard waste	30	(7.2)	11.2	(22.5)	—	—	(—)	—

1 Mill records
2 National Council for Air and Stream Improvement (NCASI) Reports 73-02 and 73-03
3 The American Paper Institute (API)
4 A short-term verification survey
5 EPA National Pollution Discharge Eliminations System (NPDES) applications
6 Literature
7 Personal interviews with recgonized authorities in the pulp and paper industry.

The majority of the data came from mill data usually taken on a daily basis for more than 1 year.

These types of data are useful for comparing a particular plant with the averages presented in Table 2-9. If the plant is above the present industry average, its technology should be examined for obvious and easy ways to improve its effluent situation. There may, of course, be good reasons for being above the industry average since, by definition, approximately half of the mills must fall in this group. However, the reasons for being above average should be determined, because the averages in 1975 certainly did not represent exemplary process water reuse and load control practices.

REFERENCES

1 W. E. Gates and S. Ghosh, *Journal of the Sanitary Engineering Division, American Society of Civil Engineers*, 97:287 (1971).

2 H. E. Orford and W. T. Ingram, *Sewage and Industrial Wastes*, 25:419 (1953).

3 C. S. Revelle, *Journal—Water Pollution Control Federation*, 37:679 (1965).

4 R. L. Woodward, *Sewage and Industrial Wastes*, 25:918 (1953).

5 J. C. Young and J. W. Clark, *Journal of the Sanitary Engineering Division, American Society of Civil Engineers*, 91:43 (1965).

6 G. J. Schroepfer, M. L. Robins, and R. H. Susag, "The research program on the Mississippi River in the vicinity of Minneapolis and St. Paul," *Advances in Water Pollution Research*, Vol. 1, Pergamon, London, 1964.

7 J. C. Young, "Chemical methods for nitrification control," *Proceedings of the XXIV Industrial Waste Conference, Part 1*, Purdue University, 1969.

8 A. E. Waugh, *Elements of Statistical Method*, 2d ed., McGraw-Hill, New York, 1943.

9 H. D. Young, *Statistical Treatment of Experimental Data*, McGraw-Hill, New York, 1962.

10 E. W. Moore, H. A. Thomas, and W. B. Snow, *Sewage and Industrial Wastes*, 22:1343 (1950).

11 E. E. Tsivoglou, "Oxygen relationships in streams," *Technical Report W-58-2*, Robert A. Taft Sanitary Engineering Center, Cincinnati, 1958.

12. J. P. Sheehy, *Journal—Water Pollution Control Federation*, 32:646 (1960).

13 H. A. Thomas, Jr., *Water and Sewage Works*, 97:123 (1950).

14 *Standard Methods for the Examination of Water and Wastewater*, 15th ed., American Public Health Association, New York, 1980.

15 D. Clifford, *Total Oxygen Demand—A New Instrumental Method*, American Chemical Society, Midland, Mich., 1967.

16 *Further studies of the sanitary significance of* Klebsiella pneumoniae *occurrence in mill effluents and surface waters*, Technical Bulletin No. 279, National Council of the Paper Industry for Air and Stream Improvement, New York, 1975.

17 J. F. Schmaltz, *Pulp & Paper Canada*, 80(7):72–76 (1979).

18 J. M. Leach and A. N. Thakore, *Identification and treatment of the toxic materials in pulp and paper wood room effluents*, CPAR Report No. 148-2, Canadian Forest Service, Ottawa, Ontario, 1974.

19 J. M. Leach, A. N. Thakore, and C. C. Walden, *Isolation of toxic constituents of Kraft pulp mill effluents*, CPAR Report No. 11-4, Canadian Forest Service, Ottawa, Ontario, 1974.

20 J. M. Leach and A. N. Thakore, *Identification and treatment of the toxic materials in mechanical pulping effluents*, CPAR Report No. 149-2, Canadian Forest Service, Ottawa, Ontario, 1974.

21 J. M. Leach and A. N. Thakore, *Identification of the toxic constituents in Kraft mill bleach plant effluent*, CPAR Report No. 245, Canadian Forest Service, Ottawa, Ontario, 1975.

22 E. P. Haydu, H. R. Amberg, and R. E. Dimick, *Tappi*, 35(12):545 (1952).

23 C. C. Walden and T. E. Howard, *Tappi*, 60(1):122 (1977).

24 W. M. vanHorn, *The causes and reduction of tastes and odors in Kraft mill wastes*, Technical Bulletin No. 32, National Council of the Paper Industry for Air and Stream Improvement, New York, 1971.

25 C. W. Dence, *The phenolic substances in pulp and paper mill effluents: Their structure, behavior, and determination.* Technical Bulletin No. 250, National Council of the Paper Industry for Air and Stream Improvement, New York, 1971.

26 W. L. Carpenter, *An investigation of improved procedures for measurement of mill effluent and recovery water color*, Technical Bulletin No. 253, National Council of the Paper Industry for Air and Stream Improvement, New York, 1971.

27 H. S. Dugal, J. W. Swanson, E. E. Dickey, and M. A. Buchanan, *Tappi*, 58(7):132 (1975).

28 H. Gehm, *State-of-the-art review of pulp and paper waste treatment*, Environmental Protection Technology Series EPA-R2-73-184, U.S. Government Printing Office, Washington, D.C., April 1973, p. 32.

29 *Methods for chemical analysis of water and wastes*, EPA-625/6-74-003, Methods Development and Quality Assurance Research Laboratory, National Environmental Research Center, Cincinnati, U.S. Government Printing Office, Washington, D.C., 1974.

30 *A study of the effect of alternate procedures on effluent suspended solids measurements*, Technical Bulletin No. 291, National Council of the Paper Industry for Air and Stream Improvement, New York, 1977.

31 *Development document for interim final and proposed effluent limitations guidelines and proposed new source performance standards for the bleached Kraft, groundwood, sulfite, soda, deink, and non-integrated paper mills segment of the pulp, paper and paperboard mill point source category*, EPA 440/1-76/047-A, U.S. Environmental Protection Agency, Washington, D.C., January 1976.

3

Basic Mathematical Analysis of the Effects of Water Pollution on Lakes and Streams

In the previous chapter, we discussed the different types of pollutants. In this chapter, we will briefly discuss the impact of these pollutants on lakes and streams. We will try to show what happens to a contaminant in a river or lake system and to describe mathematically the change in the system with time. Any mathematical relationship is simply an attempt to describe the physical phenomena that occur. Therefore, a physical understanding of a wastewater system, rather than just the mathematics, is a prerequisite to intelligent analysis. The phenomenon of interest is what happens to a pollutant after it enters the lake or stream. We are assuming that before discharge the waste stream has been subjected to some form of treatment, and we will look at its fate after discharge to a watercourse.

DISPOSAL INTO LAKES

A lake is a semiclosed system. Lakes undergo an aging process called "eutrophication." When a lake is first formed it contains little life, and is therefore termed an "oligotrophic" ("little food") lake with low nutrient levels. As the lake accumulates nutrients and develops more plant and animal life, it evolves into a "eutrophic" ("abundant food") lake with high nutrient levels. In the extreme, the lake becomes choked with plant life and evolves into a swamp. This evolution is a natural process that occurs in all lakes. The rate at which the process occurs is controlled by the amounts of nutrients that accumulate in the lake.

Aquatic plants require nitrogen, phosphorus, potassium, and other mineral nutrients. The two nutrients that most commonly act as limiting factors are phosphorus, in the form of phosphate, and nitrogen, in the form of either nitrate or ammonia.

When humans become involved in the process and increase the rate at which nutrients enter the lake, the term "cultural eutrophication" is used. The natural process may take several hundred years, but cultural eutrophication can occur in less than 20 years.

For the entire eutrophication process, it is difficult to develop a mathematical model, but for short-term effects we can develop relatively simple models that give us an understanding of the fate of pollutants in lake systems. Although there is flow both into and out of a lake, there is a considerable detention time in the lake. First, one must determine how much mixing occurs in the lake. The factors that produce mixing are currents due to inflow and currents due to wave action caused by wind. Small lakes and reservoirs can be assumed to be completely mixed, but this assumption becomes poorer as the lake becomes larger and deeper.

Complete Mix Model

For simplicity, we will analyze a completely mixed lake. Figure 3-1 diagrams the concept. Considering the concentration of only one species, the river (inlet) flows into the lake at flow rate Q_r and concentration C_r. Pollution is added to the lake at a flow rate q_p and a concentration C_p. The flow out of the lake is the sum of the inflow terms $Q_r + q_p$, and the concentration in the outflow is the same as the concentration C in the lake, because the system is completely mixed. The problem is to derive an expression for the concentration C as a function of time. Table 3-1 outlines the procedure. This solution fits what one would expect, that is, the larger the volume of the lake, the lower the concentration; the more pollutant input (increased q_p), the higher the pollutant concentration (increased W); the higher the rate of degradation of the pollutant (larger K), the smaller β and the pollutant concentration become.

This model applies only to complete mixing within the lake; another more complicated model is needed for situations of partial or no mixing. We will not cover these more complex cases, but the mathematics have been derived.

Stratification

The complete mix model does not apply to deep lakes that experience vertical stratification during certain seasons of the year. However, it might apply to the upper layer (the epilimnion) of a small, deep lake.

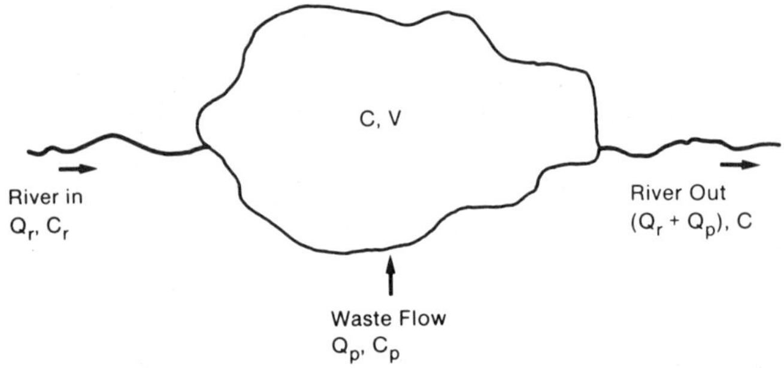

Figure 3-1 Conceptual lake. Copyright © 1981. TAPPI. Reprinted from *Introduction to Environmental Management for the Pulp and Paper Industry*, Study Guide, Volume 6, with permission

Table 3-1 Derivation of the Equation of Lake Concentration versus Time

---

Material balance on the lake:

$$\overset{1}{\text{Change in storage}} = \overset{2}{\text{inflow}} - \overset{3}{\text{outflow}} + \overset{4}{\text{sources}} - \overset{5}{\text{sinks}}.$$

Mathematically:

$$\overset{1}{V\,dC} = \overset{2}{Q_r C_r\,dt} + \overset{2}{q_p C_p\,dt} - \overset{3}{(Q_r + q_p)C\,dt} + \overset{4}{0} - \overset{5}{VCK\,dt}$$

V = lake volume
dC = change in pollutant concentration in the lake
Q_r = river (inlet) flow rate
C_r = concentration of a pollutant in the river
q_p = wastewater flow rate into the lake
C_p = concentration of a pollutant in wastewater
K = first-order rate constant (the waste is assumed to degrade by a first-order mechanism)

To simplify the equation, let

$$W = Q_r C_r + q_p C_p = \text{total inflow of waste} \qquad \text{detention time } t_0 = \frac{V}{Q}$$

$$Q = Q_r + q_p$$

Dividing through by $V\,dt$, we have

$$\frac{dC}{dt} = \frac{(Q_r C_r + q_p C_p)}{V} - \frac{(Q_r + q_p)C}{V} - CK$$

$$= \frac{W}{V} - \frac{C}{t_0} - CK$$

$$\frac{dC}{dt} + (1/t_0 + K)C = \frac{W}{V}$$

Let $\beta = 1/t_0 + K$

$$\frac{dC}{dt} + \beta C = \frac{W}{V}$$

First order linear differential equation boundary conditions

$$t = 0 \qquad C = C_0$$

Solution

$$C = \frac{W}{\beta V}(1 - e^{-\beta t}) + C_0 e^{-\beta t}$$

Concentration as function of time

$$\text{at } t = \infty \qquad C = \frac{W}{\beta V}$$

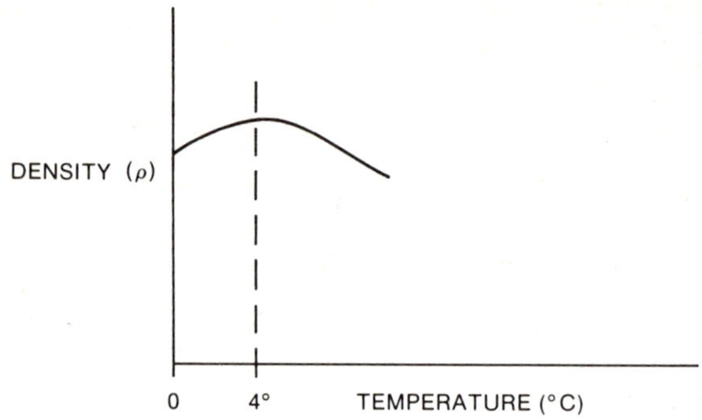

Figure 3-2 Water density as a function of temperature. Copyright © 1981. TAPPI. Reprinted from *Introduction to Environmental Management for the Pulp and Paper Industry.* Study Guide, Volume 6, with permission.

Stratification in lakes is the result of increased water density with depth (1) caused by a decrease in temperature. Figure 3-2 illustrates the density maximum of water experienced at 4°C. During spring, most lakes are uniform in temperature and are easily mixed by wind currents. As the upper waters warm in summer, their density decreases and a stable stratification results, as shown in

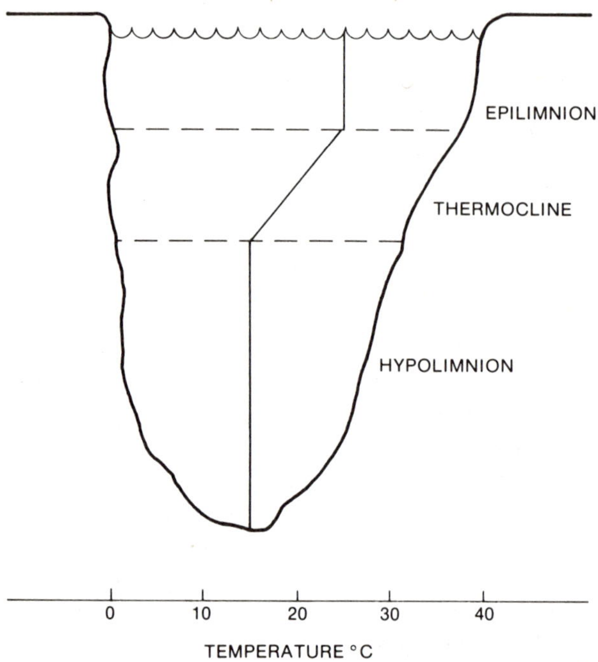

Figure 3-3 Lake stratification during summer. Copyright © 1981. TAPPI. Reprinted from *Introduction to Environmental Management for the Pulp and Paper Industry*, Study Guide, Volume 6, with permission

Figure 3-3. Three zones of stratification can be defined. The top layer (epilimnion) may be 30–50 feet deep and has a fairly uniform temperature due to mixing by wind action. The thermocline or metalimnion is characterized by its steep temperature gradient and is extremely resistant to mixing. The hypolimnion also has a fairly uniform lower temperature.

Most deep lakes in the United States experience thermal stratification during the months of May through October. In temperate climates, the decrease in air temperature that occurs with the approach of autumn causes a corresponding decrease in lake water temperature, and a mixing occurs that is called the "fall turnover," in which the lake temperature again becomes uniform. If the climate is severe enough for ice to form on the surface of the lake, a stage of winter stratification also occurs, but it is simpler than the summer one. The winter stratification consists of two zones: one just below the ice cap, in which the temperature gradient is from 0 to 4°C, and the other encompassing the rest of the lake, in which the temperature is at or near 4°C.

In deep lakes during phases of stratification, the complete mix model applies only to the epilimnion layer. Other models would have to be developed to describe the behavior of the other layers as a function of time and to predict concentrations.

DISPOSAL IN RIVERS

A stream is something more than a geographic feature represented by a line on a map. It is a complete and complex, ever-changing ecosystem. Streams are subject to much natural pollution because they drain large areas of land.

Water that does not infiltrate the ground or evaporate into the air remains on the surface of the earth and is thus termed "surface water." Surface water, as it flows off the land, collects in streams and rivers, which are the pathways for the runoff. Each river, including its branches and tributaries, drains water from a fixed geographic region called a "driange basin" or "watershed." The major U.S. watersheds are shown in Figure 3-4. These are the logical zones or regions of water management in the United States. Unfortunately, political boundary considerations make regulations on the basis of watersheds impractical.

The weather pattern in a drainage basin is a primary factor in determining the amount of flow or discharge of a river. The discharge volume is largely determined by the amount of rain received.

Most rivers have a tendency to meander rather than to follow a straight course. They bend and weave back and forth within the bounds of a broad plain. This broad plain is termed the "floodplain" of the river basin and will be under water during periods of extremely high water. It is common to include any area that can be expected to be flooded at least once every 100 years in the area defined as a floodplain.

Ideally, this floodplain should be kept free of human settlement, because it

Figure 3-4 Major U.S. watersheds

is an area that is certain to be flooded at some time. Unfortunately, human settlement is often extensive within river floodplains; people are lured there by productive soils, reliable water supplies, inexpensive transportation, and recreational potential. This human folly is the reason for so much property damage during floods.

A river tends to accumulate all of its pollution from its watershed, and then must deal with it in some manner. Because of its property of self-purification, it is capable of dealing with a certain amount of pollution introduced by humans. Microorganisms in the river decompose organic matter and suspended solids settle to the bottom, resulting in the self-purification of the river. One of the most critical parameters in a river is its dissolved oxygen content. The major source of replenishment of oxygen in a river is reaeration from the atmosphere. There are also some minor sources, such as photosynthetic activity of algae and other plants, and the oxygen dissolved in surface drainage water and groundwater infiltrates.

Algae Effects

The amount of oxygen supplied by algae depends on their concentration and on the amount of sunlight they receive. Their oxygen concentration is, therefore, a function of depth. Figure 3-5 shows the variation in algae concentration with depth. At the water surface the sunlight is intense enough to inhibit algal growth, so that the maximum concentration occurs slightly below the surface; the concentration then decreases with decreasing light intensity at lower depths (2).

Algae also experience a daily cycle of oxygen production as the sunlight comes and goes, resulting in the diurnal variation in dissolved oxygen shown in Figure 3-6.

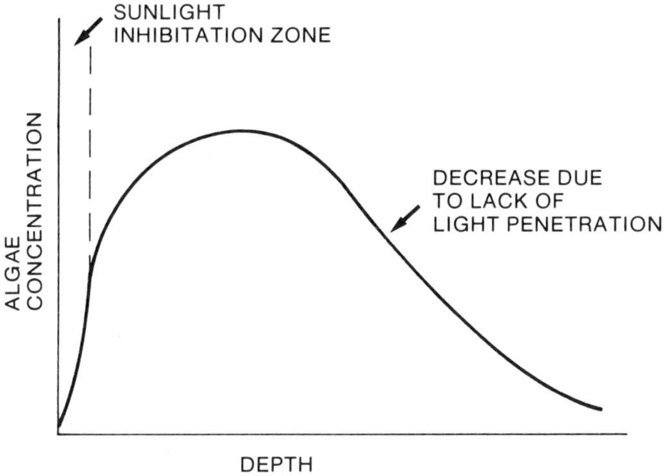

Figure 3-5 Algae concentration as a function of depth. Copyright © 1981. TAPPI. Reprinted from *Introduction to Environmental Management for the Pulp and Paper Industry*, Study Guide, Volume 6, with permission

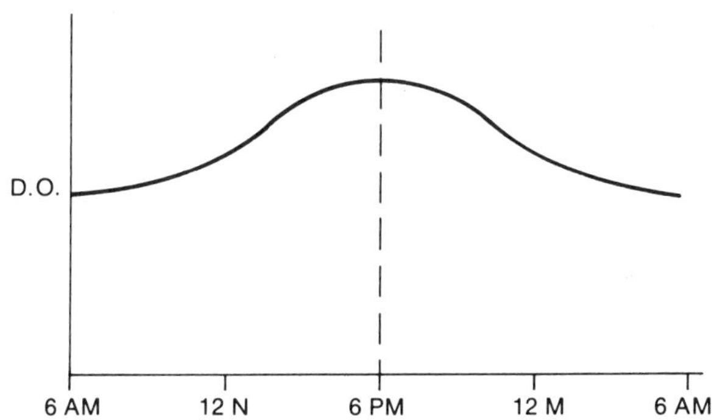

Figure 3-6 Diurnal variation of dissolved oxygen in algae. Copyright © 1981. TAPPI. Reprinted from *Introduction to Environmental Management for the Pulp and Paper Industry*, Study Guide, Volume 6, with permission

Unless a very large population of algae is present, however, algal effects are small enough to be neglected in comparison with the effect of reaeration.

Reaeration

Mathematically, reaeration is expressed by the relation shown in Table 3-2. Values of the reaeration coefficient K_2 have been estimated for various bodies of water by the Engineering Board of Review for the Sanitary District of Chicago,

Table 3-2 Reaeration Coefficients

Water Body	Range of K_2 at 20°C (Base 10)[a]
Small ponds	0.05–0.10
Sluggish streams and large lakes	0.10–0.15
Large streams of low velocity	0.15–0.20
Large streams of normal velocity	0.20–0.30
Swift streams	0.30–0.50
Rapids and waterfalls	>0.50

Reaeration equation:

$$-\frac{dD}{dt} = K_2 D$$

where D = oxygen deficit and K_2 = reaeration coefficient

Source: Copyright © 1981. TAPPI. Reprinted from *Introduction to Environmental Management for the Pulp and Paper Industry*, Study Guide, Volume 6, with permission.

[a] For base *e*, multiply by 2.3.

and these are listed in Table 3-2. The faster the stream flows, the more water comes in contact with the surface and the higher the reaeration coefficient.

Sludge Deposits

In Chapter 2 we discussed BOD and some of the mechanisms by which oxygen is depleted in the stream. A source of depletion that we have not yet discussed is benthic deposits. Heavy material that settles out forms sludge beds on stream bottoms, and these beds can exert a significant oxygen demand. Only the uppermost layer of a sludge bed, about 1–2 mm thick, is aerobic; the majority of the bed is anaerobic.

Amounts of deposition and its reverse phenomenon, scour, vary with the velocity and turbulence of the river. Sedimentation decreases the BOD of the water, while scour increases it. Each section of a river must be analyzed for benthic deposits and their importance to the oxygen balance assessed. In situations in which benthic deposits, or sludge beds, are important, their effect can be evaluated by use of an empirical equation developed by Fair, Thomas, and Moore (Table 3-3).

Mathematical Model of a River

All of the phenomena discussed above need to be combined into a mathematical model of a river. We will assume that waste is evenly distributed across the

Table 3-3 Oxygen Demand for Benthic Deposits

$$Y_m = 3.14(10^{-2}Y_0)C_T W\left(\frac{5+160W}{1+160W}\right)\sqrt{t_a}$$

Y_m = maximum daily benthic O_2 demand, g/m^2
Y_0 = BOD_5 of benthic deposit, g/kg of volatile matter at 20°C
W = daily rate of volatile solids deposition, kg/m^2
t_a = time during which settling takes place, days
C_T = temperature correction factor

$$= \frac{BOD_5 \text{ at } T}{BOD_5 \text{ at } 20°C} = \frac{(1-e^{-5K}T)}{(1-e^{-5K}20°)}$$

Source: Copyright © 1981. TAPPI. Reprinted from *Introduction to Environmental Management for the Pulp and Paper Industry*. Study Guide, Volume 6, with permission.

river, which is a poor assumption at the outfall (mill or plant discharge point) but improves farther downstream. Channelization, changing the flow pattern of a river by digging channels can also invalidate this assumption. We also assume that no mixing occurs in the direction of flow of the river, that is, we assume a type of flow in which the water in the river flows as a plug downstream and does not backmix. Other more sophisticated approaches exist to handle the fluid mechanics of river flow, but this simple one is often used. We will first deal with the problem of the concentration of a pollutant, and then use the results to deal with the dissolved oxygen problem.

The derivation of the mathematical relationships for pollutant concentration is presented in Table 3-4.

A similar approach can be used to determine the oxygen concentration in a river. This approach was developed by R. Streeter in collaboration with Earle Phelps and is called the "Streeter–Phelps equation" (5). The sources and sinks in this problem are more complex, and are listed along with the derivation of the equation in Table 3-5. The resulting equation is relatively complex, and its use requires values for S, R and P. If these values can be neglected, however, the last term drops out and the equation reduces to the classic Streeter–Phelps equation involving only BOD and reaeration effects, given in Table 3-6. This equation is most commonly used in river analysis. A plot of this equation produces the traditional dissolved oxygen sag curve shown in Figure 3-7. This curve reflects the phenomenon of natural stream purification.

The critical point on this curve is of significance, for it represents the point at which the rate of oxygen use equals the rate of reaeration, or the point at which the derivative $dD/dt = dD/d(x/v) = 0$. Table 3-7 presents expressions for the maximum oxygen deficit and its location in the river.

Table 3-4 Derivation of River Pollutant Concentration with Time or Distance

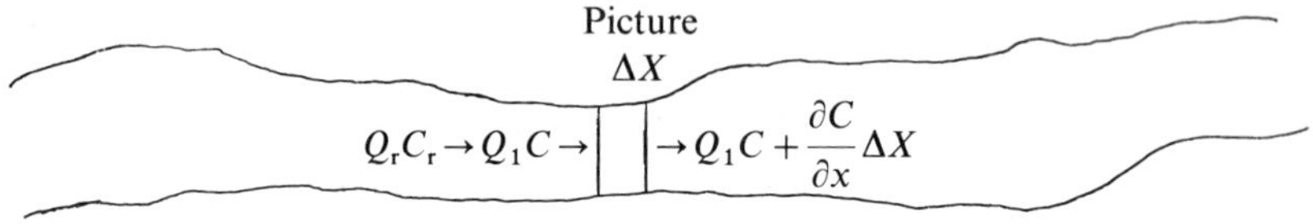

Material balance on volume enclosed by ΔX:

$$\overset{1}{\text{Changes in storage}} = \overset{2}{\text{inflow}} - \overset{3}{\text{outflow}} + \overset{4}{\text{sources}} - \overset{5}{\text{sinks}}$$

$$\overset{1}{V\Delta C} = \overset{2}{QC\Delta T} - \overset{3}{Q\left(C + \frac{\partial C}{\partial x} \Delta X \right)\Delta t} + \overset{4}{0} - \overset{5}{VK\bar{C}\Delta t}$$

Divide through by $V\Delta t$ and cancel terms:

$$\frac{\Delta C}{\Delta t} = -Q \frac{\partial C}{\partial x} \frac{\Delta x}{V} - K\bar{C}$$

$V = A\Delta x$ $\quad$ A = cross-sectional area of the stream

Take $\lim\limits_{\Delta t \to 0} \frac{\Delta C}{\Delta t} = -\frac{Q}{A} \frac{\partial C}{\partial x} - K\bar{C}$ $\quad$ recognize $\frac{Q}{A} = \upsilon$ stream velocity

$$\frac{\partial C}{\partial t} = -\upsilon \frac{\partial C}{\partial x} - K\bar{C}$$

This is general equation but it is difficult to solve, so let's look at the steady state solution.

$$\frac{\partial C}{\partial t} = 0$$

$$0 = \upsilon \frac{\partial C}{\partial x} - KC$$

$\upsilon \frac{\partial C}{\partial x} = -KC$ $\quad$ variable separable differential equation

$$\int_{C_0}^{C} \frac{dC}{C} = -\frac{K}{\upsilon} \int_0^x dx$$

$$\ln \frac{C}{C_0} = -\frac{K}{\upsilon} x$$

$$C = C_0 \exp\left(-K \frac{x}{\upsilon} \right)$$

C_0 is average river concentration at $x = 0$

$$C_0 = \frac{\text{waste input}}{\text{flow rate}} = \frac{W}{Q} = \frac{QC_r + q_p C_p}{Q_r + q_p}$$

Source: Copyright © 1981. TAPPI. Reprinted from *Introduction to Environmental Management for the Pulp and Paper Industry*, Study Guide, Volume 6, with permission.

Table 3-5 Conceptual Derivation of a River Dissolved Oxygen Profile

Sources of O_2:
  Reaeration $= K_2(C_s - C)$
  Photosynthesis $= P$
Sinks of O_2:
  BOD $= K(\mathrm{BOD_L})$
  Algal respiration $= R$
  Sludge deposits $= S$

Use an approach similar to the pollution concentration profile developed in Table 3-4 to derive the mass balance equation:

$$\frac{\partial C}{\partial t} = v\frac{\partial C}{\partial x} + K_2(C_s - C) + P - K(\mathrm{BOD_L}) - R - S$$

The time-dependent situation is complicated, so solve for the steady state case $\partial C/\partial t = 0$:

$$C_s - C = \text{dissolved oxygen deficit}$$

$\mathrm{BOD_{Lx}} = \mathrm{BOD_L}\exp(-K'x/v)$ from the development of the pollution concentration
If we insert this equation and perform the integration, the result is:

$$D = \frac{K'\mathrm{BOD_L}}{K_2' - K'}(e^{-K'(x/v)} - e^{-K_2'(x/v)}) + D_0 e^{-K_2'(x/v)} + \frac{S + R - P}{K_2'}(1 - e^{-K_2'(x/v)})$$

K_2' = reaeration coefficient (base e)
D_0 = initial oxygen deficit

Source: Copyright © 1981. TAPPI. Reprinted from *Introduction to Environmental Management for the Pulp and Paper Industry*, Study Guide, Volume 6, with permission.

Table 3-6 Streeter–Phelps Equation

$$D = \frac{K'(\mathrm{BOD_L})}{K_2' - K'}(e^{-K'(x/v)} - e^{-K_2'(x/v)}) + D_0 e^{-K_2'(x/v)}$$

Table 3-7 Expression for Maximum Dissolved Oxygen Deficit

$$D_c = \frac{K'}{K_2'}(\mathrm{BOD_L})e^{-K'\left(\frac{x_c}{v}\right)}$$

$$x_c = \frac{v}{K_2' - K'}\ln\frac{K_2'}{K'}\left[1 - \frac{D_0(K_2' - K')}{K'\mathrm{BOD_L}}\right]$$

$$t_c = \frac{x_c}{v}$$

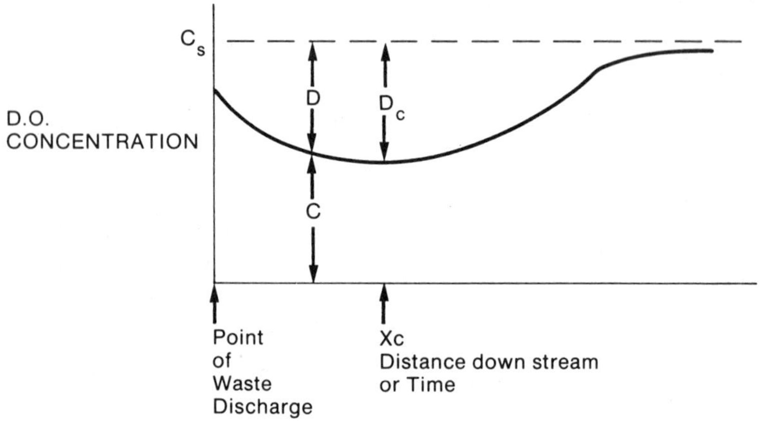

Figure 3-7 Dissolved oxygen sag curve. Copyright © 1981. TAPPI. Reprinted from *Introduction to Environmental Management for the Pulp and Paper Industry*, Study Guide, Volume 6, with permission

Aquatic Biology

The mathematical model reflects the physical parameters of pollutant concentration and dissolved oxygen but does not give an appreciation of what happens to the aquatic life in the stream at various oxygen levels. Figures 3-8 to 3-11 will aid us in understanding these interrelationships.

The stream is divided into four basic sections based on dissolved oxygen content:

1 A clean water zone (high dissolved oxygen) upstream of a pollutant discharge
2 A zone of decline (falling dissolved oxygen level)
3 A zone of damage (bottom of the dissolved oxygen sag curve)
4 A zone of recovery (rising dissolved oxygen)

In the clean water zone, the stream is populated with the species outlined in Figure 3-9. Fish, mayflies, clams, stone flies, and other organisms require a high level of dissolved oxygen. When pollutants are introduced, the dissolved oxygen begins to drop and the species present shift to those with an intermediate oxygen demand, such as those pictured in Figure 3-10. Such organisms as sowbugs, scud, blackfly larvae, leeches, damselfly nymphs, paramecia, *Vorticella*, and *Euglena* will be dominant. The only fish species present will be rough fish such as carp, buffalo, and gar.

When the dissolved oxygen drops below about 2 ppm, even these semitolerant species disappear. In this damage or danger zone only a few species of microinvertebrates such as bloodworm, sludge worm, rat-tailed maggot, and swooge

	Zones of pollution				
	Clean water	Decline	Damage	Recovery	Clean water
9 milligrams per liter Dissolved oxygen level 1 milligram per liter	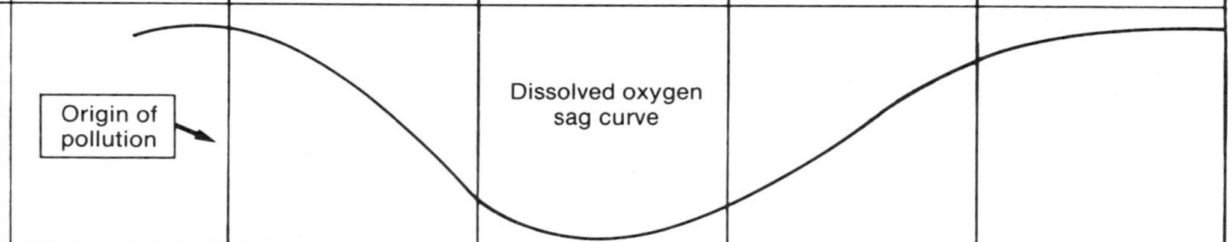 Origin of pollution		Dissolved oxygen sag curve		
Physical indices	Clear, no bottom sludge	Floating solids, bottom sludge	Turbid, foul gas, bottom sludge	Turbid, bottom sludge	Clear, no bottom sludge
Fish present	Game, pan, food and forage fish	Tolerant fishes—carp, buffalo, gars	None	Tolerant fishes—carp, buffalo, gars	Game, pan, food and forage fish
Bottom animals present	Caddisfly Stonefly	Bloodworm Snail	Sewage mosquito larvae Sewage fly larvae Rat-tailed maggot	Bloodworm snail	Caddisfly Mayfly naiads
Algae & protozoa present	Dinobryon Cladophora Ulothrix Navicula	Paramecium Vorticella	Phormidium Stigeocloneium Oscillatoria	Euglena Spyrogyra Pandorina	Dinobryon Cladophora Ulothrix Navicula

Figure 3-8 Density–diversity pattern with pollution load

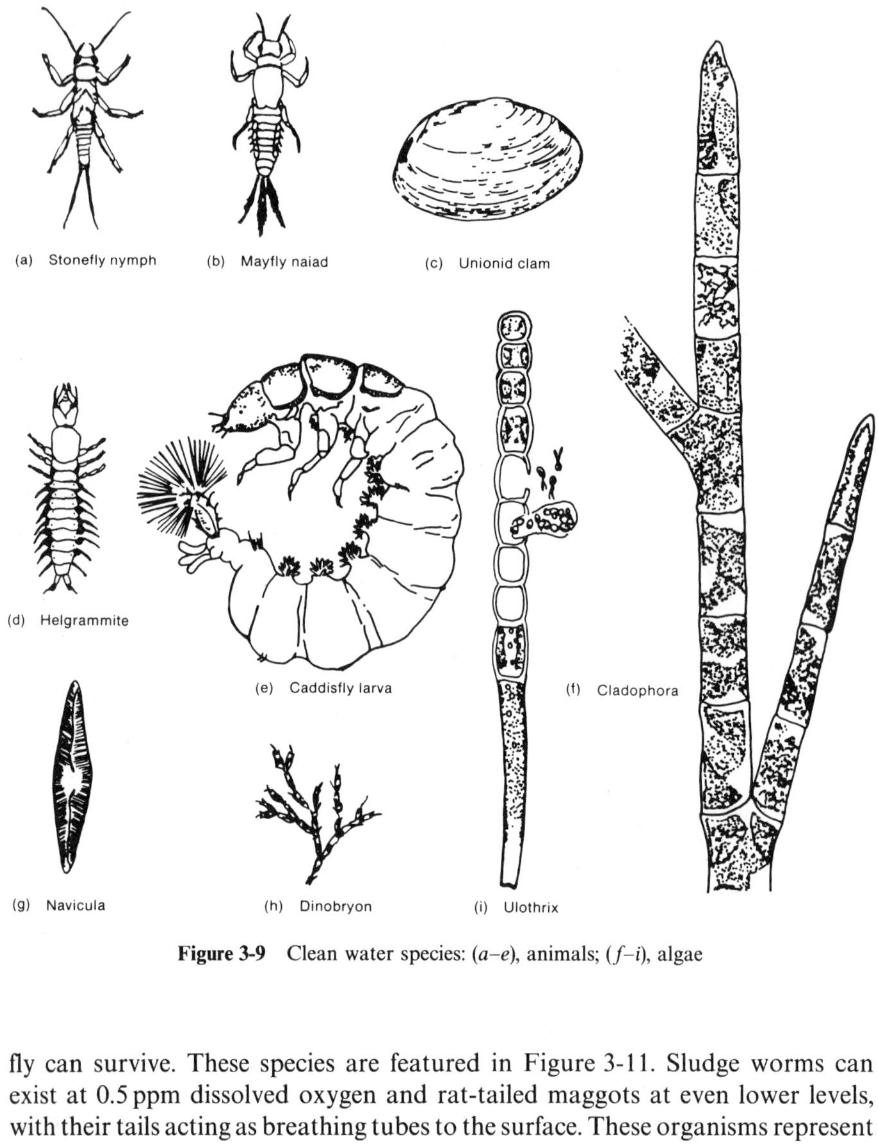

Figure 3-9 Clean water species: (*a–e*), animals; (*f–i*), algae

fly can survive. These species are featured in Figure 3-11. Sludge worms can exist at 0.5 ppm dissolved oxygen and rat-tailed maggots at even lower levels, with their tails acting as breathing tubes to the surface. These organisms represent a significant decrease in the diversity of organisms present, but the few species that do occur may exist in high density. This is the typical pattern: in clean water there are large numbers of species, each low in density, whereas in seriously polluted water there are few species in high density.

In the recovery zone, the pattern of the zone of decline is repeated. The water is likely to be clear, allowing the penetration of sunlight. Oxygen recovers to reasonable levels. Algae begin to grow, and the diurnal pattern shown in Figure 3-6 may return.

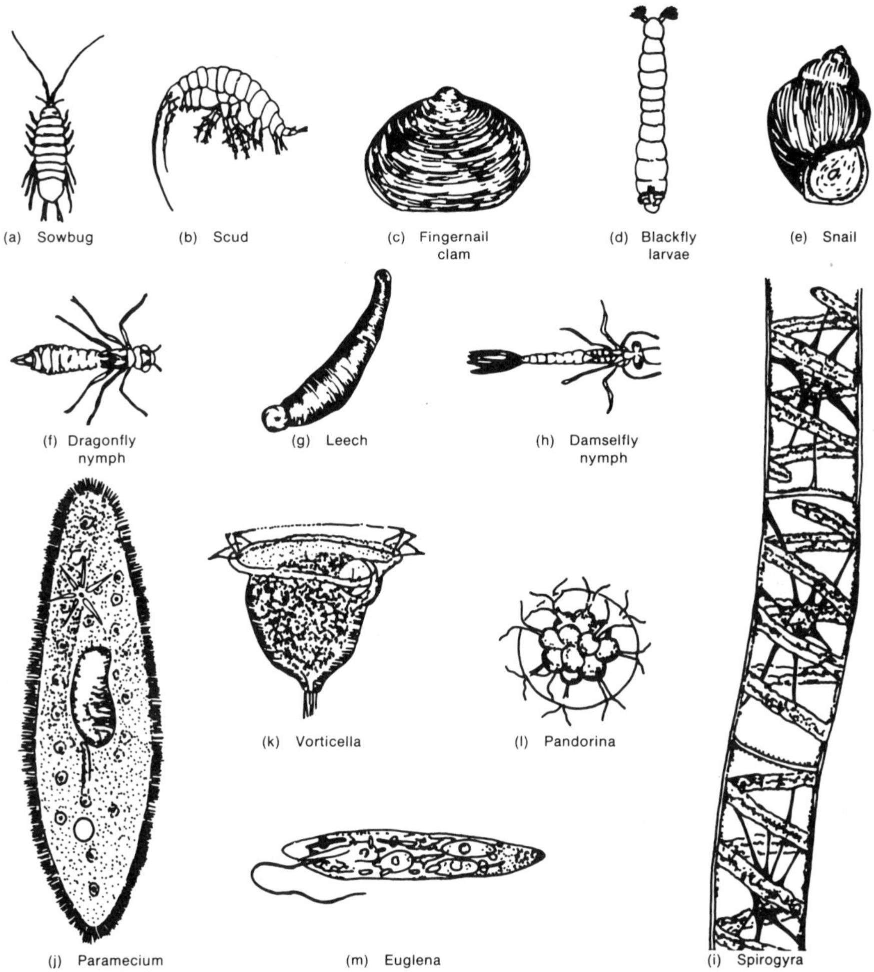

Figure 3-10 Species of intermediate tolerance: (*a–h*), animals; (*i–m*), algae and protozoa

The density–diversity pattern is manifested by fish species, macroinvertebrates, bottom animals, and microorganisms such as algae and protozoa. A study of Figure 3-8 and support Figures 3-9 to 3-11 should give insight into the changes that occur at various dissolved oxygen levels. The individual groups can be used to conduct aquatic bioassays of streams in order to assess the state of pollution of a given location in the stream. The results obtained are only semiquantitative, however, and are subject to differing interpretations.

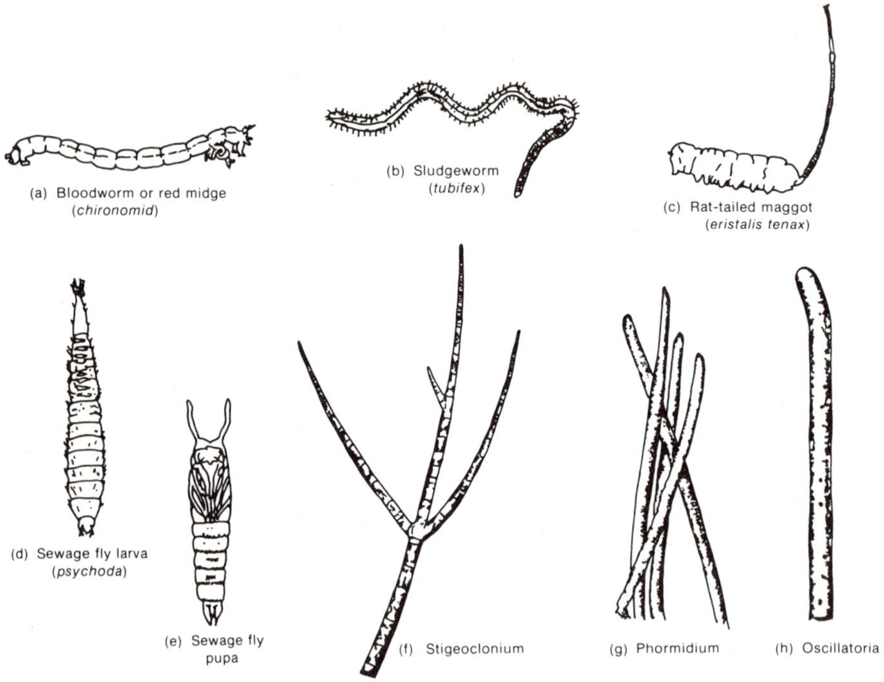

Figure 3-11 Polluted water species: (*a–e*), animals; (*f–h*), algae

APPLICATION OF MODELS

In the past, when regulations were based on stream water quality standards, models were used to estimate the stream's natural assimilative capacity and to apportion this capacity among the various waste discharges, in effect determining the effluent limits for each discharge. Since the present regulations are based on technological capabilities, as will be discussed in Chapter 4, this is no longer a prime use of models.

Models are presently used in areawide planning and in assessing the effects of locating new facilities in river basins (6). They are also used in water quality limiting situations in which dischargers may be required to do something beyond conventional treatment in order to assure stream water quality. In these critical situations, the water quality models are used to predict the effects of different discharges. The models can also be used to guide sampling programs by predicting where the most severe effects of a discharge are likely to occur.

These models are computer based, using the equations previously discussed or modifications of them. The stream is divided into sequential segments and the equation is applied to each segment in turn, while making sure that the model predictions match at the segment boundaries.

Several state regulatory agencies, as well as the EPA, have developed stream modeling packages, some of which have been evaluated by the Institute of Paper Chemistry and the National Council of the Paper Industry for Air and Stream Improvement (NCASI) (7–10). There is some controversy over how well these models actually reflect reality, because often the model predictions have not agreed with actual stream sampling data.

REFERENCES

1 J. Moor and E. Moor, *Environmental Chemistry*, Academic Press, New York, 1976, pp. 355–357.

2 G. T. Chabenoglaris (Metcalf & Eddy, Inc.), *Wastewater Engineering: Treatment, Disposal, Reuse*, McGraw-Hill, New York, 1979.

3 *Report of the Engineering Board of Review*, Part III, Appendix I, Sanitary District of Chicago, 1925.

4 G. M. Fair, E. W. Moore, and H. A. Thomas, Jr., "The natural purification of river muds and pollutional sediments," *Sewage Works Journal*, 13:270, 756 (1941).

5 H. W. Streeter and E. B. Phelps, *A study of the pollution and natural purification of the Ohio River*, Bulletin No. 146, U.S. Public Health Service, Washington, D.C., 1925.

6 *Use of mathematical modeling in the development of Areawide 208 Plans*, Technical Bulletin No. 326, National Council of the Paper Industry for Air and Stream Improvement, New York, 1979.

7 *A review of water quality mathematical models and considerations in their application to pulp and paper mill receiving water quality studies*, Technical Bulletin No. 297, National Council of the Paper Industry for Air and Stream Improvement, New York, 1977.

8 *A review of mathematical water quality model DOSAG and guidance for its use*, Technical Bulletin No. 327, National Council of the Paper Industry for Air and Stream Improvement, New York, 1978.

9 *A review of the mathematical water quality model QUALIE and guidance for its use*, Technical Bulletin No. 331, National Council of the Paper Industry for Air and Stream Improvement, New York, 1980.

10 *The mathematical water quality model QUAL II and guidance for its use—revised version*, Technical Bulletin No. 391, National Council of the Paper Industry for Air and Stream Improvement, New York, 1982.

4

Water Pollution Legislation—Past, Present, and Future

HISTORY

The first attempt, in an indirect way, to control pollution was the passage of the River and Harbor Act on March 3, 1899. This act was aimed mainly at keeping navigable rivers and harbors free enough of debris for safe navigation. The necessity for such a law tells us something about the state of some harbors at that time.

The real concern about water quality, as we understand the meaning today, had its origins in the late nineteenth century, when typhoid fever reached epidemic proportions. Typhoid was found to be a water-borne disease, and the discharge of human wastes into water supplies was a possible cause. Thus water quality became an important health issue, and in 1912 the U.S. Congress passed a Federal Water Pollution Control Act. This act mandated that the U.S. Public Health Service (PHS) research human health problems associated with water pollution. The PHS was authorized to provide technical information to the states, and a Water Pollution Control Division was established within the PHS. The Federal Water Pollution Control Act established the precedent that the individual states had responsibility for water pollution control efforts.

Until the 1940s, federal agency attention to water pollution was confined to technical research and analysis. In 1948 the Federal Water Pollution Control Act was amended to include the right of the federal government to control pollution as well as to study the problem. The major provisions of the act were:

1. A policy that the states had primary responsibility in resolving water pollution
2. Federal grants to state programs
3. Authority for interstate compacts
4. Construction loans to municipalities

Although the act contained loan provisions, money was never appropriated until the act was reauthorized in 1956.

In 1956 the Federal Water Pollution Control Act was again passed, with amendments in 1965, 1966, 1970, and 1972. The act of 1956 provided for the study of pollution problems and for matching grants for local sewage plants. The year 1956 marked the entry of the federal government into the pollution control field. At that time, the states were directed to prepare stream quality standards.

By 1970, the federal government had developed a strong desire to enforce pollution control but did not have a recent law that contained punishment provisions. Consequently, the River and Harbor Act of 1899 was resurrected to institute a permit system. In December 1970, by executive decree, the President directed that a federal permit program be set up under Section 13 of the River and Harbor Act. The permits had to be applied for before July 1, 1971, and most members of the pulp and paper industry complied. Only a handful of federal permits were issued, but this decree did signify for the first time a change in approach from receiving water quality standards to end-of-the-pipe effluent limitations. Therefore, the 1899 act became another amendment to the Federal Water Pollution Control Act (1). The River and Harbor Act of 1899 had several limitations, and subsequently fell into disuse because of several legal battles over lack of definition of "navigable waterways" and the need to establish environmental impact statements at each outfall.

The 1970 amendments to the Federal Water Pollution Control Act were named the National Environmental Policy Act (NEPA) and the Environmental Quality Improvement Act. They were significant in two respects. First, they instituted the concept of environmental impact statements. These statements document the environmental consequences of any new construction or modification, and such construction must be approved by state and federal agencies before it can be done. The second aspect of note is that the act established an environment that led the President to establish by consent decree the Environmental Protection Agency (EPA). These two aspects of the 1970 amendments are all that remain as part of the current legislation. The 1970 amendments correspond to the beginning of a time of intense environmental awareness in the United States.

ENVIRONMENTAL AGENCIES

The key federal group responsible for the environment is the EPA. The EPA is a regulatory agency. It must establish and enforce environmental standards. Establishment of standards is a central responsibility of the EPA in its pollution control effort.

Research and monitoring are also principal areas of concern for the agency. It is the purpose of the EPA to conduct research and suggest solution to environmented problems. At the same time, it has the obligation to monitor and analyze the environment. The aim is to determine cause-and-effect relationships so that control techniques can be established.

Providing technical and financial assistance is another important aspect of EPA activity. There are 10 regional offices of the EPA throughout the United States.

Each office is autonomous in that it has certain authority to deal with local matters, but each is under federal control and is directly answerable to the administrator of the EPA. EPA funds are available to independent organizations and to state and local governments to develop new technology.

Personnel development is another EPA responsibility. The agency has the task of develping the most highly skilled personnel possible to analyze and solve environmental problems. Many educational programs are available at various EPA learning centers throughout the United States.

A support group established to advise the EPA administrator on effluent limitations, water quality information, and toxic substances is the Effluent Standards and Water Quality Information Advisory Committee (ES&WQIAC). The nine-member committee is selected from the scientific community, and not from government or industry, to serve as an unbiased review group.

The Council on Environmental Quality is another federal agency with important environmental protection activities. The council consists of three members appointed by the President with the consent of the U.S. Senate. Its duty is to formulate and recommend to the President national policies that will promote the improvement of environmental quality. Trends are to be studied for the preparation of an annual environmental quality report to Congress.

A special congressional advisory group, the National Commission on Water Quality, was established to suggest possible revisions to water quality legislation, specifically PL 92–500, which will be discussed in the next section.

The key agency with responsibility for the protection of the oceans is the National Oceanic and Atmospheric Administration (NOAA). The Department of the Interior also performs several important environmental functions. At one time it was responsible for nearly all federal environmental protection operations, as well as for the supervision of the use of public lands. The Army Corps of Engineers and the Department of Agriculture are also involved with environmental matters and should be listed with this group.

FEDERAL WATER POLLUTION CONTROL ACT

The Federal Water Pollution Control Act Amendments, PL 92–500, were established on October 18, 1972. This law has been found to be free of most of the shortcomings of previous federal legislation. Before 1972, the states had allowed very liberal pollution laws in order to attract industry and commerce. This situation led to serious damage in many of the nation's rivers and lakes. The objective of PL 92–500 was to restore and maintain the chemical, physical, and biological integrity of the nation's waters. It has as its ultimate goal the elimination of all discharge of pollutants by 1985. The law's interim goal is to achieve the water quality that provides for the protection and propagation of fish, shellfish, and wildlife and for recreation in and on the water by July 1, 1983.

This legislation was unique in that for the first time it contained the power to impose substantial penalties for polluting. A polluter may be fined up to \$25,000 per day and/or receive a jail sentence of up to 1 year.

Title I of the act is a declaration of goals and policies. The goals are as stated above. The policies that it began were:

1 To designate the EPA as administrator of the law
2 To ban flow augmentation as an alternative to waste treatment
3 To award grants for pollution research.

Three areas were designated for research grants:

1 Sewer and sewage treatment
2 Advanced treatment technology
3 Industrial research

Title II of the law deals with grants for the construction of treatment works. Eight billion dollars was appropriated for this purpose; representing three-fourths of the total funding for the bill. The federal government will pay for 75% of the construction cost provided two conditions are met—that alternative waste management techniques and alternative reclamation and recycling techniques have been examined.

Section 208 deals with areawide planning. States were required to split their territories into areas and establish areawide planning commissions to handle land use planning, growth control planning, and identification and control of non-point-source discharge. Nonpoint sources must adhere to best managemnt practices criteria in order to achieve water quality goals.

Title III concerns standards and enforcement. Section 301 sets up a schedule for compliance in which by July 1, 1977, the Best Practical Control Technology Currently Available (BPCTCA) was to be installed and by July 1, 1983, the Best Available Treatment Economically Achievable (BATEA) was to be implemented. These will be discussed in depth later.

Title IV covers permits and licenses. It gives details on the National Pollutant Discharge Elimination System (NPDES) permits for any point source discharge. This section also defines the liability for oil and other hazardous substances, for which the discharger is liable for damage and clean-up costs.

Title V sets forth general provisions and describes a procedure by which private citizens can bring suit against polluters.

The component of the legislation that most affects our industry is the effluent limitation guidelines that define minimum effluent conditions for 1977 and 1983. The NPDES permit that is issued will probably be at these minimums and may be even more restrictive if local water quality conditions so dictate. In the development of the guidelines, the EPA had a consulting firm collect data on the indsutry's raw waste loadings, current state of internal and external control technology, and new technology. This information was then used to prepare a development document in which BPCTCA and BATEA were defined and used to formulate the guidelines. The guidelines were then issued as interim guidelines that are subject to a public hearing review procedure before final guidelines are issued. The EPA has issued two development documents: EPA 440/1-74-025-a in May 1974 (2), covering the unbleached Kraft and semichemical pulp segment of the paper industry, and EPA 440/1-76/047-A in January 1976 (3), covering

Table 4-1 BPCTCA Effluent Limitations in kg/kkg (lb/ton)

	Maximum 30-Day Average		Maximum/Day	
Subcategory	BOD_5	TSS	BOD_5	TSS
Dissolving Kraft	13.35 (26.7)	19.3 (38.6)	25.65 (51.3)	35.85 (71.7)
Market Kraft	7.90 (15.8)	15.85 (31.7)	15.20 (30.4)	29.40 (58.8)
BCT Kraft	6.95 (13.9)	15.10 (30.2)	13.35 (26.7)	28.05 (56.1)
Fine Kraft	5.70 (11.4)	12.40 (24.8)	10.95 (21.9)	23.00 (46.0)
Papergrade sulfite	19.60 (39.2)	24.00 (48.0)	37.60 (75.2)	44.60 (89.2)
Market sulfite	20.85 (41.7)	26.65 (53.3)	40.00 (80.0)	49.50 (99.0)
Low alpha dissolving sulfite	22.35 (44.7)	27.40 (54.8)	42.90 (85.8)	50.85 (101.7)
High alpha dissolving sulfite	26.30 (52.6)	33.65 (67.3)	52.30 (104.6)	62.50 (125.0)
GW: chemimechanical	7.05 (14.1)	10.45 (20.9)	13.50 (27.0)	19.45 (38.9)
GW: thermomechanical	5.00 (10.0)	9.20 (18.4)	9.60 (19.2)	17.05 (34.1)
GW: CMN papers	4.45 (8.9)	7.90 (15.8)	8.55 (17.1)	14.70 (29.4)
GW: fine papers	4.00 (8.0)	7.30 (14.6)	7.70 (15.4)	13.50 (27.0)
Soda	7.20 (14.4)	13.40 (26.8)	13.85 (27.7)	24.80 (49.7)
Deink	9.45 (18.9)	14.20 (28.4)	18.15 (36.3)	26.35 (52.7)
NI fine papers	4.25 (8.5)	5.90 (11.8)	8.20 (16.4)	11.00 (22.0)
NI tissue papers	6.25 (12.5)	5.00 (10.0)	11.40 (22.8)	10.25 (20.5)
NI tissue papers (FWP)	6.40 (12.8)	9.45 (18.9)	12.30 (24.6)	17.60 (35.2)

pH for all subcategories shall be within the range 5.0–9.0

Zinc[a]

Subcategory	Maximum 30-Day Average, kg/kkg (lb/ton)		Maximum/Day, kg/kkg (lb/ton)	
GW: chemimechanical	0.060	(0.12)	0.120	(0.24)
GW: thermomechanical	0.050	(0.10)	0.105	(0.21)
GW: CMN papers	0.050	(0.10)	0.105	(0.21)
GW: fine papers	0.048	(0.096)	0.095	(0.19)

[a]Applicable only to mills using zinc hydrosulfite.

Table 4-2 BPCTCA Effluent Limitations in kg/kkg (lb/ton)

	BOD_5				TSS			
Subcategory	30 Day		Maximum/day		30 Day		Maximum/Day	
Unbleached Kraft	2.80	(5.6)	5.6	(11.2)	6.00	(12.0)	12.0	(24.0)
NSSC-ammonia	4.00	(8.0)	8.0	(16.0)	5.00	(10.0)	10.0	(20.0)
NSSC-sodium	4.35	(8.7)	8.7	(17.4)	5.50	(11.0)	11.0	(22.0)
Unbleached Kraft-NSSC	4.00	(8.0)	8.0	(16.0)	6.25	(12.5)	12.5	(25.0)
Paperboard from waste paper	1.50	(3.0)	3.0	(6.0)	2.50	(5.0)	5.0	(10.0)
pH for all subcategories shall be within the range 6.0–9.0								

the bleached Kraft, groundwood, sulfite, soda, deinking, and nonintegrated paper mill segment of the paper industry.

The effluent limitations guidelines for BPCTCA, as stated in these two development documents, are presented in Tables 4-1 and 4-2. These guidelines reflect the installation of bar screens, primary clarification, and secondary treatment for most subcategories except the nonintegrated tissue papers, for which only primary treatment is suggested. Limitations are set for BOD_5, TSS, pH, and zinc.

The effluent limitations guidelines for BATEA are given in Tables 4-3 and 4-4. This technology level suggests major internal mill improvements, biological waste treatment, and physicochemical waste treatment as necessary treatment and control technologies. Limitations are provided for BOD_5, TSS, pH, zinc, and color.

The effluent limitations for new plants are given in Tables 4-5 and 4-6 as New Source Performance Standards (NSPS) and reflect internal improvements that can be achieved through effective design and layout of mill operations. The basic treatment and control processes that are suggested as a means of meeting these performance standards are similar to those identified for existing mills by 1983. Standards are set for BOD_5, TSS, pH, and zinc.

Other development documents related to the pulp and paper industry have also been issued by the EPA (4–7).

The industry's reaction to the guidelines was adverse, and legal suits were brought against the EPA under the direction of the American Paper Institute (API). All cases have been unsuccessful, and the guidelines stand as proposed.

A shortcoming of PL 92-500 was the brief way it dealt with toxic substances. This problem was corrected in June 1976 when the EPA signed a consent decree as a result of a court action that required the agency to establish revised effluent limitations and standards regulating the discharge of up to 65 classes of pollutants. These settlement agreement pollutants are listed in Table 4-7, and regulations were promulgated during 1980 covering these toxic pollutants. The

Table 4-3 BATEA Effluent Limitations in kg/kkg (lb/ton)

Subcategory	Maximum 30-Day Average BOD$_5$	Maximum 30-Day Average TSS	Maximum/Day BOD$_5$	Maximum/Day TSS
Dissolving Kraft	5.80 (11.6)	3.95 (7.9)	11.15 (22.3)	7.35 (14.7)
Market Kraft	3.35 (7.1)	2.60 (5.2)	6.80 (13.6)	4.80 (9.6)
BCT Kraft	3.00 (6.0)	2.05 (4.1)	5.80 (11.6)	3.80 (7.6)
Fine Kraft	2.55 (5.1)	1.75 (3.5)	4.95 (9.9)	3.26 (6.5)
Papergrade sulfite	8.90 (17.8)	3.00 (6.0)	17.10 (43.2)	5.60 (11.2)
Market sulfite	10.05 (20.1)	3.45 (6.9)	19.30 (38.6)	6.40 (12.8)
Low alpha dissolving sulfite	11.40 (22.8)	3.35 (6.7)	21.90 (43.8)	6.25 (12.5)
High alpha dissolving sulfite	13.80 (27.6)	4.75 (9.5)	26.50 (53.0)	8.80 (17.6)
GW: chemimechanical	3.90 (7.8)	1.65 (3.3)	3.15 (6.3)	3.10 (6.2)
GW: thermomechanical	2.10 (4.2)	1.45 (2.9)	4.05 (8.1)	2.65 (5.3)
GW: CMN papers	1.85 (3.7)	1.45 (2.9)	3.55 (7.1)	2.65 (5.3)
GW: fine papers	1.75 (3.5)	1.35 (2.7)	3.35 (6.7)	2.55 (5.1)
Soda	2.55 (5.1)	1.75 (3.5)	4.95 (9.9)	3.25 (6.5)
Deink	2.60 (5.2)	2.70 (5.4)	5.05 (10.01)	4.95 (9.9)
NI fine papers	1.35 (2.7)	0.70 (1.4)	2.60 (5.2)	1.30 (2.6)
NI tissue papers	2.15 (4.3)	1.10 (2.2)	4.15 (8.3)	2.05 (4.1)
NI tissue papers (FWP)	1.90 (3.8)	1.00 (2.0)	3.70 (7.4)	1.85 (3.7)

pH for all subcategories shall be within the range 5.0–9.0

Color

Subcategory	Maximum 30-Day Average, kg/kkg (lb/ton)		Maximum/Day, kg/kkg (lb/ton)	
Dissolving Kraft	125	(250)	250	(500)
Market Kraft	95.0	(190)	190	(380)
BCT Kraft	65.0	(130)	130	(260)
Fine Kraft	65.0	(130)	130	(260)
Soda	65.0	(130)	130	(260)

Zinc[a]

Subcategory	Maximum 30 Day Average, kg/kkg (lb/ton)		Maximum/Day, kg/kkg (lb/ton)	
GW: chemimechanical	0.0480	(0.096)	0.095	(0.19)
GW: thermomechanical	0.0415	(0.083)	0.085	(0.17)
GW: CMN Papers	0.0395	(0.079)	0.085	(0.17)
GW: fine Papers	0.0415	(0.083)	0.075	(0.15)

[a]Applicable only to mills using zinc hydrosulfite.

Table 4-4 BATEA Effluent Limitations in kg/kkg (lb/ton)

Subcategory	BOD_5		TSS		Color	
	30 Day	Maximum/Day	30 Day	Maximum/Day	30 Day	Maximum/Day
Unbleached Kraft	1.35 (2.7)	2.7 (5.4)	1.85 (3.7)	3.7 (7.4)	10 (20)	15 (30)
NSSC-ammonia	3.20 (6.4)	6.4 (12.8)	2.60 (5.2)	5.2 (10.4)	75% removal	
NSSC-sodium	2.25 (4.5)	4.5 (9.0)	2.50 (5.0)	5.0 (10.0)	75% removal	
Unbleached Kraft NSSC	1.60 (3.2)	3.2 (6.4)	2.10 (4.2)	4.2 (8.4)	12.5 (25)	18.75 (37.5)
Paperboard from waste paper	0.65 (1.3)	1.3 (2.6)	0.80 (1.6)	1.6 (3.2)		

pH for all subcategories shall be within the range 6.0–9.0

Table 4-5 New Source Performance Standard in kg/kkg (lb/ton)

Subcategory	Maximum 30-Day Average		Maximum/Day	
	BOD_5	TSS	BOD_5	TSS
Dissolving Kraft	6.10 (12.2)	8.35 (16.7)	11.75 (23.5)	15.50 (31.0)
Market Kraft	2.65 (5.3)	2.90 (5.8)	5.15 (10.3)	5.35 (10.7)
BCT Kraft	3.70 (7.4)	5.00 (10.0)	7.05 (14.1)	9.30 (18.6)
Fine Kraft	2.55 (5.1)	3.75 (7.5)	4.95 (9.9)	7.00 (14.0)
Papergrade sulfite	4.65 (9.3)	2.90 (5.8)	8.95 (17.9)	5.35 (10.7)
Market sulfite	4.65 (9.3)	2.90 (5.8)	8.95 (17.9)	5.35 (10.7)
Low alpha dissolving sulfite	11.15 (22.3)	10.0 (20.0)	21.45 (42.9)	18.6 (37.2)
High alpha dissolving sulfite	13.80 (27.6)	9.45 (18.9)	26.50 (53.0)	17.60 (35.2)
GW: chemimechnical	3.90 (7.8)	3.30 (6.6)	7.50 (15.0)	6.15 (12.3)
GW: thermomechanical	2.30 (4.6)	3.15 (6.3)	4.45 (8.9)	5.85 (11.7)
GW: CMN papers	2.00 (4.0)	3.15 (6.3)	3.85 (7.7)	5.85 (11.7)
GW: fine papers	1.90 (3.8)	3.00 (6.0)	5.60 (7.4)	5.60 (11.2)
Soda	3.15 (6.3)	4.30 (8.6)	6.00 (12.0)	7.95 (15.9)
Deink	3.90 (7.8)	4.00 (8.0)	7.50 (15.0)	7.45 (14.9)
NI fine papers	1.35 (2.7)	1.40 (2.8)	2.60 (5.2)	2.60 (5.2)
NI tissue papers	2.15 (4.4)	2.20 (4.4)	4.15 (8.3)	4.10 (8.2)
NI tissue papers (FWP)	1.90 (3.8)	1.95 (3.9)	3.70 (7.4)	3.65 (7.3)

pH for all subcategories shall be within the range 5.0–9.0

Zinc[a]

Subcategory	Maximum 30 Day Average, kg/kkg (lb/ton)		Maximum/Day kg/kkg (lb/ton)	
GW: chemimechanical	0.048	(0.096)	0.095	(0.19)
GW: thermomechanical	0.0455	(0.091)	0.09	(0.18)
GW: CMN papers	0.0455	(0.091)	(0.09	(0.18)
GW: fine papers	0.044	(0.088)	0.09	(0.18)

[a]Applicable only to mills using zinc hydrosulfite.

toxic substances are regulated under Section 307A. Industry was given until 1984 to comply with these regulations.

In the next set of BATEA guidelines the EPA looks at a total of 140 pollutants, 129 of which are classified as toxic. Of these 140, 114 are organic pollutants, 13 are metals, and 13 are resins or fatty acids. For specific information on the regulation of toxic materials, the appropriate issue of the *Federal Register* should be consulted.

Table 4-6 New Source Performance Standards in kg/kkg (lb/ton)[a]

Subcategory	BOD_5				TSS				Color			
	30 Day		Maximum/Day		30 Day		Maximum/Dy		30 Day		Maximum/Day	
Unbleached Kraft	1.55	(3.1)	3.1	(6.2)	3.75	(7.5)	7.5	(15.0)	10	(20)	15	(30)
NSSC-ammonia	3.75	(7.5)	7.5	(15.0)	3.75	(7.5)	7.5	(15.0)	—		—	
NSSC-sodium	2.60	(5.2)	5.2	(10.4)	3.85	(7.7)	7.7	(15.4)	—		—	
Unbleached Kraft-NSSC	1.90	(3.8)	3.8	(7.6)	4.00	(8.0)	8.0	(16.0)	12.5	(25)	18.75	(37.5)
Paperboard from waste paper	0.75	(1.5)	1.5	(3.0)	2.00	(4.0)	4.0	(8.0)	—		—	
pH for all subcategories shall be within the range of 6.0–9.0												

[a] The maximum average of daily values for any consecutive 30-day period should not exceed the 30-day standards shown above. The maximum for any one day should not exceed the daily maximum standards shown above.

Table 4-7 Settlement Agreement Toxic Pollutants

Acenaphthene	Haloethers
Acrolein	Halomethanes
Acrylonitrile	Heptachlor and metabolites
Aldrin/dieldrin	Hexachlorobutadiene
Antimony and compounds	Hexachlorocyclohexane
Arsenic and compounds	Hexachlorocyclopentadiene
Asbestos	
	Isophorone
Benzene	
Benzidine	Lead and compounds
Beryllium and compounds	
	Mercury and compounds
Cadmium and compounds	
Carbon tetrachloride	Napthalene
Chlordane	Nickel and compounds
Chlorinated benzenes	Nitrobenzene
Chlorinated ethanes	Nitrophenols
Chloralkyl ethers	Nitrosamines
Chlorinated napthalene	
Chlorinated phenols	Pentachlorophenol
Chloroform	Phenol
2-Chlorophenol	Phthalate esters
Chromium and compounds	Polychlorinated biphenyls
Copper and compounds	Polynuclear aromatic hydrocarbons
Cyanides	
	Selenium and compounds
DDT and metabolites	Silver and compounds
Dichlorobenzene	
Dichlorobenzidine	2, 3, 7, 8-Tetrachlorodibenzo-p-dioxin
Dichloroethylenes	Tetrachlorethylene
2, 4-Dichlorophenol	Thallium and compounds
Dichloropropane and dichloropropene	Toluene
2, 4-Dimethylphenol	Toxaphene
Dinitrotoluene	Trichloroethylene
Diphenylhydrazine	
	Vinyl chloride
Endosulfan and matabolites	
Endrin	Zinc and compounds
Ethylbenzene	

The Clean Water Act that was passed in 1977 revised and updated PL 92-500. It advanced the deadline for BPCTCA to January 1, 1979, provided a particular industry had acted in good faith in attempting to meet the previous deadline of July 1, 1977. The BATEA guidelines were divided into three subcategories:

1 Priority (toxic) pollutants
2 Conventional pollutants such as BOD_5, TSS, and pH
3 Nonconventional pollutants, which is a catch-all category that includes color

The Clean Water Act also allows waver of the BATEA guidelines for conventional pollutants if the costs of meeting the guidelines exceed the associated benefits. Thus this legislation introduced the concept of cost–benefit analysis to environmental legislation.

The November 18, 1982, issue of the *Federal Register* contained the latest version of the Federal Water Pollution Control Act, which took effect on January 31, 1983. Its structure is basically no different from that of PL 92-500, but some specific limitations have changed. It established limits for conventional pollutants (BOD_5, TSS, pH) and priority pollutants (zinc, pentachlorophenol, trichlorophenol). It did not regulate other priority pollutants because there were no reasonable data upon which to promulgate such regulations. The regulations also did not include any limits on nonconventional pollutants because, for example, in the case of color there was no uniform national concern; thus color was left to the states to regulate on a case-by-case basis. For ammonia the cost impact was unjustifiable, and for resin and fatty acids the data were not sufficient for meaningful regulations.

The law contains limits for all pulp and paper manufacturing categories and covers Best Practical Control Technology Currently Available (BPT), Best Available Technology Economically Achievable (BAT), Performance Standards New Source (PSNS), and Pretreatment Standards for Existing Municipal Facilities (PSES). These standards, where duplicated, are very similar to those given in Tables 4-3 to 4-6. Because these recent minor changes are considered less important than the data from the time of enactment of PL 92-500, a benchmark piece of legislation that reflected a whole new approach to the regulation of pollution, the original standards are given in these four tables and the reader is referred to the November 18, 1982, issue of the *Federal Register* if the current data are required.

TOXIC SUBSTANCES LEGISLATION

In the 1980s, toxic substances have become a major issue of concern. There are several pieces of legislation that deal with the regulation and management of toxic materials, such as the Safe Drinking Water Act, the Resource Conservation and Recovery Act of 1976, the Comprehensive Environmental Response,

Compensation, and liability Act, the Federal Insecticide, Fungicide, and Rodenticide Act, and the Toxic Substances Control Act (8). Each of these acts will be discussed briefly regarding their implications for the paper industry.

Safe Drinking Water Act

The Safe Drinking Water Act was passed by Congress in 1974 in an attempt to safeguard public drinking water supplies by providing both primary and secondary standards. The primary standards deal with health factors involving drinking water and the secondary standards with its aesthetic qualities. The act defines a public water supply as any system that serves 25 or more individuals. Therefore, many mill drinking water systems, which are often private wells, would qualify for regulation under this act.

The act sets the acceptable level of nitrates at 10 mg/1 or less. The maximum level of organic and inorganic contamination is specified. The level of turbidity should be 1 JTU (Jackson turbidity unit) or less. A waiver for turbidity up to 5 JTU may be granted if:

1 Turbidity does not interfere with disinfection.
2 Turbidity does not hinder the persistence of disinfection.
3 Turbidity does not interfere with microbiological determinations.

A set of complex microbiological criteria based on the coliform test must also be met, which are too involved to deal with in this text.

The frequency with which drinking water must be tested for these various potential contaminants is defined by state agency regulations and varies from state to state. Such information must be obtained from state EPA or equivalent agencies. If a drinking water system fails to meet the requirements, an employer is mandated by law to so inform those employees served by the system.

Contamination of drinking water supplies by toxic chlororganics is of major concern, and such compounds are covered under the organics regulations of the Safe Drinking Water Act.

Resources Conservation and Recovery Act (RCRA)

Historically, solid waste was the last type of pollution to be regulated. It was not considered at all until the passage of the Solid Waste Disposal Act of 1965, which only provided funds for research and for the demonstration of new and improved methods of disposal. It took the Resource Conservation and Recovery Act of 1976 to effectively regulate the land disposal of waste. RCRA does more than just regulate land disposal, however; it controls the disposal of all solid wastes. In the terms of this act, solid wastes are "any garbage, refuse, sludge from a waste treatment plant, water supply treatment plant or air pollution control facility, and other discarded material, including solid, liquid, semisolid or contained gaseous materials resulting from industrial, commercial, mining or agriculture activities and from community activities unless such is specifically excluded from the regulations." This definition is very comprehensive and is meant to include all forms of potentially toxic materials, including liquids and gases.

The objectives of the law can be stated simply as protection of health and the environment and conservation of valueable resources. The law is adminstered by the Office of Solid Wastes within the EPA.

The critical section of RCRA is Subtitle C, which regulates hazardous wastes. They are defined as "solid waste or combination of solid wastes which because of its quantity, concentration, or physical, chemical or infectious characteristics may a) cause or significantly contribute to an increase in mortality or an increase in serious irreversible or incapacitating reversible illness or b) pose a substantial present or potential hazard to human health or the environment when improperly treated, stored, transported or disposed of, or otherwise managed." The EPA publishes a list of hazardous wastes and criteria for determining their characteristics, which are to be revised periodically. The list first appeared in the *Federal Register* of May 19, 1982. As might be expected, this is a very controversial list.

Generators of hazardous wastes are required to keep records of amounts and types and to use appropriate storage containers. This and other information must be provided to those who transport, store, and dispose of hazardous wastes. Similar records must be kept by those who store, transport, and dispose of such wastes. Treatment, storage, and disposal facilities must obtain an appropriate permit from the EPA in order to operate.

Discharges that are regulated under the NPDES system of the Clean Water Act are exempt from RCRA requirements. The RCRA requirements do not apply to plants producing less than 1,000 kg per month of combined total hazardous wastes. Such plants are considered small generators. This exemption does not apply, however, if the plant generates acutely toxic wastes. The exemption limit for acutely toxic waste is 1 kg/month.

The RCRA regulations have been designed to be consistent with the Hazardous Materials Transportation Act administered by the Department of Transportation.

Comprehensive Environmental Response, Compensation, and Liability Act

The comprehensive Environmental Response, Compensation, and Liability Act was passed at the end of 1980 and has been nicknamed the "superfund." The act has three main purposes:

1 To require the reporting of the release of hazardous substances into the environment
2 To compile a list of all sites at which hazardous wastes have been disposed
3 To provide funds to be used for the cleanup of dangerous waste disposal sites

This act was designed to correct an already severe problem with disposal sites that existed in 1980. It covers release to all segments of the environment—air, water, and land—and all forms of hazardous substances.

The action that brings the act into force is the release of a reportable quantity of toxic and/or hazardous waste. A "reportable quantity" is defined as 1 pound unless specified otherwise in other regulations. The facility that released the material must report the release to the National Response Center and must notify parties potentially injured by the release. This usually means publication in a local paper. Any person in charge of a facility from which a reportable quantity of a hazardous substance is released who fails to adhere to the reporting requirements is subject to a $10,000 fine and to imprisonment for up to 1 year. The act imposes stringent record-keeping requirements and mandates that such records be retained for 50 years. Any destruction or alteration of such records carried a possible fine of $20,000 or imprisonment of up to 1 year or both.

Damages that are recoverable under the provisions of the act include the cost of removal or remedial action by government personnel; the cost of the response incurred by others; damages for injury of material resources; and costs incurred in assessing the state of injury. The act does not impose liability for personal injury, medical claims, property damage, loss of marine harvest, or a number of other types of claims. However, some of these types of claims can be brought under other existing state and federal laws. The act contains a liability limit of $50,000,000 for damages.

One of the main purposes of the superfund was to create two pools of money to deal with past and present toxic waste disposal sites. The Hazardous Substances Response Trust Fund is generated from taxes on petroleum and chemicals to be imposed until September 30, 1985, or until the fund reaches $1,380,000,000. These funds are to be used to clean up the existing problems at various sites resulting from poor disposal practices in the past. The second fund created by the act is a Past Closure Liability Trust Fund, to assume liability for damages due to any releases from permitted hazardous waste disposal facilities. This fund is financed by a tax of $2.13 per dry weight ton of hazardous waste received at permitted disposal sites that will remain at the site after site closure. Materials that will be disposed of by incineration are therefore exempt.

Federal Insecticide, Fungicide, and Rodenticide Act (FIFRA)

FIFRA originated in 1947 to register pesticides moving in interstate commerce. The basic statute was substantially amended in 1972, 1975, and 1978, and still deals mainly with the manufacture and transport of pesticides. The slimicides used in the paper industry qualify under this act. All new pesticides proposed for use in the United States must be registered and must meet certain safety and efficacy criteria. The material must be certified for specific applications. Registration may be canceled if any question of safety to humans or to the environment ever arises.

The act applies to all pesticides used in the United States, whether or not manufactured there. It does not apply, however, to pesticides destined for export. Toxic materials not considered pesticides are regulated under TSCA.

Toxic Substances Control Act (TSCA)

TSCA, passed in 1976, established an extensive test program for both existing and new chemicals to assess their effects on human health or on the environment. It was designed to fill any gaps existing in other acts (Clean Water Act, Clean Air Act, FIFRA, RCRA) regarding toxic materials. It is intended mainly to affect producers of chemicals. Its impact is secondary on pulp and paper mills, because they do not directly manufacture toxic chemicals for sale, but mills do use chemicals regulated under TSCA. Chemical divisions of paper companies are directly affected.

TSCA requires the EPA to publish a chemical substances inventory. In 1983 this list consisted of 59,224 substances (9), and new substances are being added at a phenomenal rate. From this inventory a priority list of the 50 most dangerous chemicals was compiled. This priority list is a subject of considerable controversy.

Some of the tests required are for carcinogenicity, mutagenicity, teratogenicity, behavior modification, synergism, and toxicity. The cost of these tests for any specific chemical is substantial.

As of 1984, the EPA's major focus under TSCA was on PCBs, asbestos, and chlorofluorocarbons (CFCs). However, implementation of the act is still in a developmental stage.

REFERENCES

1 *Federal Water Pollution Control Act Amendments of 1972*, Conference Report, 92nd Congress, 2nd Session, House of Representatives Report No. 92-1456, U.S. Government Printing Office, Washington, 1972

2 *Development Document for Effluent Limitations Guidelines and New Source Performance Standards for the Unbleached Kraft and Semi-chemical Pulp Segment of the Pulp, Paper, and Paperboard Mills Point Source Category*, EPA U.S. Environmental Protection Agency, Washington, D.C. 440/1-74-025-a, May 1974.

3 *Development Document for Interim Final and Proposed Effluent Limitations Guidelines and Proposed New Source Performance Standards for the Bleached Kraft, Groundwood, Sulfite, Soda, Deink, and Non-Integrated Paper Mills Segment of the Pulp, Paper and Paperboard Mill Point Source Category*, Vols 1 and 2, EPA 440/1-76/047-A, U.S. Environmental Protection Agency, Washington, D.C. January 1976.

4 *Development Document for Effluent Limitations Guidelines for the Bleached Kraft, Groundwood, Sulfite, Soda, Deink, and Non-Integrated Paper Mills Segment of the Pulp, Paper, and Paperboard Point Source Category*, EPA 440/1-76/047-b, U.S. Environmental Protection Agency, Washington, D.C. December 1976.

5 *Final Development Document for Effluent Limitations Guidelines, New Source Performance Standards, and Pretreatment Standards for the Pulp, Paper, and Paperboard and the Builders' Paper and Board Mills Point Source Categories*, EPA 440/1-82/025, U.S. Environmental Protection Agency, Washington, D.C. October 1982.

6 *Development Document for Effluent Limitations Guidelines and New Source Performance Standards for the Builders' Paper and Roofing Felt Segment of the Builders' Paper and Board Mills Point Source Category*, EPA 440/1-74-026a, U.S. Environmental Protection Agency, Washington, D.C. May 1974.

7 *Development Document for Proposed Effluent Limitations Guidelines and Standards for Control of Polychlorinated Biphenyls in the Deink Subcategory of the Pulp, Paper, and Paperboard Point Source Category*, U.S. Environmental Protection Agency, Washington, D.C., October 1982.

8 Eugene T. Holmes, *Introduction to Environmental Law for pulp and Paper Managers*, Tappi Press, Atlanta, 1982.

9 *RTECS; Registry of Toxic Effects of Chemical Substances*, 1981–1982 ed., 3 vols. (Rodger L. Tatken and Richard J. Lewis, Sr., eds.), DHSS (NIOSH) Publication No. 83–107, U.S. Government Printing Office, Washington, D.C., June 1983.

5

The Process Modification Approach versus External Treatment

In meeting the requirements imposed by law, two alternative methods are used—modification of the process to make it nonpolluting or application of add-on external treatment technologies. To meet the 1979 requirements of BPCTCA, only external treatment was required; this method was chosen by most mills. The mill owners believed that, although the in-process route existed, it was nonproven technology and therefore of higher risk. The industry spends in excess of $500 millions per year to operate and maintain existing treatment facilities. To meet the requirements of BATEA, a combination of internal modification and external treatment will probably be required. The development of nonpolluting papermaking processes is one of the engineering challenges facing the industry in the 1980s.

COST OPTIMIZATION

Paper manufacturers must evaluate their positions with regard to wastewater treatment and energy consumption, and then adopt policies that make the best long-term economic sense. For existing and new mills, the best available internal and external control technologies will be required to meet the proposed EPA effluent standards. Figure 5-1 illustrates the type of capital and operating cost optimization that will be required to meet the BATEA EPA standards at various pulp and paper production facilities. As shown, there exists some optimum combination of in-plant control technology and external treatment to minimize the environmentally related capital and operating costs. Only careful examination of the alternative options in internal and external effluent control will permit optimum solution of the capital and operating cost function illustrated in Figure 5-1.

Exactly where the optimum lies is a question that must be answered on an individual mill basis, but the industry as a whole has probably not moved as far as it could in process water conservation. At present the industry is in the process of examining potential areas for increased water conservation. Among the factors giving impetus to greater water economy have been:

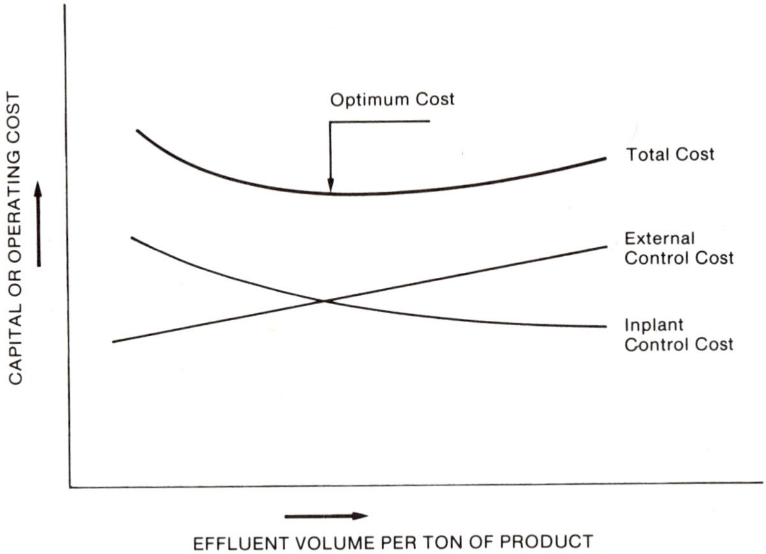

Figure 5-1 Treatment cost optimization

1 Reduced process losses of fiber, fines, and filler materials
2 Reduced cost for purchase or processing of fresh water
3 Reduced cost for heating of process streams

Heating costs assume even greater prominence in an era of increasing energy conservation efforts. It has been suggested (1) that the cost of production can be described by the following relationship:

$$\text{Cost} = k\left[\frac{\begin{matrix}\text{annual production} \\ \text{(tons)}\end{matrix} \times \begin{matrix}\text{flow/ton} \\ (\text{m}^3/\text{ton})\end{matrix} \times \begin{matrix}\text{raw waste load} \\ \text{(kg/ton)}\end{matrix}}{\begin{matrix}\text{final effluent requirements} \\ \text{(kg/ton)}\end{matrix}}\right]^{\alpha}$$

in which k is a constant and α is approximately two-thirds. This formula, based on factors that will be of increasing importance in the future, illustrates the importance of developing more efficient, less polluting processes.

Five parameters will be useful in evaluating future process development:

1 Water use per ton of product
2 Net energy consumption per ton of product
3 Residuals generated per ton of product
4 Number and complexity of the unit operations involved
5 The extent to which the use of exotic construction materials is required to counter corrosion

It may be possible with innovation to meet the BATEA requirements by the process modification route; certainly, the groundwork for such a step is being laid.

The largest single source of wastewater in a fully integrated operation is the paper machine. Process modifications that are possible in this area will be discussed in Chapter 7. The perspective to be adopted is that of a total water reuse scheme for a nonintegrated operation, but the concepts and principles apply equally well to the paper machine stock preparation area of an integrated operation. The largest source of BOD load is the bleach plant, which will be discussed in Chapter 13.

DISCHARGE HISTORY OF THE PULP AND PAPER INDUSTRY

The paper industry has a long history of reducing process water consumption per unit of production. For instance, in the southern Kraft industry, water usage in the manufacture of bleached Kraft pulp was reduced from 100,000 gallons per ton of production to approximately 40,000 gallons per ton over the period 1946–1956. More recent industrywide data for all categories suggest a decrease from the 1959 level of water consumption, which was 39,000 gallons per ton, to approximately 27,000 gallons per ton in 1972 (2). For nonintegrated fine paper manufacture in 1970 the average process water use was 36,207 gallons per ton, and by 1972 this was lowered to 23,834 gallons per ton (3); by 1976 this had decreased to 15,200 gallons per ton. For combination paperboard manufacture,

Table 5-1 Comparison of Average Water Consumption versus the Lowest for the 10 Major Production Categories

	1981 Average Consumption (gal/ton)	1980 Best Consumption (gal/ton)
Dissolving Kraft	44,200	35,500
Market bleached Kraft	35,000	17,000
BCT bleached Kraft	37,600	24,800
Alkaline fine paper	44,900	19,500
Unbleached Kraft	15,800	5,600
Unbleached Kraft and Semi Chemical	11,400	7,300
Dissolving sulfite	70,600	66,700
Papergrade sulfite	51,400	9,300
Miscellaneous integrated	27,500	5,500
Nonintegrated fine paper	18,400	5,600

the average discharge went from 8,700 gallons per ton in 1970 to 7,100 gallons per ton in 1972 to 7,000 gallons per ton in 1976. Nonintegrated tissue manufacturers reported an average discharge of 21,600 gallons per ton in 1972, which was decreased to 14,480 gallons per ton by 1975 (4).

Results of a recent survey (5) of paper industry water use are shown in Table 5-1 for various production categories. This table shows both average water use and the lowest reported use for each category.

There are examples in the industry of nonintegrated mills with zero effluent discharge in the manufacture of combination paperboard or tissue. This trend is continuing at an accelerating rate due to two factors: environmental legislation and rising energy costs.

REFERENCES

1 P. E. Wrist, *Tappi*, 61 (11): 41 (1982).

2 D. Marshall, A. Springer, and I. Gellman, "The relation between water use, economy and waste water management for environmental quality protection in the paper industry," *Proceedings of the WWEMA Industrial Pollution Conference and Exposition*, Houston, Jexus, March 30–31, 1973.

3 A. Springer, D. Marshall, and I. Gellman, *The relation between process water quality characteristics and its reuse potential in fine paper mills*, Technical Bulletin No. 287, National Council of the Paper Industry for Air and Stream Improvement, New York, 1976.

4 A. Springer, D. Marshall, and I. Gellman, *The relation between process water quality characteristics and its reuse potential in non-integrated manufacture of tissue and toweling*, Technical Bulletin No. 289, National Council of the Paper Industry for Air and Stream Improvement, New York, 1976.

5 J. C. Wyvell, J. C. Adams, G. Valentine, and T. Shelnutt, *An assessment of the potential for water reuse in the pulp and paper industry*, Final Report RU-84-1, Bureau of Reclamation, U.S. Department of the Interior, Washington, D.C., March 1984.

6

Basics of Sampling and Conducting Mill Pollution Load Surveys

In this chapter we will summarize two of the key components of a mill pollution load survey: monitoring flow rates and conducting a reasonable sampling program.

SAMPLING BASICS

The basis for the pollution abatement program in any plant or mill depends upon the information that is obtained from some form of sampling program. If a poor procedure is used in obtaining a sample or if the sample is not taken in the proper place, there is no possibility of obtaining accurate information on which to base a decision.

In obtaining good samples, three criteria must be met (1):

1 The sample must be representative of the waste stream
2 Proper sampling technique must be followed in obtaining the sample
3 The sample must be properly preserved or protected from degradation until it is analyzed

Obtaining a representative sample (the first criterion) is not simple due to fluctuations in the flow rate and composition of the water stream. The velocity distribution is also variable in a sewer, so it is important to sample where the flow is well mixed. The outlet of a well-mixed tank is an excellent sampling point, as is a turbulent section of a straight sewer stream. The normal method used to obtain a representative sample, despite variations in flow and waste composition, is to take a series of samples and either analyze them separately or to mix them to form a composite sample for analysis. Automatic composite samplers will be discussed later, but in general the greater the frequency of sampling for the composite, the more accurate the result.

Batch processes are difficult to sample representatively because their discharges appear in slug form. A batch system must be sampled frequently over the entire discharge cycle to provide a representative composite sample. Alternatively, a series of flow proportional samples can be taken and combined in order to characterize the process.

After taking whatever precautions are necessary to ensure a representative sample, proper sampling technique must be used. Proper technique varies with the type of sample and will be discussed in detail in the next section. Some obvious general precautions are to avoid contamination from dirty sample bottles and to mark each sample properly. The sample label should contain such information as (1) the location where the sample was taken, (2) the date and time of collection, (3) the type of sample (that is, grab or composite), and (4) a notation of information that may change before laboratory tests can be made, such as temperature, pH, or appearance. It is surprising how often proper labeling is neglected and the amount of wasted effort this causes.

TYPES OF SAMPLES

Basically, there are two types of samples—grab and composite. A composite sample is made up of a series of grab samples mixed together. Usually but not always, composite samples are preferred. For instance, if the waste stream is at constant flow and does not vary in composition, a grab sample provides all necessary information. Alternatively, composite samples obscure the extremes of composition that can be provided by a series of grab samples; these extremes are necessary if the variation in composition over time of the waste stream must be determined.

Grab Samples

For grab samples, the quantity of sample taken must be related to the tests to be performed. Table 6-1 lists the volume of sample required to perform various tests. The required sample volume is the sum of the volumes required for all tests to be performed; double this amount should be taken in case of questionable test results that require repetition of one or more tests. A common sample size is 1 l. The sample is usually collected in a wide-mouth bottle to minimize collection time and to make the sample more representative of a specific instant in time.

Composite Samples

In order to minimize the number of samples to be analyzed, it is often desirable to mix several grab samples together to form a composite sample. The total amount of composite sample required is also related to the number and types of tests to be performed, but in general the minimum amount should not be less than 2 l and the minimum size of each grab sample making up the composite

Table 6-1 Volume of Sample Required for Determination of the Various Constituents of Industrial Water

Test	Volume of Sample (ml)[a]
Physical Tests	
Color and odor	100–500[b]
Corrosivity	Flowing sample[b]
Electrical conductivity	100[b]
pH, electrometric	100[b]
Radioactivity	100–1,000
Specific gravity	100[b]
Temperature	Flowing sample[b]
Toxicity	1,000–20,000[b]
Turbidity	100–1,000[b]
Chemical tests	
Dissolved gases	
Ammonia, NH_3	500[c]
Carbon dioxide, free CO_2	200[c]
Chlorine, free Cl_2	200[c]
Hydrogen, H_2	1,000[c]
Hydrogen sulfide, H_2S	500[c]
Oxygen, O_2	500–1,000[c]
Sulfur dioxide, free SO_2	100[c]
Miscellaneous	
Acidity and alkalinity	100
Bacteria, iron	500
Bacteria, sulfite-reducing	100
Biochemical oxygen demand	100–500
Carbon dioxide, total CO_2 (including CO_3^{2-}, HCO_3^-, and free)	200
Chemical oxygen demand (dichromate)	50–100
Chlorine requirement	2,000–4,000
Chlorine, total residual Cl_2 (including OCl^-, HOCl, NH_2Cl, $NHCl_2$, and free)	200
Chloroform—extractable matter	1,000
Detergents	100–200
Hardness	50–100
Hydrazine	50–100
Microorganisms	100–200
Volatile and filming amines	500–1,000
Oily matter	3,000–5,000
Organic nitrogen	500–1,000
Phenolic compounds	800–4,000
pH, colorimetric	10–20
Polyphosphates	100–200
Silica	50–1,000
Solids, dissolved	100–20,000
Solids, suspended	50–1,000
Tannin and lignin	100–200

Table 6-1 *(Cont'd.)*

Test	Volume of Sample (ml)[a]
Cations	
Aluminum Al^{3+}	100–1,000
Ammonium, NH_4^+	500[c]
Antimony, Sb^{3+} to Sb^{5+}	100–1,000
Arsenic, As^{3+} to As^{5+}	100–1,000
Barium, Ba^{2+}	100–1,000
Cadmium, Cd^{2+}	100–1,000
Calcium, Ca^{2+}	100–1,000
Chromium, Cr^{3+} to Cr^{6+}	100–1,000
Copper, Cu^{2+}	200–4,000
Iron, Fe^{2+} and Fe^{3+}	100–1,000[c]
Lead, Pb^{2+}	100–4,000
Magnesium Mg^{2+}	100–1,000
Manganese, Mn^{2+} to Mn^{7+}	100–1,000
Mercury, Hg^+ and Hg^{2+}	100–1,000
Potassium, K^+	100–1,000
Nickel, Ni^{2+}	100–1,000
Silver, Ag^+	100–1,000
Sodium, Na^+	100–1,000
Strontium, Sr^{2+}	100–1,000
Tin, Sn^{2+} and Sn^{4+}	100–1,000
Zinc, Zn^{2+}	100–1,000
Anions	
Bicarbonate, HCO_3	100–200
Bromide, Br^-	100
Carbonate, CO_3^-	100–200
Chloride, Cl^-	25–100
Cyanide, Cn^-	25–100
Fluoride, Fl^-	200
Hydroxide, OH^-	50–100
Iodide, I^-	100
Nitrate, NO_3^-	10–100
Nitrite, NO_2^-	50–100
Phosphate, ortho, PO_4^{2-}, HPO_4^{2-}, $H_2PO_4^-$	50–100
Sulfate, SO_4^{2-}, HSO_4^-	100–1,000
Sulfide, S^{2-}, HS^-	100–500
Sulfite, SO_3^{2-}, HSO_3^-	50–100

[a] Volume specified in this table should be considered as a guide for the approximate quantity of sample necessary for the particular analysis. The exact quantity used should be consistent with the volume prescribed in the standard method of analysis whenever the volume is specified.

[b] Aliquot may be used for other determinations.

[c] Samples for unstable constituents must be obtained in separate containers, preserved as prescribed, completely filled, and sealed against all exposure.

should be about 200 ml for sampling intervals of about 1 hour and 25 ml for intervals of 3–5 minutes.

Either time or flow can be used as the basis for composite sampling. When time is the basis, a fixed volume of sample is drawn at specific time intervals. Time-based composite samples are more common. When flow is the basis, a fixed volume of sample is drawn whenever a certain volume of flow has passed the sampling point. For automatic sampling, this type of apparatus is more difficult to construct, since it must be capable of monitoring the flow past the sampler. Flow measurement techniques will be discussed later in this chapter.

The length of time over which a sample can be composited is limited by the

Table 6-2 10 gal Composite Sampler Characteristics

1	Capability for AC-DC operation with adequate dry battery energy storage for 120-hour operation at 1-hour sampling intervals
2	Suitable for suspension in a standard manhole while providing access for inspection and sample removal
3	Total weight including batteries under 18 kg (40 lb)
4	Sample collection interval adjustable from 10 minutes to 4 hours
5	Capability for flow-proportional and time-composite samples
6	Capable of collecting a single 9.5 l (2.5 gal) sample and/or collecting 500 ml (0.13 gal) discrete samples in a minimum of 24 containers
7	Capability for multiplexing repeated aliquots into discrete bottles
8	One intake hose with a minimum internal diameter (ID) of 0.64 cm (0.25 in.) and a weighted, streamlined intake screen that will prevent accumulation of solids
9	Intake hose liquid velocity adjustable from 0.61 to 3.0 m/sec (2.0–10.0 fps) with dial setting
10	Minimum lift of 6.1 m (20 ft)
11	Explosion proof
12	Watertight exterior case to protect components in the event of rain or submersion
13	Exterior case capable of being locked, and with lugs for attaching steel cable to prevent tampering and provide some security
14	No metal parts in contact with the waste source or samples
15	An integral sample container compartment capable of maintaining samples at 4–6°C for a period of 24 hours at ambient temperatures upto 38°C
16	With the exception of the intake hose, capable of operating in a temperature range between − 10 and 40°C
17	Purge cycle before and after each collection interval, and a sensing mechanism to purge in the event of plugging during sample collection and then collect a complete sample
18	Capable of being repaired in the field

Table 6-3 Inventory of Automatic Composite Wastewater Samplers

Sampler	Power Supply	Type of Sample	Type of Pump	Intake Tube ID (mm)[a]	Liquid Intake Velocity (cm/sec)[b]	Purge Cycle
Sigmamotor WA-2	AC	Time	Peristaltic	3.17	7.9	No
Sigmamotor WD-2	AC-DC	Time	Peristaltic	3.17	7.9	No
Brailsford EV-1	AC-DC	Time or flow	Vacuum pump	4.76	0.45	No
Brailsford DU-1	DC	Time or flow	Piston	4.76	0.45	No
Brailsford EP-1	DC	Time	Piston	4.76	0.45	No
Hants Mark 3B	Manual vacuum	Time	Manual vacuum	6.35	75[c]	No
ISCO 1391-X	AC-DC	Time or flow	Peristaltic	6.35	21	Yes
ISCO 1392	AC-DC	Time or flow	Peristaltic	6.35	61	Yes
Sirco MKVS7	AC-DC	Time or flow	Piston	9.52	98	Yes
Pro-Tech CG-125P	Gas	Time or flow	Gas lift	3.17	207	Yes
QCEC CVE	AC	Time	Piston	6.35	61–152	Yes
N-Con Scout	DC	Time	Peristaltic	6.35	7.6	Yes
N-Con Surveyor	AC	Time or flow	Impeller	12.70	36	Gravity
N-Con Sentinel	AC	Time or flow	Optional	NA	Variable	NA

Source: Reference 2.
[a] Multiply by 0.0394 to obtain inches.
[b] Multiply by 0.0328 to obtain feet per second.
[c] Mean
[d] NA, not applicable.

time the sample can be stored without changing the characteristics of the waste to be analyzed.

Composite sampling is usually done with an automated sampler. Continuous samplers are marketed commercially, and must be carefully examined to ensure that they are suitable for the characteristics to be sampled and the sampling location. The ideal characteristics of a composite sampler (2) are listed in Table 6-2.

No sampler yet devised meets all of these criteria. Some available automatic composite samplers are listed in Table 6-3.

In some cases, it is possible and even desirable to construct a composite sampler rather than buy it. A simple one for open sewers consists of a contoured scoop mounted on a rotating shaft. The scoop dips into the sewer and collects a sample; then, as the shaft rotates, a hose connection to the center of the shaft makes the sample flow out. This type of sampler, sometimes called a "calm shell sampler," can be made to be flow proportional by adjusting the shape of the scoop. Its sampling rate is controlled by the rate of rotation of the shaft. The scoop and drain can be made large to avoid plugging. The clam shell sampler is inexpensive and relatively trouble free.

Some measurements for which composite samples are commonly collected are listed in Table 6-4.

Composite samplers are not 100% reliable. A multiple sampler study (2) indicated an average overall failure rate of approximately 16%. The major cause of sampler malfunction is plugging of intake lines and is thus related to the solids content of the waste stream being sampled.

It is necessary to choose the frequency at which to set the sampler. This frequency is dependent on two factors: the variability of the waste stream and the principal test to be performed on the sample. Table 6-5 provides a rough guide for selecting the compositing schedule. Compositing samples over a period

Table 6-4 Common Wastewater Parameters for Which Composite Samples Might Be Considered

Specific conductance
Five-day biochemical oxygen demand
Chemical oxygen demand
Total suspended solids
Ammonia nitrogen
Total Kjeldahl nitrogen
Nitrite-nitrate nitrogen
Total phosphorus
High-temperature coliform
Toxic substances

Table 6-5 Suggested Sampling or Compositing Schedule

Characteristic	High Variability	Low Variability
BOD[a]	4 hours	12 hours
COD or TOC[a]	2 hours	8 hours
Suspended solids	8 hours	24 hours
Alkalinity or acidity	1 hour grab	8 hour grab
pH	Continuous	4 hour grab
Nitrogen and phosphorus[b]	24 hours	24 hours
Heavy metals	4 hours	24 hours

Source: Reference 3.

[a] The compositing schedule used when continuous samplers are not employed depends on variability, that is, 15 minutes for high variability to 1 hour for low variability.

[b] Does not apply to nitrogen or phosphorus wastes (e.g., fertilizer).

of less than 2 hours is seldom necessary because the sewer system exerts approximately this much of an equalizing effect.

SAMPLE PRESERVATION

After collection, a composite or grab sample must be properly stored and preserved or the best sampling program is made worthless. Preservation techniques employed include refrigeration, biocides, and radical pH adjustment. Refrigeration is the most commonly used technique because it alters the characteristics of the sample the least. The preservation technique chosen must be compatible with the test parameters to be monitored. Tables 6-6 and 6-7 list recommended storage or preservation procedures that are compatible with the various tests of interest. Also given in these tables is the maximum period of time the sample can be kept before testing if the test result is to be valid.

Table 6-6 Recommended Storage Procedures

	Sample Storage	
Analysis	Refrigeration at 4°C	Frozen
Total solids	OK	OK
Suspended solids	Up to several days	No
Volatile suspended solids	Up to several day	No
COD	Up to several days	OK
BOD	Up to 1 day in composite sampling systems	Lag develops; fresh sewage seed must be used

Source: Reference 1.

Table 6-7 Sample Preservation

Parameter	Preservative	Maximum Holding Period
Acidity-alkalinity	Refrigeration at 4°C	24 hours
Biochemical oxygen demand	Refrigeration at 4°C	6 hours
Calcium	None required	7 days
Chemical oxygen demand	2 ml H_2SO_4/l	7 days
Chloride	None required	7 days
Color	Refrigeration at 4°C	24 hours
Cyanide	NaOH to pH 10	24 hours
Dissolved oxygen	Determine on site	No holding
Fluoride	None required	7 days
Hardness	None required	7 days
Metals, total	5 ml HNO_3 per litre	6 months
Metals, dissolved	Filtrate: 3 ml 1:1 HNO_3 per litre	6 months
Metals, ammonia	40 mg $HgCl_2$[a]/l, 4°C	7 days
Nitrogen, Kjeldahl	40 mg $HgCl_2$[a]/l, 4°C	Unstable
Nitrogen, nitrate-nitrite	40 mg $HgCl_2$[a]/l, 4°C	7 days
Oil and grease	2 ml H_2SO_4/l, 4°C	24 days
Organic carbon	2 ml H_2SO_4/l (pH 2)	7 days
pH	Determine on site	No holding
Phenolics	1.0 g $CuSO_4$/1 + H_3PO_4 to pH 4.0, 4°C	24 hours
Phosphorous	40 mg $HgCl_2$[a]/l, 4°C	7 days
Solids	None available	7 days
Specific conductance	None required	7 days
Sulfate	Refrigeration at 4°C	7 days
Sulfide	2 ml zinc acetate/l	7 days
Threshold odor	Refrigeration at 4°C	7 days
Turbidity	None available	7 days

Source: Reference 4.

[a] Disposal of mercury-containing samples is a recognized problem; research investigations are underway to replace it as a preservative.

The reason for prompt application of preservation techniques is to prevent various changes that can occur over time in the sample. For instance, metal cations can precipitate as hydrides; the valance state of ions may change due to oxidation or reduction; or metal cations may be adsorbed on the surface of glass, plastic, or quartz containers. In addition to chemical reactions or physico-chemical changes, microbiological changes can occur in the sample. Cells may lyse to increase the BOD and COD, or the BOD and COD may be decreased by the conversion of organic matter to microbiological mass with conversion losses to CO_2 and H_2O. The generation or breakdown of microbiological mass can change the form of nitrogen and phosphorus found in the sample. As a general policy, air should be excluded from stored samples to minimize both chemical oxidation reactions and microbiological activity. Thus it can be seen that proper sample preservation is a very important consideration.

CHOICE OF SAMPLING SITES

Half of the problem of conducting a pollution load survey concerns the collection, preservation, and testing of the samples. The other half of the problem involved the selection of sampling sites and the determination of the flow rate at the various sites chosen.

To choose sampling sites, a flow diagram of the entire mill sewer system is needed as a basis. Care must be taken to determine that the diagram represents the current sewer configuration rather than a past situation. It is best to walk through the system with the diagram in hand in order to check that it is accurate or to note where changes have been made that are not represented on the diagram.

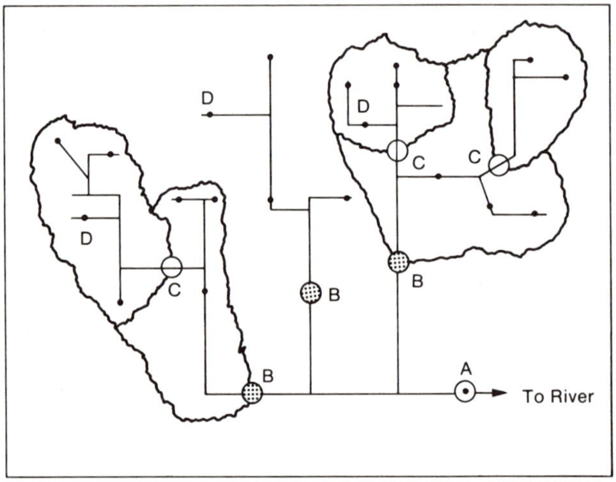

OPERATIONS	PERFORMED AT STATION A	B	C	D
Continuous Flow Recording	X	X	—	—
Recorded - pH, Cond., or Other Process Variable	X	X	X	—
Temporary Flow Meter	—	—	X	X
Composite Sampler	X	X	—	—
Grab Sampler	—	—	—	X

Figure 6-1 mill process sewer diagram. (Courtesy NCASI.)

Figure 6-1 is a schematic diagram of a portion of a mill process sewer system; on it are suggested locations for flow recording, composite samples, and grab samples. Point A represents the main sewer to the treatment plant or from the treatment plant to the river. Points B are sites on main trunk systems leading to the main sewer. In both locations A and B, it is desirable to record the flow continuously and to install composite samplers. The B locations are monitored so that, when an upset occurs, it is possible to determine which area of the mill is responsible. Locations C and D represent troublesome subsystems that may require temporary monitoring during difficulties; thus temporary flow meters and grab samples or monitoring of a specific variable is suggested.

MEASUREMENT OF FLOW

An important aspect of any mill effluent survey program is to determine the flow rates in the different mill sewers. Measurement of flow rate is covered in *Standard Methods* (6) and depends on the type of sewer in which the flow is measured, pipes or open channels.

Flow in Pipes

A commercially available magnetic flow meter is the lowest pressure drop device for permanent measurement of flow in pipes. Magnetic flow meters operate on the principle of Farraday's law, which stated that a magnetic field disturbed by the flow of a fluid through it generates a voltage proportional to the mass of the fluid. Magnetic flow meters are accurate and relatively trouble free; their one drawback is that they are expensive.

Venturi meters are the logical option for applications in which the cost of a magnetic flow meter cannot be justified and large pressure drop losses cannot be tolerated. For Reynolds numbers greater than 10,000, friction losses through a venturi meter are usually in the range of 10% of the available head. The coverging venturi section is designed to recover most of the head loss. The basic equation for a venturi meter is

$$V = C_v \sqrt{2g_c(\Delta P)/\rho(1 - B^4)} \tag{6-1}$$

where V is the velocity in the venturi throat, ΔP is the pressure drop upstream of the throat, usually monitored by a manometer, g_c is a gravitational conversion factor, ρ is density, B is the ratio of the venturi throat diameter to the inside diameter of the pipe, and C_v is the venturi coefficient to compensate for friction loss, which is 0.98 for Reynolds numbers greater than 10,000.

If there is excess pressure in a line and cost is a major consideration, the logical choice is an orifice meter. The same basic equation as stated above applies to an orifice meter, except that the venturi coefficient (C_v) is replaced by an orifice coefficient (C_o) that is a strong function of B and the Reynolds number. Perry's

Chemical Engineering Handbook (7) can be consulted for the detailed relations and also for a description of how best to construct and install an orifice meter. It is one of the most popular techniques used for monitoring flow in a pipe on a continuous basis.

In recent years, ultrasonic flow meters have been developed that operate by detecting the shifts in an ultrasonic energy field transmitted through the wall of the pipe caused by the flow through the pipe. These instruments are available in portable versions. They must be used with care because the ultrasonic energy often does not penetrate completely across the pipe, thus making them very sensitive to velocity profile effects, pipe size, and type of flow.

Rotometers are the instruments of choice for very small fluid flow rates in small pipes. The area available for the flow to pass varies with the height of the float, and float position is usually directly proportional to flow rate.

In situations in which there is no available place to put a flow meter or in which pipes are buried in concrete and therefore inaccessible, it may still be possible to measure the flow rate through the use of a tracer technique called the "salt dilution approach." In this technique, a small stream of known flow and concentration is fed into the stream whose flow rate is to be determined. Then, an appreciable distance downstream, where complete mixing can be assumed to have taken place, the concentration of the total stream is determined. A simple material balance provides the flow rate:

$$Q = C_T q/(C_F - C_b) \tag{6-2}$$

where Q is the large pipe flow rate, q is the flow rate of the added tracer, C_T is the concentration in the tracer stream, C_F is the final concentration in the large pipe, and C_b is the concentration in the blank, that is, in the large pipe upstream of the tracer addition point.

A variety of materials can be used as tracers, but dyes or salts are most popular. If a dye is used, its concentration is usually measured colorimetrically. If sodium chloride is used, flame photometry or differences in conductivity can be used to measure the concentration. A rotometer is commonly used to measure the flow of the tracer.

The salt dilution technique is not limited to the measurement of flow in inaccessible pipes but can be used to measure flow in any situation, including open channels.

Flow in Open Channels

Parshall Flume. The open-sewer equivalent of the venturi meter is the Parshallflume. Figure 6-2 shows the top and side view of a Parshall flume. This device is ideal for use in open sewers where continuous monitoring of flow is justified. It is often used on the final mill effluent from the plant. Parshall flumes have the advantages of causing minimum head loss, being self-scouring, having reasonably high accuracy, and being able to function over a wide range of flow

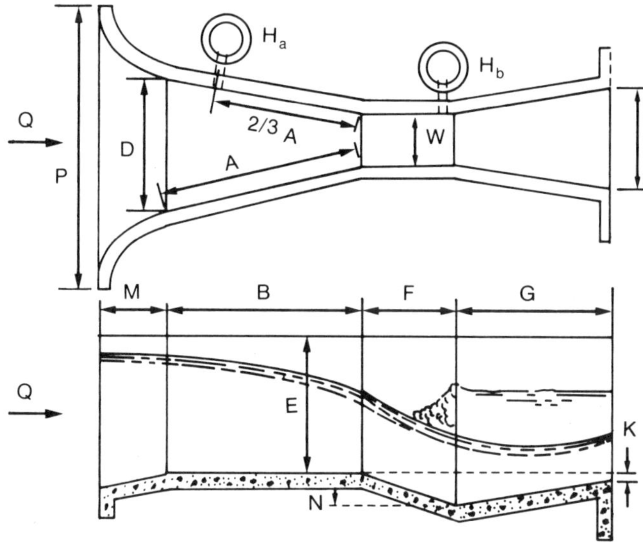

TYPICAL DIMENSIONS:

W	A	B	C	D	E	F	G	K	N	M	P	CAP-CFS
6"	2'-7" /16	2'-0"	1'-1 1/2"	1'-3 5/8"	2'-0"	1'-0"	1'-6"	3"	4 1/2"	1'-0"	2'-11 1/2"	0.05-3.9

APPROX. EQUATIONS

WIDTH	FORMULA
3"	$Q=0.992\ H_a^{1.54}$
6"	$Q=2.06\ H_a^{1.58}$
9"	$Q=3.07\ H_a^{1.53}$
12-24"	$Q=4W\ H_a^{1.52}$

Figure 6-2 Parshall flume. (Courtesy NCASI.)

rates. Figure 6-3 illustrates the last advantage, showing, for instance, that for a 3-in throat width the flume is capable of monitoring flows from 0.6 to 35 million gallons/day (MGD). For such a 3-in. flume, the flow rate is related to the hydraulic head by the relation

$$Q = 0.992H_a^{1.54} \tag{6-3}$$

The functional form is the same for other flume widths, but both the coefficient and the exponent vary, as given by the approximate formulas at the bottom of Figure 6-2. The general formula is

$$Q = 4WH^n \tag{6-4}$$

where Q is the discharge in cubic feet/second, W is the throat width in feet, H is the

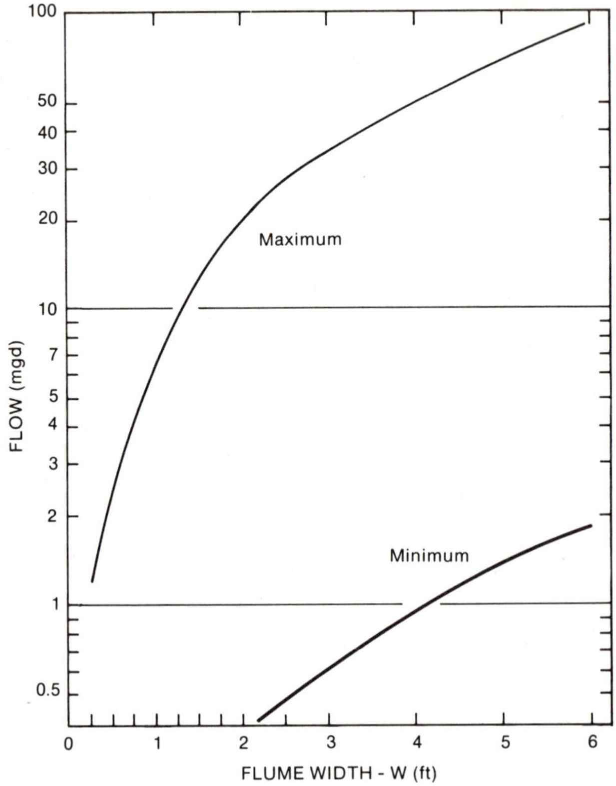

Figure 6-3 Parshall flume design limits. (Courtesy NCASI.)

head of water above level floor in the converging section in feet, and n is $1.522W^{0.026}$.

Palmer–Bowlus flume. A Palmer–Bowlus flume may be used when installation of a Parshall flume is not justified. Figure 6-4 shows the various shapes of Palmer–Bowlus flumes. Such flumes may be nothing more than a level section of floor placed into the sewer. The length of the floor should be approximately the same as the diameter of the conduit. This type of flume is easily installed in existing sewers, since no drop in head is needed. The following equation applies to the Palmer–Bowlus flume (8):

$$Q^2/g_c = A_c^3/b \tag{6-5}$$

$$V_c^2/2g_c = A_c/2b = d_c/2 \tag{6-6}$$

where A_c is the area at the critical depth in square feet, d_c is the critical depth in V_c is the critical velocity, and b is the width of the flume.

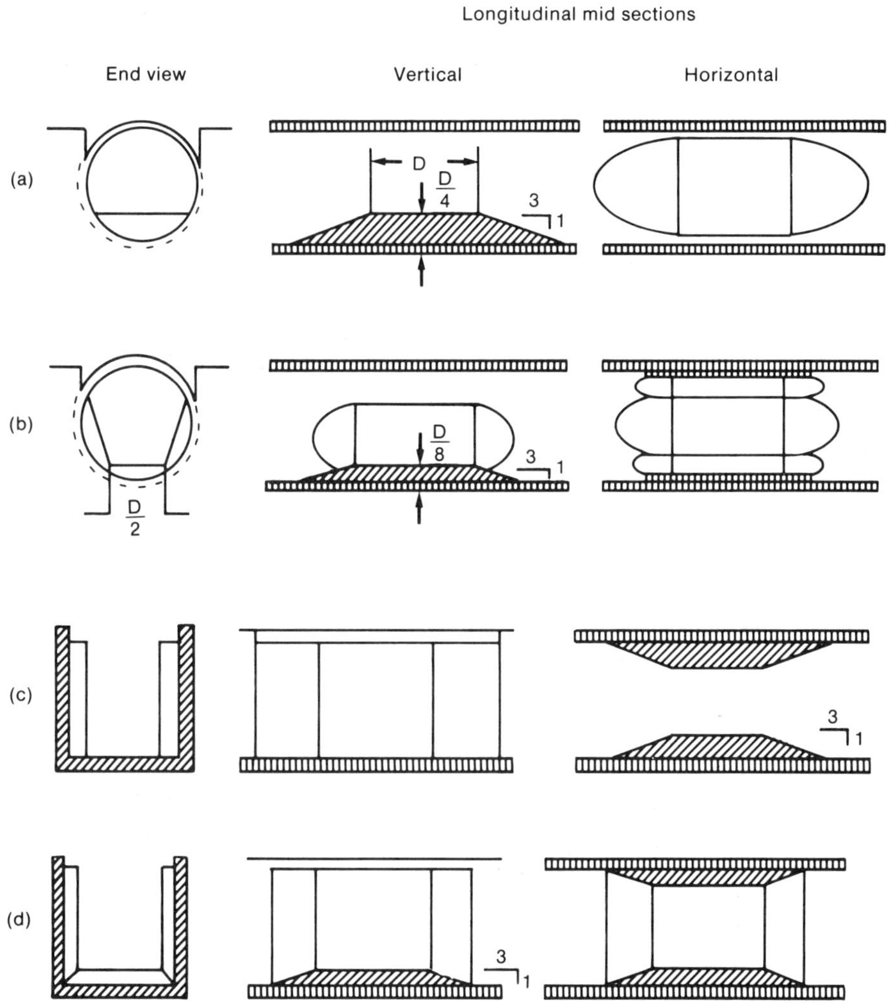

Figure 6-4 Palmer–Bowlus flumes

These flumes work quite well, and reasonable accuracy can be obtained at up to 0.95 of the sewer depth.

Weirs. The weir is the device most commonly used to measure open channel flow because it is low in cost and easy to install. A weir is essentially a specially shaped dam that is placed across a channel. Figure 6-5 shows some common types of sharp-crested weirs, and Figure 6-6 shows the profile of a sharp-crested weir and how it would be installed across a channel. Proper form of the crest is important to accurate measurement. The water flowing over the crest is called

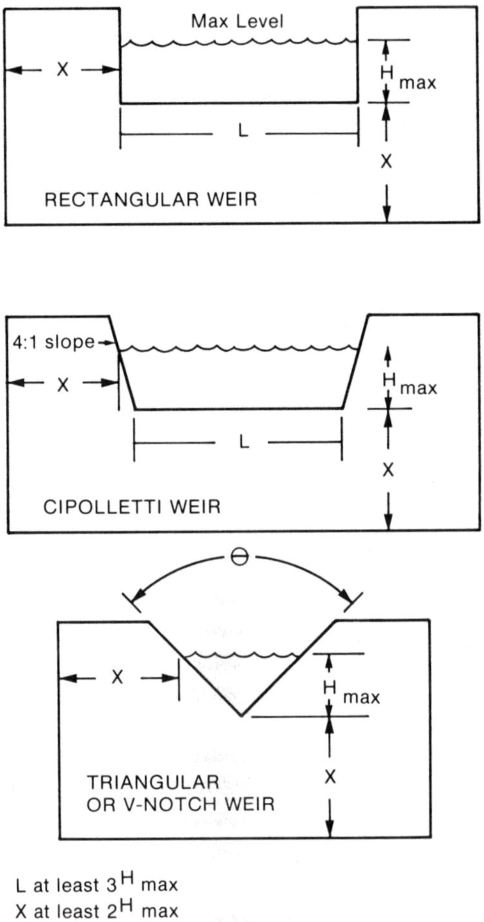

Figure 6-5 Sharp-crested weirs

the "nappe." The main problem with weirs is that they constrict the flow as it passes over the crest. The Cipolletti weir was designed to minimize this effect.

In order to obtain good flow measurement with a weir, the following general requirements should be met:

1 The weir should be installed perpendicular to the flow.
2 The weir should be exactly level to ensure uniform depth of flow.
3 The head should be measured at a distance upstream at least three times the depth of the weir crest in a quiet section away from all disturbances.
4 The weir should be located in a straight stretch of sewer with a low slope.
5 The approach velocity should be less than 0.3 ft/sec if possible and uniform in profile.

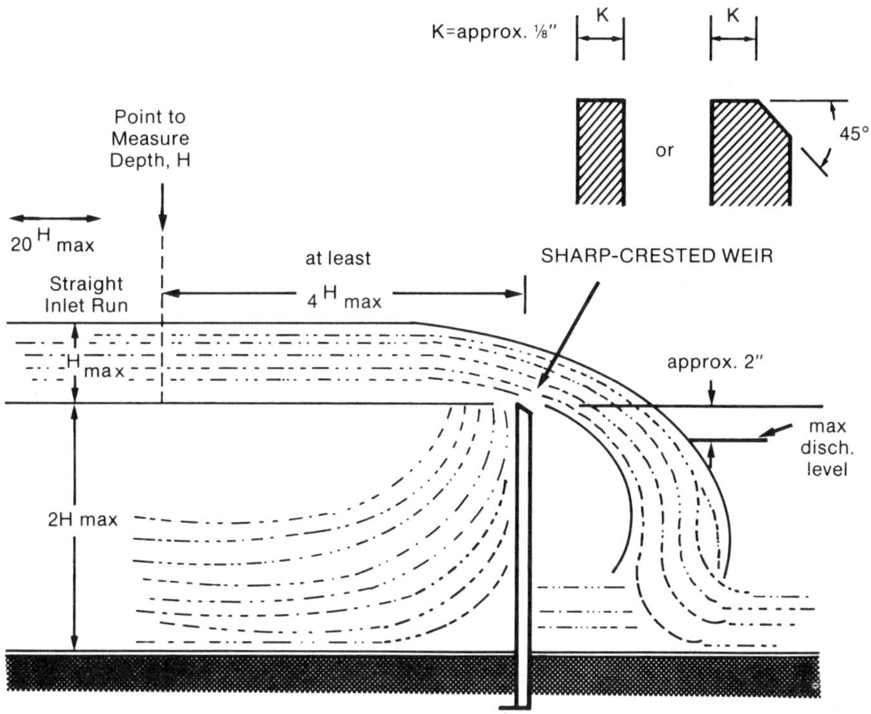

Figure 6-6 Profile a sharp-crested weir

6 The weir should be made of 1/8- to 1/4-in. plate material and the notch edge should be shaped to approximately a 45° angle, with the high portion of the bevel facing upstream. The sharp edge is important for preventing the nappe from adhering to the crest. The edge of the weir must be kept clean.
7 The approach channel should be clear of debris.
8 The pocket formed by the crest must be ventilated to prevent the formation of a vacuum.
9 The connection of the weir to the channel should be waterproof so that all flow passes over the weir.

One way to meet these requirements is to construct a weir box for insertion into the sewer. Figure 6-7 illustrates the important features of a weir box, such as the point from which to measure the head and the screen or baffle to give even flow distribution.

The flow rate is determined in a properly installed weir from the depth of the crest. The general formula is

$$Q = CH^{3/2} \tag{6-7}$$

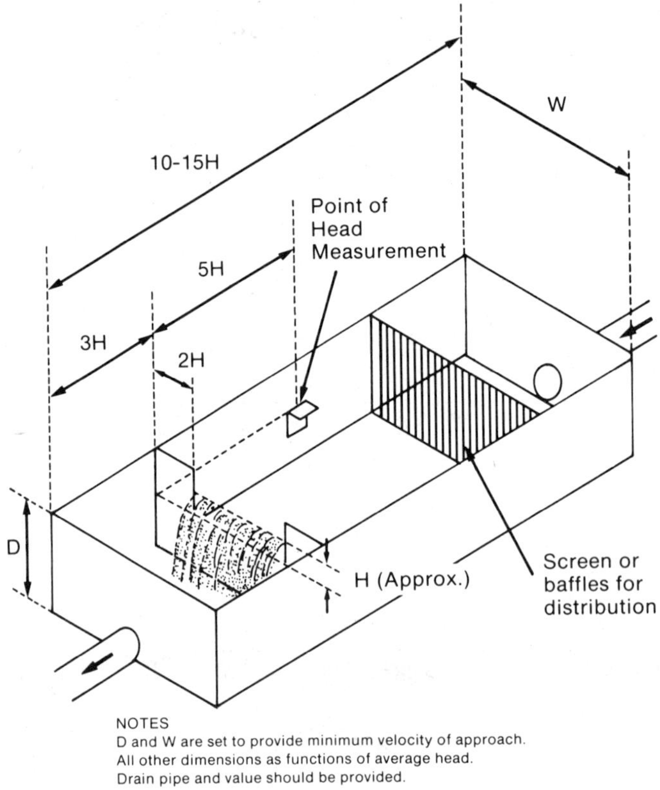

Figure 6-7 Weir box

where Q is the flow rate in cfs, H is the head above the crest in feet, and C is a coefficient that depends on the type of weir.

Rectangular Weirs. Rectangular weirs are the most common type of weir, useful in most applications. The rectangular weir does, however, have the problem of end contraction. For this weir, the flow rate can be calculated from the Francis formula

$$Q = 3.33LH^{3/2} \tag{6-8}$$

where Q is the flow in cfs, L is the effective width of the weir in feet, and H is the head in feet. Figure 6-8 is a nomograph of this formula that is useful in determining the width of weir required for a given application. Caution should be taken in using the formula or nomograph at low flows because very low heads cause the nappe to cling to the weir face and make the calculation inaccurate.

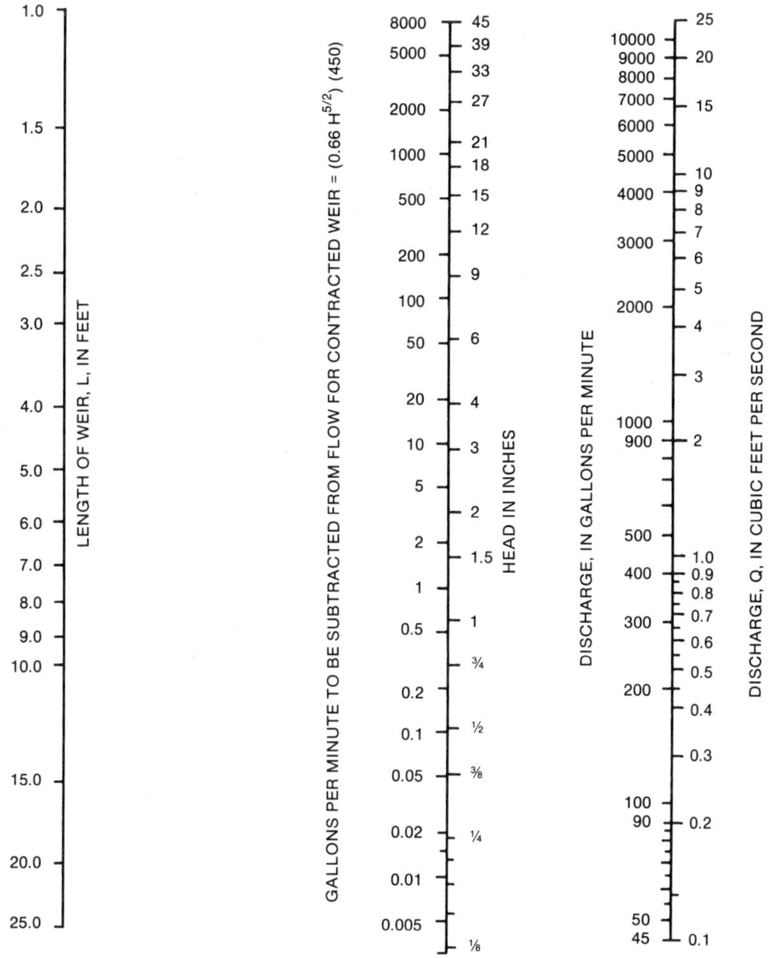

Note: Based on Francis weir formula as follows:

$Q = 3.33\ LH^{3/2}$ for suppressed weir)

or $Q = 3.33\ (L-0.2H)\ H^{3/2}$

$= 3.33\ LH^{3/2} - 0.66H^{5/2}$ (for contracted weir with two end contractions)

Where:

Q = discharge, in cubic feet per second.

L = length of weir, in feet.

H = head in feet.

Figure 6-8 Nomograph for the capacity of a rectangular weir

Cipolletti Weirs. The trapezoidal or Cipolletti weir was invented to aid in solving the problem of end contraction. The usual slope of the ends is one horizontal to four vertical. The end contraction makes measurements with rectangular weirs slightly inaccurate because the amount of contraction changes with H. The general equation for a Cipolletti weir is

$$Q = 3.367LH^{3/2} \tag{6-9}$$

where Q is the discharge flow rate in cfs, L is the length of the weir opening at the base in feet, and H is the head in feet.

When the approach velocity is greater than about 0.3 ft/sec, the velocity head must be compensated for. The velocity head is defined as $V^2/2g_c$. The adjustment

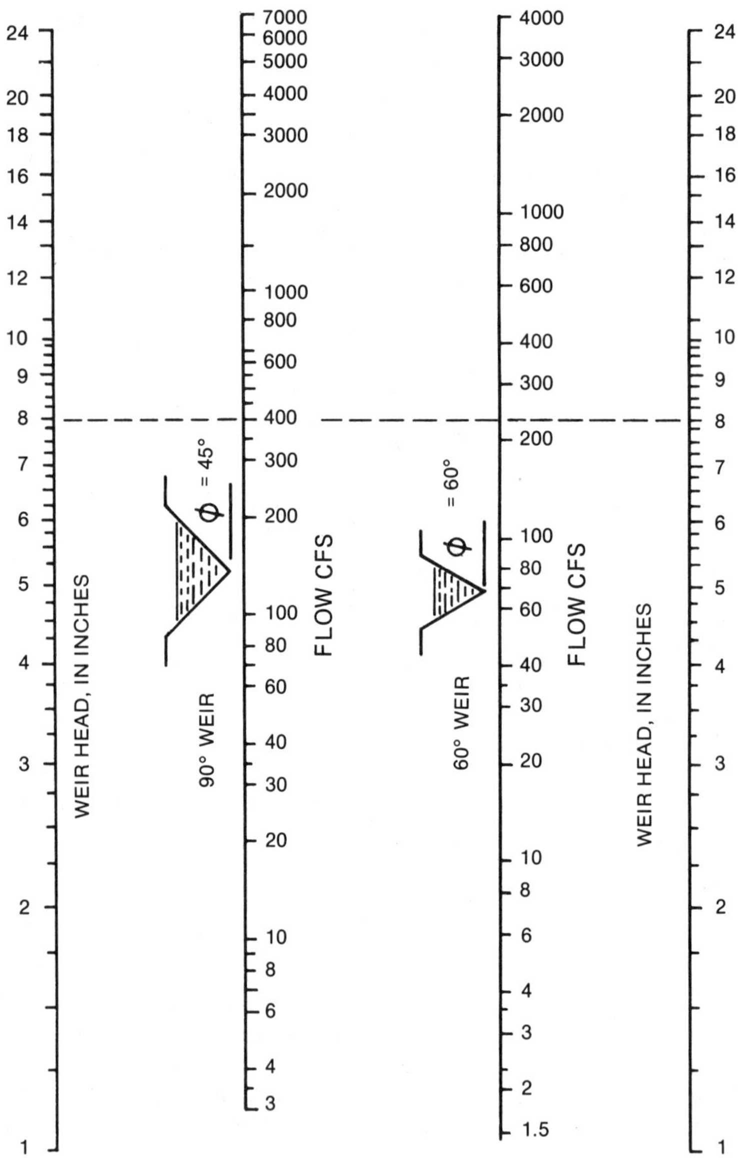

Figure 6-9 Nomograph for the capacity of a triangular weir

is made by substituting the following expression in the flow formula for $H^{3/2}$:

$$H^{3/2} = (H + V^2/2g_c)^{3/2} - (V^2/2g_c)^{3/2} \tag{6-10}$$

Triangular Weirs. For measuring small flows with a weir, the logical choice is a V-notch weir. V-notch or triangular weirs are good for flows up to 2 cfs (900 gpm). The two most popular notch angles are 90° and 60°. Figure 6-9 is a nomograph that can be used to calculate the flow rate from the measured head. The formula for a 90° notch weir is

$$Q = 2.49H^{2.5} \tag{6-11}$$

and for a 60° notch weir is

$$Q = 1.44H^{2.5} \tag{6-12}$$

where Q is the flow in cfs and H is the head in feet. V-notch weirs should not be used if the head is less than 0.3 ft because the nappe will cling to the weir face.

Open-Pipe Discharges

Another situation that is often encountered in a plant is the need to measure flow from an open pipe. Figures 6-10 and 6-11 illustrate two possible cases. The

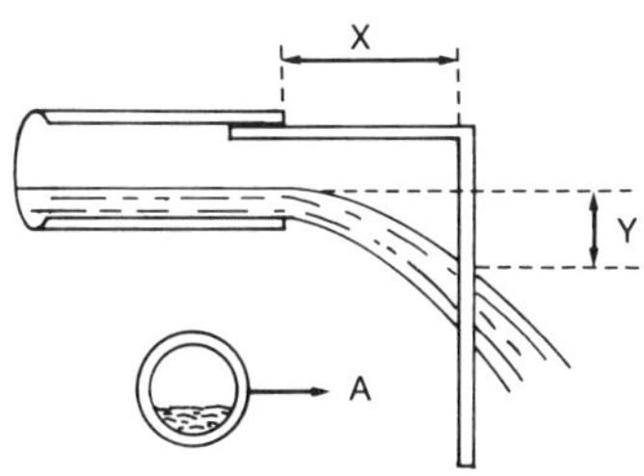

Figure 6-10 Open-pipe discharge. (Courtesy NCASI.)

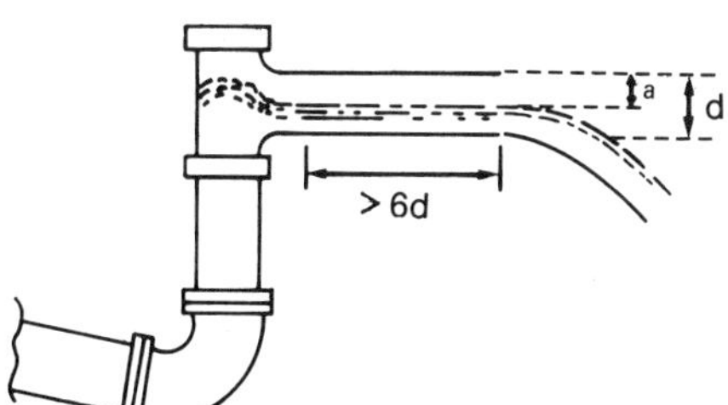

Figure 6-11 Standpipe overflow weir. (Courtesy NCASI.)

flow rate can be determined more accurately from a standpipe, as shown in Figure 6-11. It it is important to monitor discharge flow on a regular basis, open discharges should be converted to this configuration.

For the open-pipe discharge shown in Figure 6-10, the flow rate can be calculated by

$$Q = 1{,}800Ax/\sqrt{y} \tag{6-13}$$

where Q is the discharge in gpm, A is the cross-sectional area of liquid in the pipe, and x and y are measured as shown in Figure 6-10. For the standpipe overflow shown in Figure 6-11,

$$Q = 3{,}900[1 - (a/d)]^{1.88}d^{2.48} \tag{6-14}$$

where a is the distance in feet to the liquid surface and d is the inside diameter of the pipe in feet. Q is the volumetric discharge rate in gpm.

REFERENCES

1 *Handbook for Monitoring Industrial Wastewater*, Technology Transfer Division, U.S. Environmental Protection Agency, Washington, D.C., August 1973.

2 D. J. Harris and W. J. Keffer, *Wastewater sampling methodologies and flow measurement techniques*, EPA 907/9-74-005, U.S. Environmental Protection Agency, Region VII, Surveillance and Analysis Division, Technical Support Branch, Field Investigation Section, Chicago, Illinois, June 1974.

3 W. W. Eckenfelder, *Water Quality Engineering for Practicing Engineers*, Barnes and Noble, New York, 1970.

4 *Methods for chemical analysis of water and wastes*, EPA 16020-07/71, National Environmental Research Center, Cincinatti, Ohio 1971.

5 J. McKeown, *Procedures for conducting mill effluent surveys*, Technical Bulletin No. 183, National Council of the Paper Industry for Air and Stream Improvement, New York, 1965.

6 *Standard Methods for the Examination of Water and Wastewater*, 15th ed., American Public Health Association, New York, 1980.

7 R. H. Perry and C. Chilton, *Chemical Engineers Handbook*, 5th ed. McGraw Hill, New York, 1975.

8 *Design and construction of sanitary and storm sewers*, ASCE Manuals and Reports on Engineering Practice no. 37, WPCF Manual of Practice no. 9, *Water Pollution Control Federation*, New York, 1969.

7
Possible Process Modifications in the Stock Preparation and Paper Machine Areas

When papermaking stock is run on the paper machine wire, a certain amount of the solid matter is not retained, but is passed through the wire together with most of the water used for suspending the fibers. The water removed from the stock on the wire is known as "white water."

Fifty years ago, it was generally believed that white water could not be reused without sacrificing too much of the quality of high-grade papers. Today, however, all mills practice some degree of white water reuse and consider it advantageous to do so.

COMPOSITION OF WHITE WATER

The composition of the white water depends on the type of machine, type of stock, type of nonfibrous furnish, size of the wire, and amount of suction used on the wet sheet. Ordinarily, white water contains fiber debris, small fibers, soluble matter, and a high percentage of nonfibrous suspended matter such as pigments, starches, and dyestuffs. Usually the ratio of finely divided matter (filler and fiber debris) to fiber is higher in white water than in the original furnish. For papers containing filler, the ash content of the stock passing through the wire may be twice to three times that in the original stock.

The total amount of solid matter in the white water varies greatly from mill to mill, depending on the grade of paper and method of white water treatment. The through fraction, or percentage of fiber leaving the slice that passes through the wire, may range from less than 5% for heavyweight papers to over 50% for lightweight tissues on a closed system (1). For newsprint paper, the total losses through the wire may run 35–40% or higher, depending on the machine speed (2). Losses are greater at higher speed. Table 7-1 shows the amount of solids present in white water from a mill making filled grades of printing and writing papers (3). White water from unfilled papers contains much lower percentages of solid matter than indicated in this table.

The composition of white water depends on its source on the paper machine.

Table 7-1 Solids Present in White Water from Machines Making Printing and Writing Papers (Machine Speed 700 fpm)

Paper Grade	Solids (lb/1,000 gal)
Book paper (19% ash)	34.3
Lightweight bond (6.5% ash)	13.0
Offset (11% ash)	26.4
Coating base stock (8–10% ash)	12.6
Mimeograph paper (12% ash)	18.3

The richest white water is obtained from the trays under the first part of the Fourdrinier wire. This rich tray water is generally returned directly to the mixing box, refiners, or other points as makeup water. A less rich white water is obtained from the pit under the wire. In some mills, the pit is divided into sections to collect water of different richness, with the richest used directly as makeup and any excess or less rich water sent to saveall s. Thick stock from breaks at the couch roll is generally collected separately in the couch pit and sent directly to the machine chest as makeup.

WHITE WATER RETURN SYSTEMS

A great deal of effort has gone into closing up machine systems and recirculating the maximum amount of white water. Some of the systems in use are quite complicated. A fairly simple white water system taken from a nonintegrated fine paper mill is shown in Figure 7-1 and a water balance in Figure 7-2. Integrated pulp and paper mills have more complicated water recirculation systems.

Deciding how to design a closed water system is a matter of getting water of sufficient quality and quantity to the right point in the process.

Equipment required to close a nonintegrated operation depends first on the existing equipment at the plant and second on how the plant chooses to recycle. Let's consider a model system and propose a rational scheme for closure.

The first logical step is to use rich white water from the wire pit and couch pit in as many applications as possible. These would include hydropulper dilution water for all stock systems (long fiber, short-fiber hardwood, and broke) and consistency regulation in all stock systems.

The second step is to make maximum use of saveall clarified clear water. This water should be used for all machine showers that can tolerate this quality water. The water quality will probably be adequate for wire showers, headbox showers, grooved roll showers, trim knockdown showers, wire knockoff showers, and breast roll showers. These shower applications presuppose a well-controlled

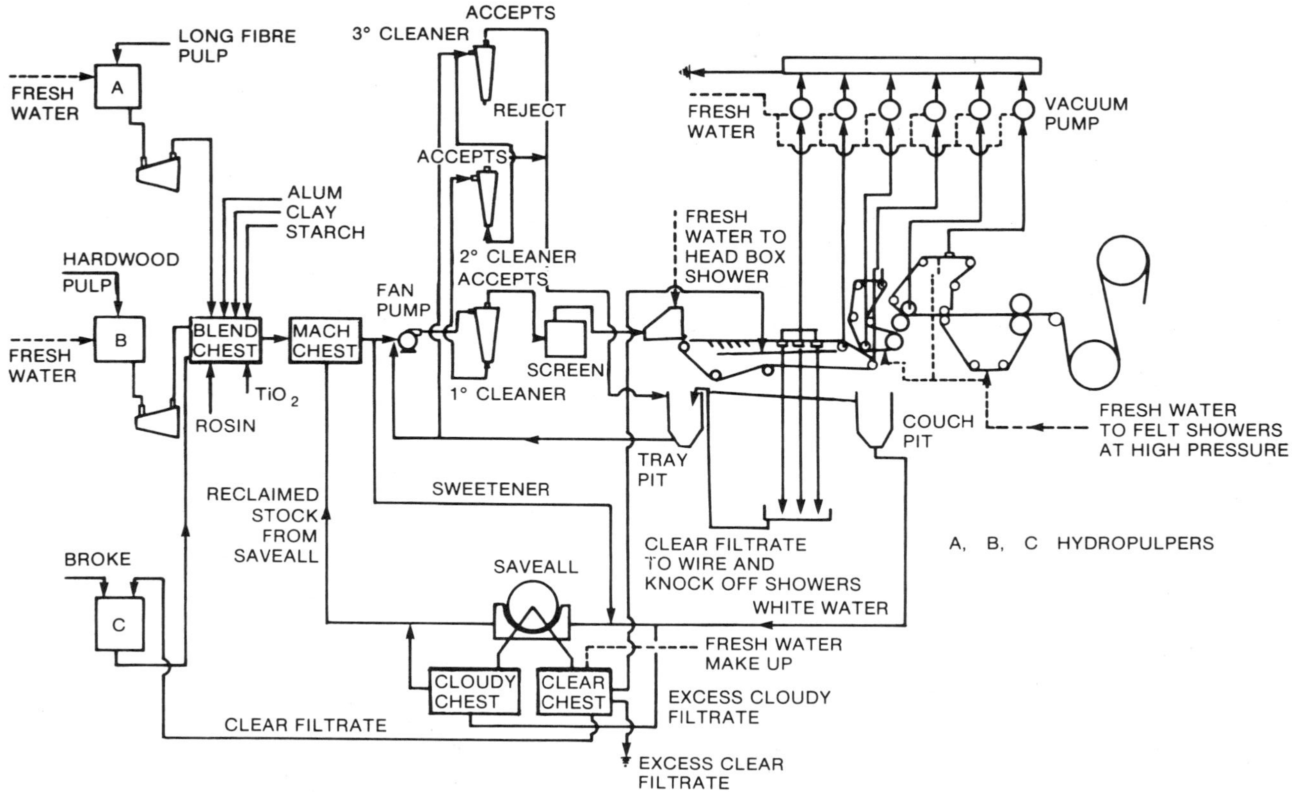

Figure 7-1 Model of a nonintegrated mill

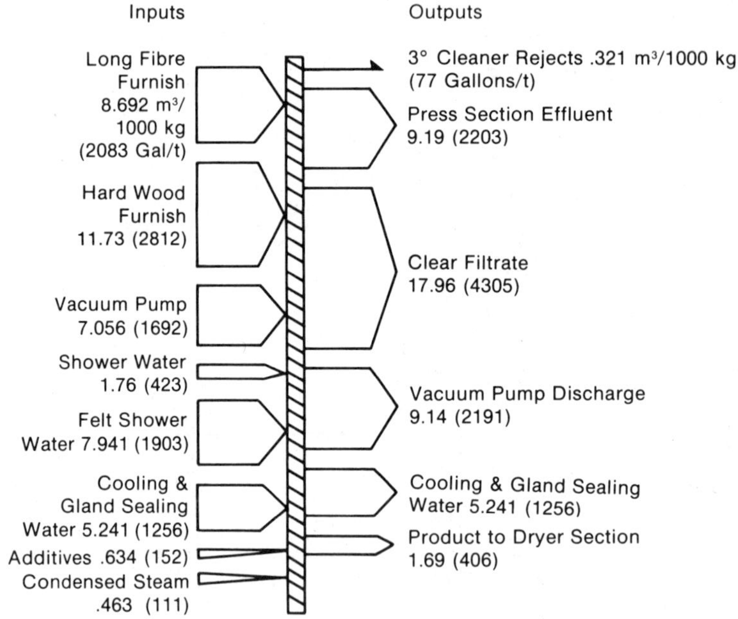

Figure 7-2 Total water balance for the model paper machine

saveall producing high-quality clarified effluent. It is often desirable to protect the showers with an inline strainer in case a saveall upset occurs.

The next item to consider is how to handle the large volume of seal water used on water ring vacuum pumps. One possibility would be to use saveall clarified water in this application. The saveall water quality might be satisfactory if the pH of the system is kept sufficiently high. However, if rosin sizing is practiced and a pH of around 4.5 is maintained in the system, serious corrosion problems with the mild steel housing of the pump could be expected. The Nash Company recommends the following quality: abrasion less than 50 mg/l, soft fibrous suspended matter less than 360 mg/l, pH above 5.5, total dissolved solids less than 500 mg/l, total hardness less than 200 mg/l, chlorides less than 100 mg/l, sulfides less than 50 mg/l, and sulfates less than 100 mg/l. This quality is much more restrictive than is found in practice. Because a large volume of water is involved, it seems reasonable to consider creating a separate recirculation system for the vacuum pumps.

The basic idea in a recirculation system is to collect the seal water from all pumps in a pit located below the pumps, cool this water by some means, and return it to the pumps. A recirculation system would require the installation of vapor-liquid separators before the vacuum pumps.

There are several alternative schemes for such recirculation systems. Figure 7-3 shows that the maintenance of stable vacuum conditions with the water ring pump is highly dependent on the temperature of the sealing water. All schemes take into account this fact and try to maintain a sealing water

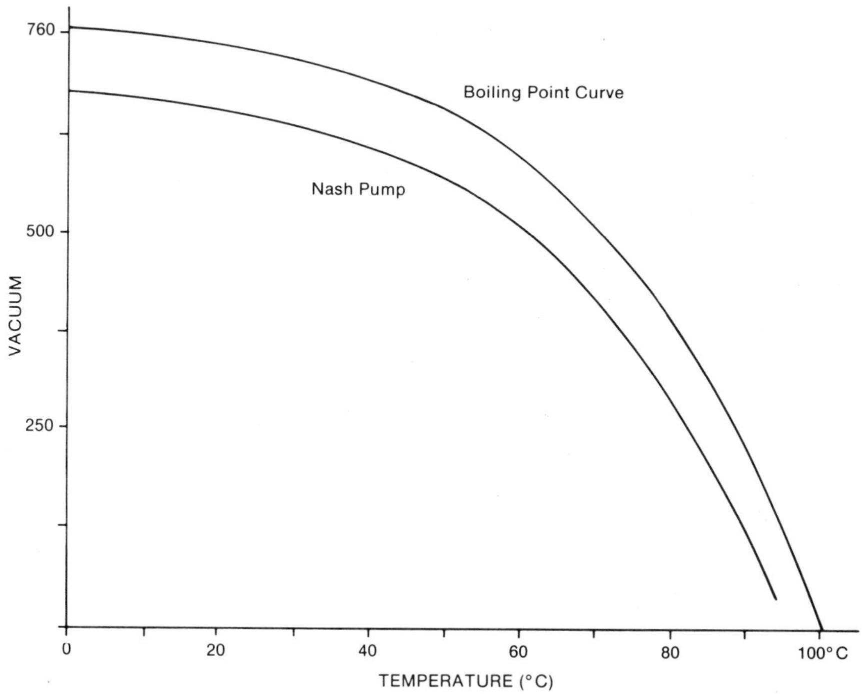

Figure 7-3 A comparison of the boiling point curve and the mean vacuum from a water ring pump at different temperatures. Copyright © 1981. TAPPI. Reprinted from *Introduction to Environmental Management for the Pulp and Paper Industry*, Study Guide, Volume 6, with permission

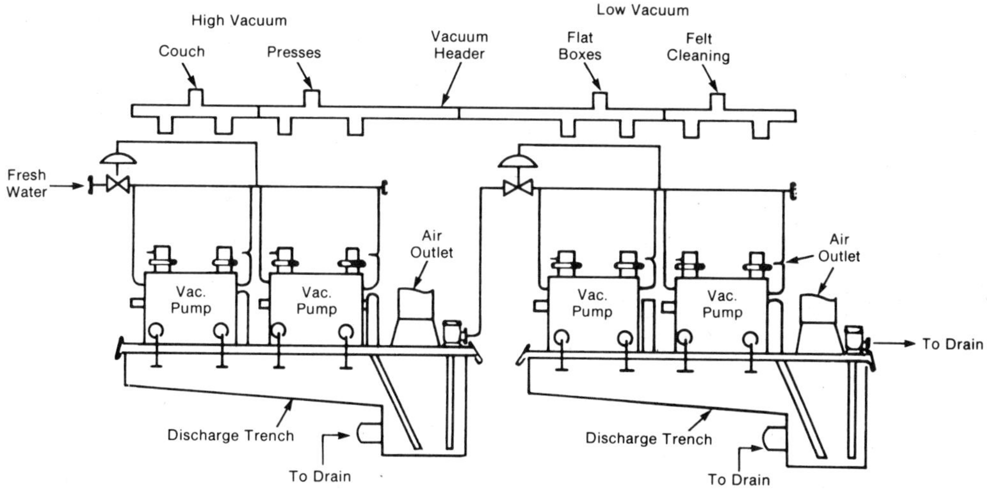

Figure 7-4 Cascading of vacuum pump seal water system

temperature below 125°F (4). Since a lower vacuum is developed at higher temperatures, one scheme recommends using fresh cold water on the pumps requiring a high vacuum and then using this same water to provide sealing water for the lower-vacuum pumps. Figure 7-4 illustrates such a cascade system. Figure 7-5 shows another approach in which a cooling tower is used for the recycled water; alternatively, a heat exchanger could be used instead of the cooling tower. A fourth alternative is illustrated in Figure 7-6, in which the pumps are put in a recirculation system and the temperature of the recirculated water is controlled by bleeding in cold fresh water at a thermostatically controlled rate.

From the perspective of water conservation, the cooling tower system is preferred, since it requires no fresh water except for makeup and blowdown. There are several cooling tower systems operating in the paper industry.

Vacuum pump seal water use can be eliminated entirely by replacing the water ring pumps with high-speed exhausters or lobe-type pumps (5).

The next item to consider for the model system is the felt shower water. It would be ideal to use saveall clarified water. However, to my knowledge, no fine paper mill has yet successfully used this approach. Success in this area has been observed in combination paperboard and in tissue manufacture. Felt showers are a sensitive application requiring consistently high-quality water.

The best method for handling felt shower water is to minimize the volume

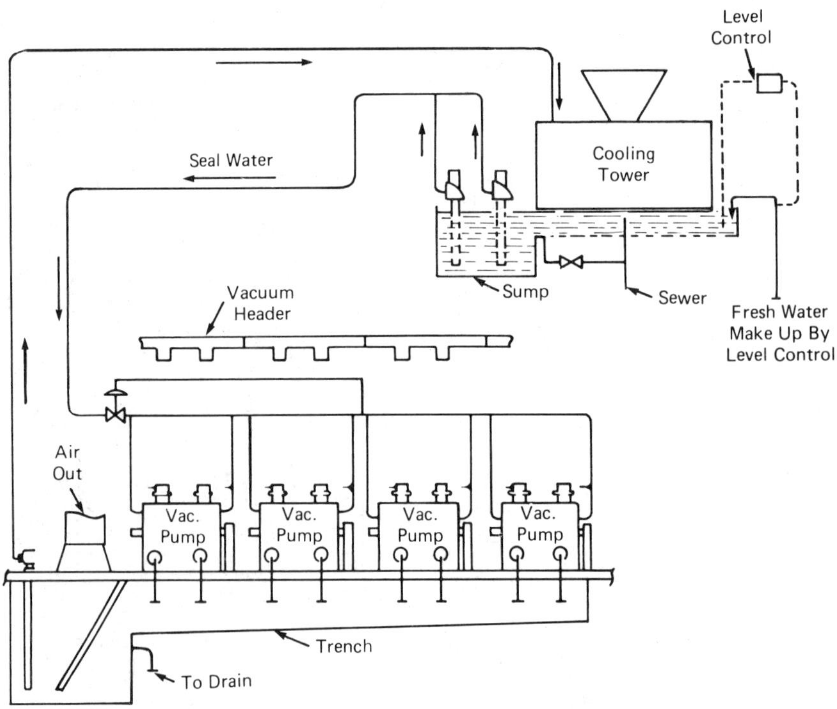

Figure 7-5 Closed-seal system with a cooling tower

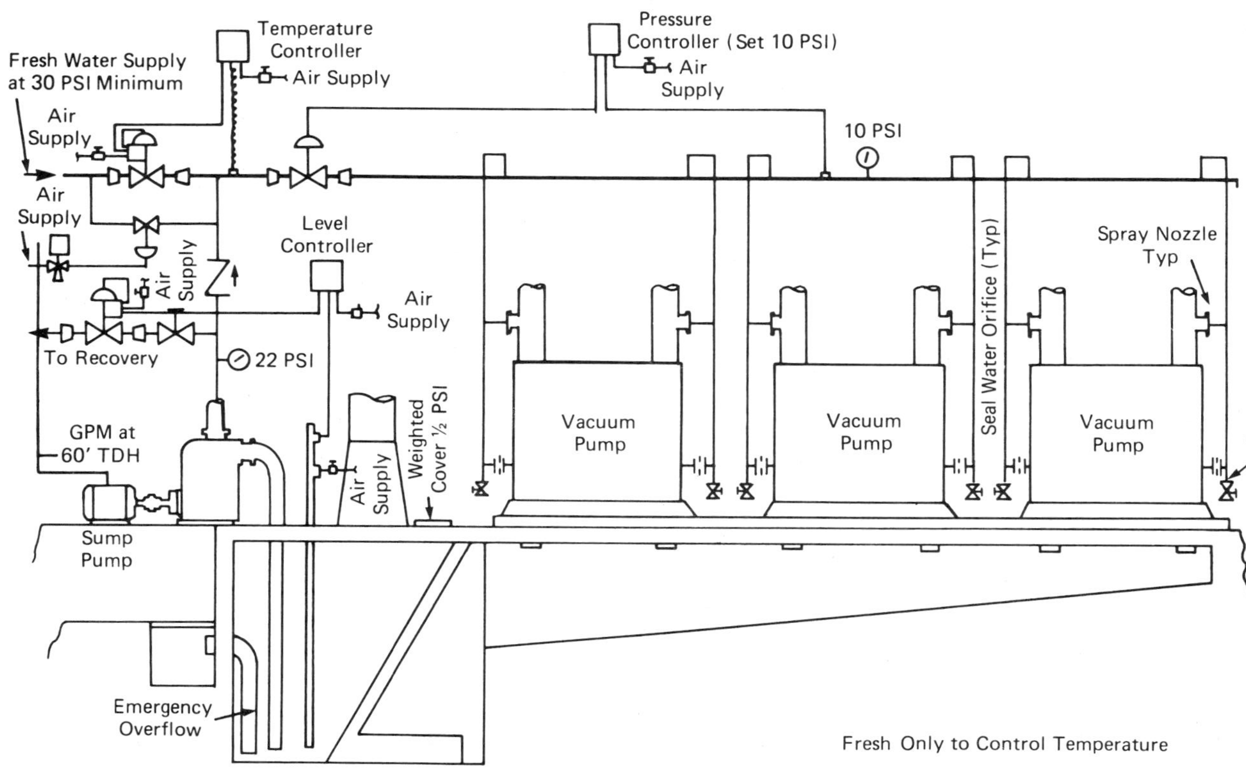

Figure 7-6 Seal water system controlling temperature by a freshwater bleed

by using high-pressure, low-volume showers. Substantial water savings are possible with the application of these devices. Consider an example from the literature of a typical modern press section converting from low- to high-pressure showers (6). Table 7-2 indicates a possible 94% reduction in shower water volume. Some manufacturers would disagree with this example, believing that a low-pressure leveling shower might be necessary in addition to the high-pressure cleaning showers in order to achieve adequate cleaning. In addition, they might disagree with the length of time the high pressure shower is required. Still, a very significant reduction in volume can be achieved. If each low-pressure cleaning shower were replaced by a high-pressure intermittent cleaning shower, approximately a 90% reduction in volume would result. In such a change from low- to high-pressure showers, however, a whole new felt cleaning system would be designed that would further reduce the number of cleaning showers as well.

For purposes of our model system, we will assume that high-pressure felt showers are applicable and that a 90% reduction in volume is possible, thereby leaving only 190 gal/ton involved in showering felts. It is further assumed that this volume of fresh water can be tolerated.

The next item to consider is gland seals and cooling water. The cooling water is best handled by creating a clear water sewer for the mill in which all uncontaminated cooling water is collected and discharged from the mill untreated. It contains only thermal energy as a contaminant. Many of the gland seals can be replaced by mechanical seals. Where mechanical seals cannot be employed, the seals can be left on fresh water.

Table 7-2 Analysis of a Representative Felt Shower System

1. Purge-type grooved roll shower	270 gpm
2. Felt wetting shower	104 gpm
3. Low-pressure felt cleaning shower	40 gpm
4. Felt lube shower	30 gpm
Total	444 gpm
Or total usage 444 (1,440) = 639,360 gallons/day	
Recommendation to Correct the Above Situation	
1. Use a high-pressure, low volume felt cleaning shower	40 gpm
2. Redesign the felt lube shower to minimize water use	6 gpm
3. Replace purge-type grooved roll cleaning shower with intermittent high-pressure cleaning shower 40 gpm on an intermittent basis or 20 gpm continuously	20 gpm
Total	66 gpm
Or total usage 66 (1,440) = 41,040 g/d, which is 6.05% of the above or a 94% reduction in water use	

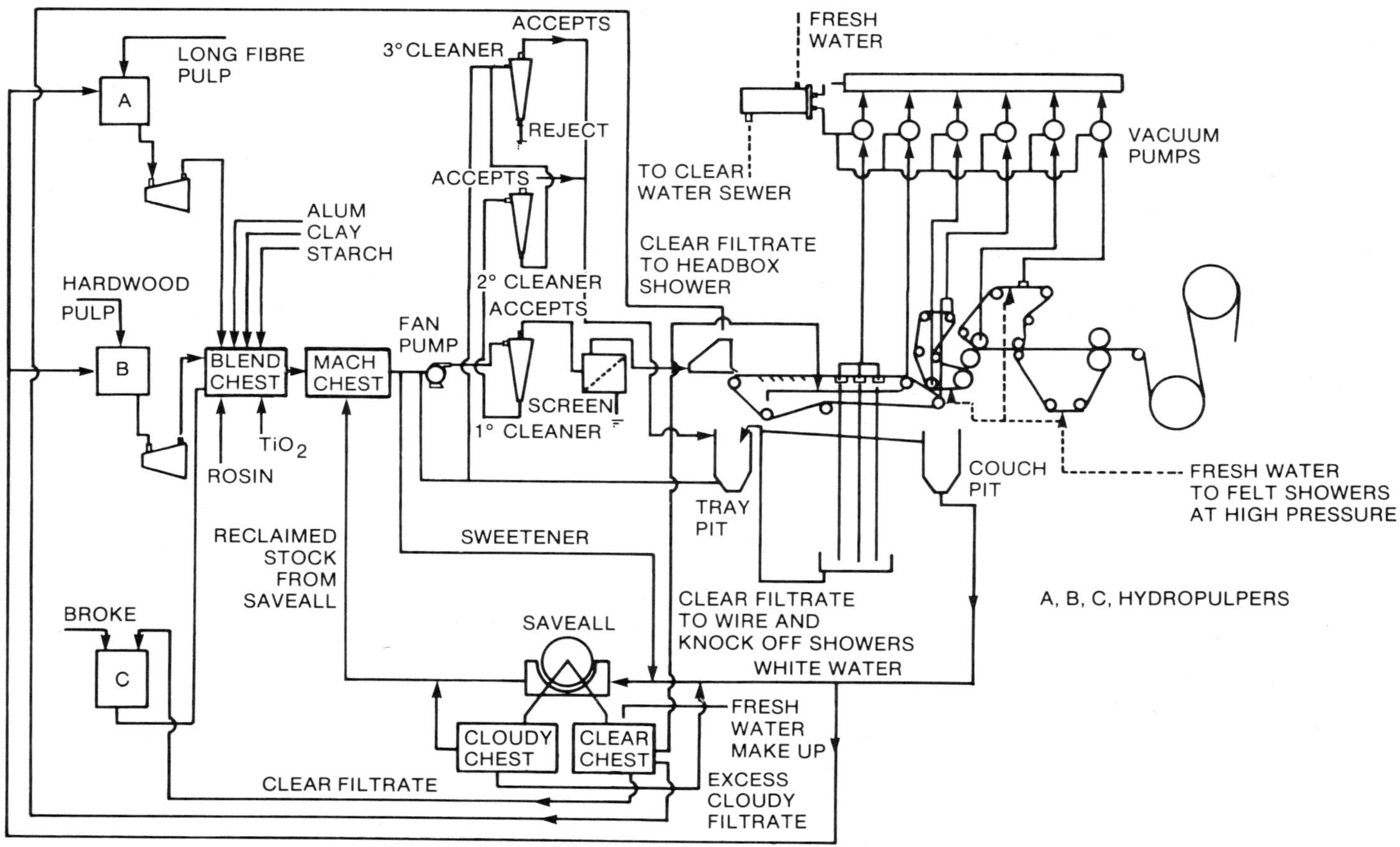

Figure 7-7 Complete water recycling strategy for a model nonintegrated mill

The total fresh water used for the model system must equal the volume discharged, which is the volume of water leaving with the paper web to the dryer section (406 gallons per ton) plus the volume of cleaner rejects (77 gal/ton), a total of 483 gal/ton. Fresh water at 152 gal/ton of product will be used for additives makeup and 190 gal/ton for felt showers. This leaves 141 gal/ton for gland seals and edge squirts.

Figure 7-7 shows these changes implemented on the model nonintegrated mill. The diagram is not complete in all details of the mill, and consistency regulators, gland seals, bearing cooling water, edge deckles, edge squirts, and the clear water sewer are not shown.

SAVEALLS

A key element in the system is the saveall. Five different saveall alternatives are commonly found in paper mills: inclined screens, settling tanks, drum savealls, flotation savealls, and polydisc savealls.

The inclined wire screen is a very crude device that has been used as a saveall. It is not very effective in removing suspended matter, and the effluent from the screen cannot be reused except for stock dilution. Mills using this type of saveall generally run the excess effluent to the sewer.

The revolving screen saveall may be of two types, the simple screen or the vacuum type. The simple screen saveall consists of a wire-covered cylinder that revolves in a vat containing the white water. A fiber mat is formed on the wire due to a difference in head between the water levels inside and outside the cylinder. The efficiency of this type of saveall is low, except where the water contains long fibers or where a fiber mat is used as a precoat on the filter. In a special drum-type submerged-wire saveall (Jones-Vacuum), the drum is equipped with 48 suction troughs; the wire containing the recovered fiber leaves the drum and carries the recovered fiber to a conveyor chute, after which the wire is cleaned by a series of showers and returned to the saveall (7).

The vacuum filter saveall is more efficient than the simple screen type. It produces high-clarity water that contains about 120–240 mg TSS per liter (1–2 lb/1,000 gal), none of which is long fiber (8, 9). According to Rudolfs and Axe (10), the average reduction of suspended solids by vacuum filtration is about 50%, although this varies from 20 to 70% depending on the length of the fiber in the white water. In filtering white water containing a high filler content, it is advantageous to add some long-fiber stock to the water in order to produce a more efficient fiber mat. As a rule, filter savealls are limited to the treatment of white water of low filler content.

A special type of filter, known as the "waco filter," is widely used in European mills and is also found in a few U.S. mills as a paper mill white water saveall (3). A Waco filter operates on a strictly mechanical basis, and no pH adjustment or added chemicals are required. A preformed cellulose mat is used as the filter. The mat is continuously formed on a traveling wire before the filtering operation, after which the wire carrying the preformed mat (containing about 0.025 lb fiber per square foot of exposed wire surface) is passed around the main cylinder (diameter 78 in.) in a vat. The white water flows through the mat and wire into the interior of the cylinder, while the suspended matter is retained on the fiber mat. As the wire leaves the cylinder, the mat and recovered material are doctored off and returned to the system. An average efficiency of about 92–97.5% has been reported on filled papers in a mill making bond and book papers containing 6.5–19% ash (3).

The recovered fiber from a filter-type saveall is usually sent to the stock chests. Part of the clarified water is used for shower water on the paper machines, and the rest is sent to hot water storage from which water is drawn to the refiners and chests. The drum type savealls are not as efficient as disc savealls, which are their modern replacement (11).

Sedimentation savealls, or clarifiers, operate on the same principles used in the sedimentation of raw water. The white water is first treated with coagulant to flocculate the finely suspended matter, which is then settled out in basins equipped for the continuous removal of sludge. Alum and alkali are used as coagulants in sedimentation savealls. This type of saveall is much more efficient than the filtration types, particularly when the white water contains a high filler content, for example, in book mills. However, the efficiency of sedimentation savealls is more affected by changes in consistency.

Sedimentation savealls are rarely used in modern mills because they require high residence times. This is a major disadvantage if multiple grades are run.

Flotation savealls are based on the principle that solid particles with rise to the surface of a liquid if the liquid is loaded with air. Thus, in this type of saveall, air is mixed with the waste water so that free air bubbles are produced when the pressure is reduced. When the air is properly introduced, the bubbles attach themselves to the floc and cause it to float to the surface. The floated fiber is then removed by skimming. To operate properly, air bubbles of the appropriate size must be formed. The fibrous and filler particles must be coagulated or flocculated and have their surfaces reduced in wettability before they can be removed by flotation (12), cations must be present, and the velocity must be between 3 and 5 cm/sec in order not to break up the flocs (13).

There are two types of flotation savealls, those that operate under reduced pressure and those that operate at atmospheric pressure. There is one type that is a combination sedimentation and flotation type. Only the atmospheric pressure units are used at present in the industry.

In general, flotation savealls have the advantage of requiring very little space because of their low retention times. They have the disadvantage of not working well on waters containing a high percentage of fillers or other heavy materials. Flotation savealls have a greater thickening capacity than the sedimentation type (14).

The Sveen–Pedersen saveall operates at atmospheric pressure to a retention chamber where air at 46–60 psi, about 1–2% by volume, is introduced. From there, the water is pumped into the saveall, where the release of pressure at this point produces air bubbles that carry fiber and filler particles to the surface for

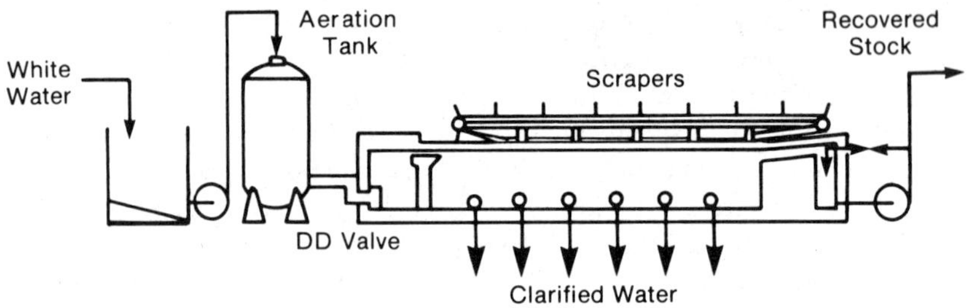

Figure 7-8 Seven–Pedersen saveall

removal by skimming, while the clarified water is removed from the bottom. Figure 7-8 sketches a Sveen–Pedersen saveall. Before entering the saveall, a special colloidal glue is sometimes added to stabilize it and provide greater air buoyancy. For lightweight sheets, a glue dosage of 2 ppm (0.5 lb/ton) is typical, and for heavy sheets a dosage of 4 ppm (1.0 lb/ton) (15). Activated silica sol or a polymeric flocculant may be used in place of glue. Aluminum ion is essential to the proper operation of the Sveen–Pedersen saveall.

In an electrophoretic investigation of the Sveen–Pedersen process, Hukki and Rinne (16) showed that the combined effect of alum and Sveen glue was to stabilize the zeta potential at very near the zero value, thus providing the necessary condition for flocculation. It is also claimed that alum gives the air bubbles a positive charge so that they will attach themselves more readily to the negatively charged agglomerates (17). The rosin size–alum complex present in the white water tends to coat the fibers and clay particles, giving them a nonpolar air–solid surface that makes them flotable (16). White water containing 15–17 ppm aluminium ion is satisfactory. Additional aluminum ion can be added if necessary, but must be added well before the water enters the saveall in order to be effective (15). The saveall operates satisfactorily at pH values between 4.2 and 7.3 (18).

Effluents as low as 12–24 ppm in solids content (0.1–0.2 lb/1,000 gal are obtained on water from unfilled papers and about 24–48 ppm (0.2–0.4 lb/1,000 gal) on water from filled papers. Typical results on many different grades of paper are shown in Table 7-3 (19). As a general rule, the recovered fiber and filler (at 2–3% solids) are returned to the stock system, preferably as close to the paper machine as possible. The clarified water is withdrawn through tubes in the bottom of the unit into a storage chamber from which it is pumped to the showers, regulators, and beaters.

The Sveen–Pedersen saveall design has been improved by the addition of lamellae to the flotation chamber, which permit higher throughput rates while maintaining efficiency. This is a German development, termed a "Favair device," and is marketed in the United States by the Permutit Corporation.

Through the use of a circular configuration and a variable influent position that moves with the float skimming mechanism, Krofta has developed a near zero velocity influent condition in a flotation saveall that should lead to improved performance (20).

Table 7-3 Typical Results With the Sveen–Pedersen Saveall

	Solid Matter lb/1,000 gal	
Type of Paper	Into Saveall	Out of Saveall
Chipboard	1.6–2.5	0.08–0.30
Lineboard	2.6–3.6	0.08–0.25
Book, clay	7.2–17.9	0.10–0.31
Book, carbonate	12.5–38.2	0.19–0.40
Bond, wood	4.0–12.4	0.12–0.29
Newsprint	7.0–9.2	0.17–0.38
Tissue, toilet	2.6–5.8	0.14–0.27

In a polydisc saveall, the disc filter consists of a number of ribbed, pie-shaped sectors covered with a filtering medium and held together in a complete disc, A series of these discs are mounted on a centershaft with axial sections called "cores" (11). The water flows through the filtering medium into the sector, and inward in the sector to an opening at the bottom leading to the hollow space in the middle of the centershaft. The water flows along this space and out the end of the centershaft. There is a sliding wear plate seal arrangement at the end of the centershaft that controls the distribution of the water into "cloudy" or "clear" portions.

The polydisc saveall operates in the following manner (Figure 7-9):

1 Vacuum on: mat forming, discharge cloudy. The water and fines start to flow through the filtering medium as the sector becomes submerged. Some fiber and fines go through the medium, and this flow is kept seperate as 'cloudy' water by the valve arrangement. During this time, the longer fibers staple or form on the medium into a mat of fibers. The flow of this water through the centershaft, sealed valve, and drop leg piping induces a hydraulic vacuum to form the mat, usually at 7–10 in. of mercury

2 Mat filtering, discharge clear. The mat formed now acts as the filtering medium that traps and collects the fines as the water passes through, resulting in clear water, usually having a solids content of less than 50 ppm. As the clear water flows down the drop leg pipe, the induced vacuum should be 10–12 in. of mercury to draw the maximum volume through the mat. If the vacuum becomes too high, mat removal is difficult

3 Vacuum off, mat removed. As the disc rotates further, the mat with the collected fines comes out of submergence and the valve blanks off the end of the core under that portion of the disc, which shuts off the vacuum. A knockoff shower peels the mat from the sector, and the mat with the fines and filler drops into a recovered stock trough. This recovered stock can then be directed into a chest to be metered back into the machine system

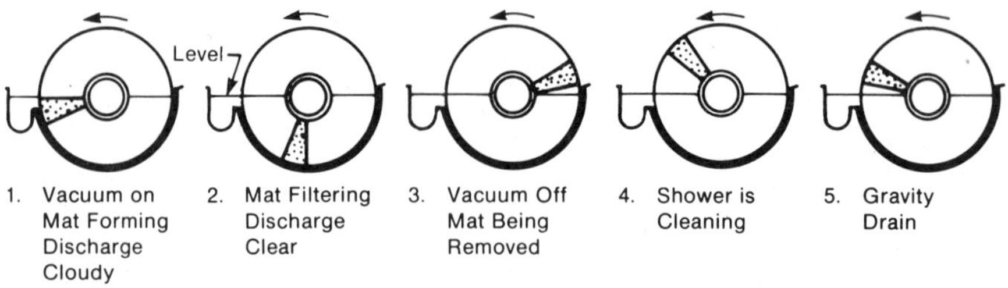

Figure 7-9 Disc saveall operating sequence

4 Shower cleaning, gravity drain. After the mat discharges, a cover cleaning shower washes the sector, and some of the wash water goes into the centershaft and out through the valve (gravity drain). The remainder of the shower water splashes off the sector and into the vat. The gravity drain water has some fiber content and is therefore usually directed into the cloudywater seal pit

The piping arrangement for a disc saveall generally consists of the inlet flow, which is a mixture of white water, sweetener stock, and recycled cloudy water; cloudy and clear drop leg piping and gravity leg into the seal pits; recovered stock to a stock chest; and three shower water headers—sheet knockoff, sluice, and cover cleaning showers. The cover cleaning and sheet knockoff showers have been previously described and require clear water at 80 psi, while the sluice shower is used to ensure that the recovered stock does not hang up in the chutes. This water can be taken from the cloudy water compartment at 40 psi. Installation of a disc saveall requires a minimum of 15 ft from the centerline of the saveall to the surface of the water in the seal pits to ensure proper vacuum during operation. Sweetener stock is the source of long fiber used to form the filter mat on the disc; the recovered stock consists of a mixture of sweetener and white water solids.

When water recycling is an issue, the choice usually must be a disc saveall or a flotation saveall. Disc savealls can produce good white water clarity and fiber/filler recovery with minimum lag time. This permits the saveall to become an integral part of the stock preparation and machine white water system with good flow control of the circuits. Also, disc savealls are much more stable in their operating characteristics than are flotation savealls.

One advantage of the flotation saveall over the disc filter is that it does not require sweetener stock, and thus its capacity is not dependent on stock freeness. Another advantage is lower initial capital costs. The disadvantages are the larger floor space it occupies and its reputation for requiring significant operator attention (21). Flotation savealls have generally been used

1 In fine paper mills where a drum or disc unit is not suitable because of the particular system's stock characteristics

2 In small mills where the floor space requirement does not offset the lower capital costs

Flotation savealls require coagulant addition and a relatively constant feed rate. Operating costs for service, chemicals, and electrical energy are comparatively high for flotation (22).

The choice between flotation and disc savealls is a matter that each mill must decide. Both produce adequate-quality water for recycle to all uses, with the exception of felt showers and edge squirts.

In a well-run paper mill, the water leaving the mill should contain no more than 0.05 g/l (0.5 lb/1,000 gal) of suspended material, and in most cases less, since this amount is only a moderate saveall performance. Fiber loss from paper mills should not exceed 1% of total production.

WASH-UP WATER

In the model system of Figure 7-7, it is easy to overlook a significant item—the water needed for washing and general cleaning. It comes as a surprise to find that this use can amount to 25% of the total water demand of the mill (23) and is a major item in a closed system.

The way to handle wash-up water is to use saveall clarified water as much as possible for cleaning. Effluent from the cleaning operation is then sent to the mill's primary clarifier and on to a surge basin external to the primary water system of the mill. This water can then be further clarified by processing through the freshwater accelerator system and returned to the main process water system. This approach assumes that a freshwater coagulation precipitation treatment system is available at the plant. Any excess fresh water added during a wash-up will have to be slowly worked out of the system as water to the dryer section, cleaner rejects, screening rejects, or primary clarifier rejects.

In totally closed paperboard mills, the surge basin size used is one to two times the total tankage capacity of the mill. It is best to size the basin using data on water loss and frequency of wash-ups. A dynamic computer simulation of the mill would be very useful in sizing the surge basin and the white water storage tank used to contain spills.

It is advisable to minimize the number and volume of wash-ups as much as possible without jeopardizing production or product quality.

POTENTIAL PROBLEMS FROM WATER REUSE

Table 7-4 lists potential operating problems that arise from extensive water reuse. This information comes from the literature and from discussions with mill representatives. The problems are broken down into three general classes—those arising from the presence of higher levels of suspended solids, those arising from

Table 7-4 Potential Problems Encountered in Water Reuse in Paper and Board Manufacturing

Dissolved Solids Buildup	Suspended Solids Buildup	Thermal Energy Buildup
Slime	Dirt	Temperature
Foam	Erosion	Sizing problems
Pitch	Fines	Machine room temperature
Corrosion	Felt plugging	Reduced vacuum pump capacity
Sizing	Wire plugging	
Product mottling	Wire life	
Color	Felt life	
pH control	Reduced drainage rate	
Precipitation	Shower plugging	
Scale		
Odor		
Retention		

the accumulation of dissolved solids, and those arising from the increased retention of thermal energy.

Of the 25 potential problems listed in Table 7-4, only 7 were cited to be of real concern by the operating personnel surveyed. These problems are listed in Table 7-5 and can be broadly categorized as plugging, corrosion, scale, foaming, slime, sizing, and color. Difficulties that seemed to arise first were those associated

Table 7-5 Problems Encountered in Low-Water-Use Paper Mills

Suspended Solids Buildup	Dissolved Solids Buildup	Thermal Energy Build up
Plugging Problems	Corrosion	
Wire showers	Scale	
Headbox showers	Color	
Felt filling	Slime	
Bearing cooling lines	Foaming	
	Sizing	
Color		
Deposits on headbox lip		
Deposits on wire return rolls		

Source: Copyright © 1981. TAPPI. Reprinted from *Introduction to Environmental Management for the Pulp and Paper Industry*, Study Guide, Volume 6, with permission.

with suspended solids buildup and the subsequent plugging of shower nozzles, small lines, and felts. Suspended matter may also lead to deposit formations that may be of biological and/or nonbiological nature. Some mills experienced an intensification of biological slime difficulties and some a diminishing as a consequence of more intensive water reuse practices.

Dissolved and colloidal materials present a more serious problem than do suspended solids (21, 24–26). After long periods of running under conditions of low freshwater use, the problems of scale and corrosion emerge. With very high degrees of recirculation, the concentration of dissolved material increases to a point where the wet-end chemistry of the papermaking system may be affected and difficulties with foam and sizing begin to occur. Table 7-5 lists all documented problems that have been reported by mills, but no single mill has experienced all of these problems. With even greater water reuse in paper operations than now occurs, there may be additional difficulties that have not yet been experienced. Such problems might be related to the increased fines content of the sheet (21). In many cases, water reuse yields positive benefits as well, such as improved drainage due to increased white water temperature (21, 27).

Plugging Problems

One of the first and most common problems that mills encountered in using reclaimed water was plugging, specifically plugged lines, wire showers, headbox showers, wires, and felts. Plugging has generally been attributed to the presence of long fiber in the water. A useful test for the presence of long fiber involves observations of a water sample under a microscope at 80 power to determine the presence of particles longer than 0.3 mm. Suspended solids analysis has not provided sufficient delineation of the problem, since tissue mills have reported suspended solids levels of over 3,000 mg/l, with no plugging difficulties on self-purging showers as long as there was no long fiber present (21).

Difficulties with long fiber have been avoided if water clarification devices (clarifiers, saveall, microstrainers, A-B Floatwash fractionators, Sweco concentrators) are used to remove long fiber. In critical applications, further protection may be provided with in-line slotted strainers. Successful applications have used slotted strainers approximately five to seven times finer than the opening being protected. Thus, for wire showers with a nozzle diameter of 0.05–0.06 in., the strainer size would be 60 mesh. A strainer is also used to protect nozzles from scale flakes in a freshwater shower system.

In addition to the slotted in-line strainers, which are manufactured by a wide variety of companies, several special in-line protection devices have been developed and marketed specifically for use in closed white water systems. Among these devices are the Sweco concentrator, the Albany Engineered Systems 3600 strainer, the Sinclair float wash fractionator, and the Celleco Microsorter. These specialized devices may be of value in solving a particular problem (27, 28), but for most applications the slotted strainer is sufficient.

The C-E Bauer Hydrosieve and the Dorr Oliver DSM screen have been used as postscreening devices, but usually they are not 100% efficient in rejecting long-fiber material and therefore are not the best choices for this application.

Corrosion

Corrosion is the deterioration of metal by either electrochemical or bacterial processes. The corrosion rate is affected by the complex interactions of numerous factors, including dissolved oxygen level; pH; dissolved solids, chlorides, sulfates, and so on; hardness; alkalinity; acidity; temperature; flow velocity past the surface; carbon dioxide concentration; and contact with dissimilar metals. The intensity of the dissimilar metal problem is related to the relative positions of the two metals in the galvanic series. The number of factors involved in corrosion makes it a complex and difficult problem to control.

Corrosion results in increased downtime and higher maintenance costs. However, the direct association of these effects with corrosion over a period of time is difficult to establish because routine equipment changes and modifications play such an important role in maintenance costs. At some fine paper and board mills, it has been implied that it was necessary to remove all dissimilar metal from the system or to insulate such metal from neighboring equipment. Most mills routinely replace mild steel piping as it deteriorates with stainless steel or fiberglass. A study that related water quality to corrosion conditions (29) implied that for nonintegrated systems 316 stainless steel would be satisfactory for closed system conditions but that 304 stainless steel would be borderline in its protective characteristics.

A prime candidate for corrosion problems is the iron casting of Nash vacuum pumps. One tissue mill reported a corrosion problem when clarified water at pH 4 was used as pump seal water (4). Combination paperboard manufacturers have also reported difficulties in this area (30). However, an Israeli fine paper mill reported the use of clarified white water at pH 5 on Nash vacuum pumps from 1964 to 1972 with no corrosion or gumming (31).

One mill reported that a pump using straight white water of pH 4.2–4.9 and solids content of 1,200–1,800 ppm (10–15 lb/1,000 gal) for 1 year was found to be eroded beyond repair. The efficiency of the pump was measured several times during the year and showed steady deterioration (32). To help relieve the erosion problem, Nash has developed lobe purges to bleed off erosive material that accumulates at the outer rim of the pump body.

Techniques available for controlling corrosion include pH adjustment, cathodic protection, protective coating, and addition of chemical inhibitors. The inhibitors available include calcium bicarbonate, chromates and dichromates, nitrates, polyphosphates, silicates, borax, sodium benzoate, and organic inhibitors. No universal corrosion inhibitor for white water systems exists. An inhibitor that is satisfactory for one system may be ineffective or even harmful in another. Those that appear most promising are the polyphosphates, silicates, and borax. However, so far, chemical inhibitors have not been used in paper mill water systems.

The only techniques employed in combination paperboard, fine paper, and tissue mills for corrosion control have been

1 pH control
2 Replacement of piping with more resistant material
3 Replacement of bronze pump impellers and valves in order to remove dissimilar metals from the system

Under aerobic conditions electrolytic corrosion tends to dominate, but under anaerobic conditions bacterial corrosion becomes important. In the presence of sulfate, which accumulates under conditions of extensive water reuse, sulfate-reducing bacteria contribute to increased corrosion rates. The bacteria are able to use the cathodic hydrogen that would otherwise slow or stop the corrosion process. There is also a chance of odor from reduced sulfides that might be produced. Bacterial corrosion is controlled by the use of an effective biocide. One combination paperboard mill made filler from the accumulated sulfate by forming insoluble barium sulfate through the daily addition of 200–300 kg barium carbonate (33). It was controlling sulfate levels to about 300 mg/l.

A practical way to forecast the severity of corrosion problems to be expected from process water recycling is to use metal coupon tests to determine the difference in corrosion rates before and after implementation of a recycling program. As a rule of thumb, mild steel corrosion rates of less than 35 milli-inches per year (MPY) are considered normal, 35–50 MPY are questionable, and rates greater than 50 MPY signal the need for some method of corrosion retardation.

Deposition

Problem with deposition, or scale, are the opposite of those with corrosion. Scale results from the crystallization, precipitation, or coagulation of nonresinous substances, as differentiated from pitch or slime. The accumulation of scale is usually associated with one of four anions—carbonate, silicate, sulfate, or oxylate. In combination with the cations calcium, magnesium manganese, iron, aluminum, or barium, these anions are the source of most scale difficulties. The most prevalent scale is that resulting from calcium and magnesium carbonates.

The usual way to control scale is to control the hardness level of the water or, in other words, the cation level. Hardness values greater than 400 ppm have been observed in a tissue mill practicing a high level of water reuse with no adverse effects on the system (4).

Water containing soluble ferrous and manganous compounds will favor the growth of iron and manganese using bacteria that may contribute to deposition problems. High concentrations of alum can lead to the precipitation of aluminum hydroxide, which may form a chemical slime (34, 35). Aluminum compounds have also caused felt clogging (36).

The traditional method of scale control is the use of sequestering or dispersing agents. Common ones used are polyphosphates, ethylene diamine tetraacetic acid (EDTA), and synthetic organic polyelectrolytes.

Slime and Odor

Slime and odor are appropriately considered together since, in all cases of mill problems with product odor, the source has been traced to biological growth. Most problems with biological growth in white water systems have been due to poor housekeeping, lack of an effective biocide program, or stagnant areas in the water or stock preparation system.

As water use in decreased, the concentration of dissolved solids increases in the system. The associated higher carbohydrate and nutrient levels constitute a favorable environment for bacterial growth. A potential inhibitor of biological growth in this enriched environment is the longer detention and increased concentration of any biocide added to the system. As the water reuse intensifies, the temperature may shift from the mesophyllic range of 20–35°C (68–95°F) to the thermophyllic range of over 50°C (over 122°F), causing a corresponding shift in bacterial populations. The thermophyllic organisms tend to form spores and thus are more difficult to kill or control. It is not practical to operate a mill water system at high enough temperatures to suppress bacterial growth for reasons of human safety and comfort (37). Under closed white water conditions, a shift toward more anaerobic organisms has been seen (37).

The interaction of the enriched environment, biocide concentration, detention time, and temperature shift is likely to be responsible for the presence or absence of slime problems when white water reuse is increased.

Either slime boards or total bacterial counts are used to determine the level of bacterial activity in a system. If the method of total bacterial count is used, counts greater than 10^6 bacteria per milliliter usually signal potential difficulties, although this number will vary somewhat from mill to mill. If the slime board technique is used, wet slime accretion of more than 10–15 $g/100\,cm^2/day$ usually signals potential difficulties.

Mills cope with slime principally by using more frequent wash-ups and by the occasional addition of single or multiple slimicides at several points in the system. Among the biocides used are chlorine, chlorinated phenols, some amines, quaternary ammonium compounds, organosulfur compounds, and silver compounds. Sometimes the combination of a biocide and a dispersing agent is the most effective and economical approach to slime control. Some mills operating under low water use add chlorine or chlorine dioxide directly to the fan pump as an effective slime control procedure.

Color

Coloring of the recycled water is a problem, particularly when white and colored grades are sequentially manufactured. To cope with the problem, it is common to begin with white grades and fresh water, running with fresh water until enough white water becomes available for the hydropulpers and for consistency regulation. Mills develop a color compatibility sequence, beginning with white grades, then producing light colored grades, and finishing with the darkest grades, after which it is necessary to wash up before repeating the cycle. A mill manufacturing 150 tpd of tissue saved 130,000 gall water per dump by adopting the color compatibility production sequence of white—pink—yellow—blue—green before dumping water from the system (38). For this mill's production schedule this amounted to a saving of only half a million gallons per month, but the benefits could be much greater in a mill with many short runs in its production schedule.

Water recycling capability is improved if the dyes used have a high preference for the fiber and are easily bleached. White water systems must be separated for extensive water reuse practices in multimachine mills if incompatible colors or colors and whites are made at the same time.

Felt Hairs

Some believe that the problem of felt hairs is minimal with felts of high synthetic content and that there is no longer any justification for sewering the water from the felt conditioning system (23). Another approach is to return contaminated press section effluent to the couch pit of the machine on which it occurs. In the course of subsequent refining, felt hairs reportedly are shortened to a harmless length (32). Press effluents processed in this way apparently have not caused any problems.

A more conservative approach involves screening the press effluent before returning it to the white water system. A horizontal rotating center-fed cylindrical screen has been used successfully. One mill found a 140-mesh nylon screen adequate (32), while another mill used a 120-mesh stainless steel filter medium (31). In the latter case, the press effluent was sent to the saveall after screening. Several other types of screens have been tried and found to be ineffective. Circular vibrating screens suffered from stapling of felt hairs and blocked rapidly. Side hill screens blocked frequently. Slotted vibrating pressure screens plugged rapidly, and the backwash was ineffective in clearing them (33).

Dissolved Salt Buildup

As effluents are reduced and increased amounts of paper machine white water are recirculated throughout the mill, the concentration of electrolytes increases substantially. The potential for buildup will be discussed quantitatively later in this chapter. Many of the normal wet-end processes, especially those depending on the ionic forces of attraction and repulsion, will be altered in such high-salt environments.

Trials conducted under totally closed conditions on the model paper machine at Western Michigan University (WMU) led the investigators to conclude that, with adjustments, a 70% groundwood/30% Kraft catalog paper virtually identical to that now produced commercially can be made under totally closed conditions (39). It was learned that salt levels up to 4,000 mg/l had little effect on all physical and optical properties, since these measures in the trial papers were equal to or better than those of the base line samples. These results were confirmed by trials conducted at Miami University (MU), where a high-opacity offset sheet was manufactured on a model paper machine under conditions of complete water reuse (40). This study found potential problems with brightness reversion, sizing, and retention that can be controlled. Therefore it appears possible to manufacture fine paper under high degrees of recycling

One problem that has occurred in the manufacture of combination paperboard at dissolved solids levels greater than 4,000 mg/l is a mottled appearance of the product. This problem was not observed in the model machine trials at WMU or MU. It could be caused by dissolved organic components. In the manufacture of tissue, a high dissolved salt content could not be tolerated if it caused a mottled appearance or a harsh feel in the product.

Dirt Content

In the manufacture of toilet and facial tissue and fine papers, the dirt level in the product is critical and affects consumer preference for the product. One tissue mill reported problems due to dirt specks when the process water discharge was less then 4,255 gal/ton in the manufacture of toilet tissue and less than 3,750 gal/ton in the manufacture of white facial tissue (38).

MATERIAL BALANCE CONSIDERATIONS

As a water system is closed, the concentrations of various components in the system have the potential to build to intolerable levels. To illustrate this problem, a detailed mass balance for various states of closure was performed on a system similar to that shown in Figure 7-7. The concentration of a component i in the headbox water was calculated assuming a fixed addition rate of component i to the water system. Component i can be anything in the system—fiber, wet-strength resin, alum, biocide, dye, chloride, sulfate, and so on. The rate of buildup depends on five factors:

1. The state of closure of the system
2. The rate of retention of the substance in the sheet (Ri)
3. The manner in which the material fractionates in the saveall (Fi), that is, the degree to which it remains with the stock returned to the system or leaves with the clarified water
4. The manner in which the material fractionates in the primary clarifier (Ei)
5. The fraction of material rejected in the stock cleaning system (Si)

The last factor is relatively unimportant because it usually represents a small quantity of material. Retention values (*Ri*) can range from 1 to 0.0582. The lower limit assumes that the only factor contributing to retention is the dissolution of the substance in the water going with the sheet to the dryer section. Some writers use the term "substantive" to refer to material that has a preference for fiber and "nonsubstantive" to refer to material that does not. The lower limit for retention values is for nonsubstantive materials. Saveall fractionation *Fi* ranges from 0.0242 to 1 and primary clarifier fractionation *Ei* from 0.0153 to 1, assuming that the lower limit is fractionation due only to solubility in the water leaving with the solids. Representative values of *Ri*, *Fi*, and *Ei* are listed in Table 7-6, but this table is based on limited data; actual values could vary considerably from those listed. The actual values for a particular mill should be determined from data taken on site.

The results of the mass balance calculations are shown in Figures 7-10 and 7-11. Figure 7-10 illustrates a water system that is closed by using a clarifier as the main water clarification element, and Figure 7-11 shows a system using a saveall similarly. Figure 7-10 shows that if a material *i* is completely retained in the sheet, its concentration in the headbox approximately doubles at zero effluent discharge. Thus the sludge purge from the system can act as a significant purge for certain substances. The top curve of Figure 7-10 shows the worst case for the system using a clarifier, in which the material is retained only in the water going with the sheet and in the sludge.

The system employing a saveall is different because saveall sludge is recycled into the system. High fractionation values in the saveall lead to a further concentration buildup. A much more severe concentration buildup can occur when the saveall fractionation value is high and the sheet retention value is low. This situation is illustrated by the top curve of Figure 7-11, for which $Fi = 0.5$ and $Ri = 0.0592$. Increasing the sheet retention value to 0.5 greatly dampens the high overall fractionation effect and leads only to a doubling of the concentration. For

Table 7-6 Representative Values for *Ri*, *Fi*, and *Ei* in Tissue Manufacturing

Component	*Ri*	*Fi*	*Ei*
Fiber	0.95	0.97	0.95
Suspended solids	0.90	0.95	0.92
Dissolved solids	—	0.5	—
Total solids	—	0.8	—
Chloride	0.06	0.1	—
Sulfate	0.06	0.1	—
Chemical oxygen demand	—	0.93	0.85
Magnesium	—	0.6	—
Calcium	—	0.35	—
Aluminum	0.3	0.9	—
Silica	—	0.8	—
Lead	—	0.8	—

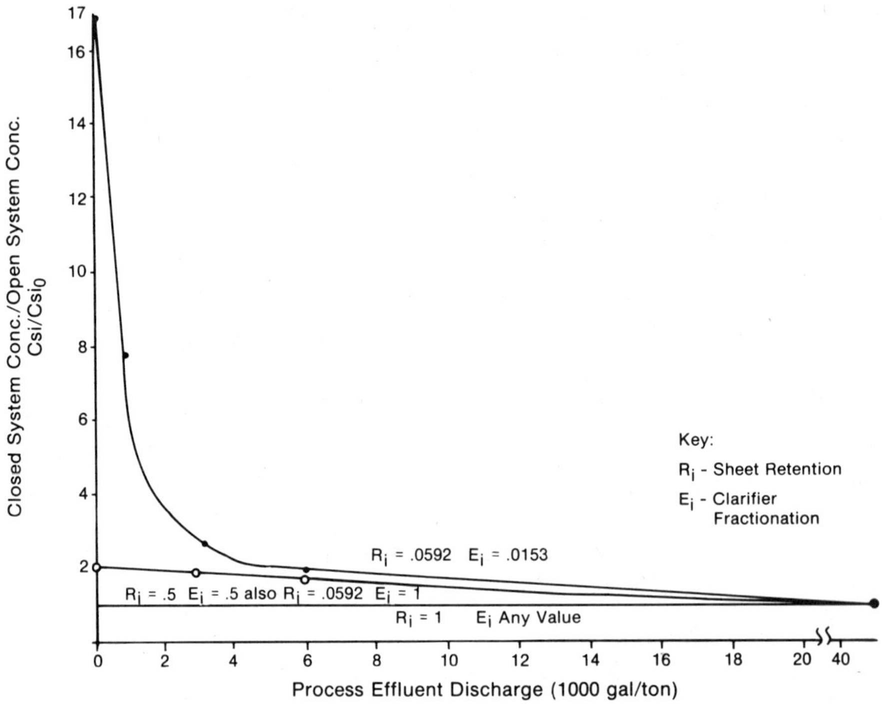

Figure 7-10 Concentration buildup as a function of process effluent discharge for an illustrative mill employing a clarifier as its water clarification element

any given state of process water discharge, the level to which a component will build is a function of both *Ri* and *Fi*.

A system could employ both a saveall and a clarifier to achieve water conservation. Normally, all water in the system would flow to the saveall before going to the clarifier. The level to which a component would build is a function of *Ri*, *Fi*, and *Ei* at a given level of discharge. The sheet retention value (*Ri*) and the saveall fractionation value (*Fi*) would be of prime importance in determining the level of buildup. However, the clarifier fractionation value would moderate the effect of the saveall fractionation value.

Since retention is such a dominant factor in controlling the level of buildup of materials under conditions of low water use, work has begun on the development of a retention control strategy for closed systems. The basic background information on retention behavior has been described (24, 25), as have various elements of such a strategy (26, 41). In order to implement this strategy, appropriate instrumentation and control algorithms must be developed.

The determining factor in concentration buildup is the amount of purge from the system, or the amount of process effluent discharge. When attempting to predict the level of concentration buildup for decreased effluent discharge, one does not simply move down one of the curves shown in Figures 7-10 and 7-11

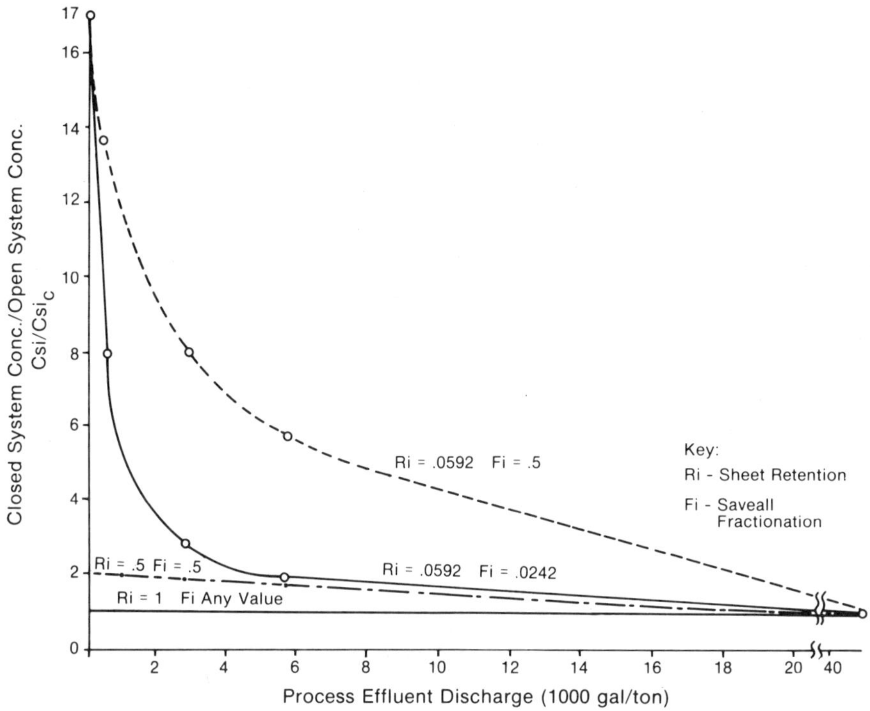

Figure 7-11 Concentration buildup as a function of process effluent discharge for an illustrative mill employing a saveall as its primary water clarification element

to obtain different concentration ratios. Rather, one may shift from one curve to another because the degree of closure affects the values of *Ri*, *Fi*, and *Ei*. However, in practice, these effects would probably be small and *Ri*, *Fi*, and *Ei* would remain relatively constant until a very small purge volume high level of system closure is reached.

Figures 7-10 and 7-11 can be used in one of two ways—to estimate the level of buildup of a component, given the above limitation, or to determine the amount by which the addition of component *i* to the system should be reduced to maintain constant headbox conditions.

It is interesting to note that the curves suggest, for the specific situations evaluated, that the maximum concentration a component will reach is 17 times the open system concentration, and that just a slight preference for retention in the sheet will significantly reduce a component's buildup. Similar experience has been confirmed by other investigators (37).

The shape of the curves for a fixed retention value and a fixed fractionation value are significant in that there seems to be a breakpoint at about 3,000 gal/ton. This would imply that a mill with a water system similar to that illustrated in Figure 7-7 could expect to close its water system to 3,000 gal/ton with few difficulties from concentration buildup. Only when a system was closed to below 3,000 gal/ton would problems begin to appear.

WATER QUALITY CONSIDERATIONS

The buildup of various materials in a system is manifested as a change in the water quality, resulting in operational problems experienced by the plant. An early attempt to identify the process water quality requirements for different types of paper manufacturing is given in Table 7-7. Certain grades of paper, such as cigarette paper, facial tissue, condenser paper, filter paper, and photosensitive paper, may require water of greater purity than that listed for fine papers. These are exceptions, however, and most of the criteria listed in Table 7-7 are more restrictive than necessary for the given grade. The choice of level for a particular parameter was often based on what a well-operating water treatment system could achieve rather than on a level actually shown to be necessary.

In an attempt to better identify the water quality required in a specific application for a given type of paper manufacturing, the National Council of the Paper Industry for Air and Stream Improvement has established a computer water quality data repository in which actual data on samples of process water are stored by application. These data are used to generate maximum observed operational water quality characteristics (MOOWQCs), which represent an attempt to identify the lowest-quality water that can successfully be used in a given application without causing any operational difficulties. Data are available for combination paperboard, fine paper, and tissue mills. The appropriate Technical Bulletin can be consulted for details (30, 38, 42). The data are given in terms of functional water quality tests as well as chemical water quality parameters, and these data indicate much more liberal requirements than those specified in Table 7-7.

Table 7-7 Standards for Process Water (Maximum Permissible Limits, ppm)

Substance	Fine Papers	Ground wood Papers	Kraft Papers	
			Bleached	Unbleached
Turbidity as SiO_2	10	50	40	100
Color in platinum units	5	30	25	100
Total hardness as $CaCO_3$	100	200	100	200
Calcium hardness as $CaCO_3$	50	—	—	—
Alkalinity to methyl orange as $CaCO_3$	75	150	75	150
Iron as Fe	0.1	0.3	0.2	1.0
Manganese as Mn	0.05	0.1	0.1	0.5
Residual chlorine as Cl_2	2.0	—	—	—
Silica (soluble) as SiO_2	20	50	50	100
Total dissolved solids	200	500	300	500
Free carbon dioxide as CO_2	10	10	10	10
Chlorides as chlorine	—	75	—	—

Several mills have implemented the principles discussed in this chapter, and their experiences have been reported (37, 43–46).

REFERENCES

1 *Paper Maker and British Paper Trade Journal* (London), 128:208 (September 1954).

2 G. F. Underhay, *Tappi*, 37 (11):547 (1954).

3 W. P. Nesbitt, *Pulp & Paper Magazine of Canada*, 51:82 (March 1950).

4 A. Springer and R. Garcia, "An analysis of the potential consequences of minimizing vacuum pump seal water discharge," *Tappi*, 64 (2):67–70 (1981).

5 J. G. Coats, Jr., *Tappi*, 51 (8):95A (1968).

6 D. J. Morrissey, *American Paper Industry*, 54 (5):15 (1972).

7 *Paper Trade Journal*, 142:20 (1958).

8 H. A. Morrison, *Paper Trade Journal*, 122 (15):41 (1946).

9 R. S. Hawley, *Sewage Works Journal*, 11:512 (1939).

10 W. Rudolfs and E. J. Axe, *Paper Trade Journal*, 125 (3):53 (1947).

11 W. V. Stevens, "White water systems engineering," paper presented at the Conference on Retention of Fine Solids in Paper Manufacture, State University of New York, Syracuse, N.Y., June 17, 1975.

12 W. Friend, *Tappi*, 39 (6):172A (1956).

13 *Paper Maker Annual Review* Number 61–62, 1957.

14 F. Kück, *Wochenblatt für Papierfabrikation*, 79:3–6, 33–36 (1951).

15 H. R. Clodfelter, *Pulp & Paper*, 21:54 (September 1947).

16 R. T. Hukki and R. Rinne, *Paperi ja Puu*, 36:129 (April 22, 1954).

17 W. D. Harrison, *Paper Trade Journal*, 109 (7):41 (1939).

18 P. Easton and R. Baum, *Tappi*, 33 (6):301 (1950).

19 O. F. Hutchinson, *Paper Trade Journal*, 127 (23):44 (1948).

20 Personal communication, Dr. Milos Krofta, Lenox Institute for Research, Inc., 101 Yokun Ave., Lenox, mass. 01240.

21 T. Frette, *Paper Technology*, 15:271 (1974).

22 J. Velinski, *Papier* (Darmstadt), 26:619 (October 1972).

23 J. J. Wilkinson, *PPI, Pulp & Paper International*, 15 (5):59 (1973).

24 M. Strutz, *The effect of total water reuse and alum control on first pass retention*, M.S. thesis, Miami University, 1981.

25 J. Boylan, *Evaluation of anionic and nonionic retention aids for closed mill systems*, M.S. thesis, Miami University, 1982.

26 M. Strutz and A. Springer, "A wet-end optimization strategy for total water reuse," *Tappi*, 65 (3):139–142 (March 1982).

27 V. M. Mehta, *Pulp & Paper Canada*, 80 (9):T270 (1979).

28 B. Buchanan, *Pulp & Paper*, 54 (3):60 (1980).

29 D. F. Bowers, "Changes in water properties and corrosivity with closure," Paper presented at the TAPPI Environmental Conference, Atlanta, 1983.

30 A. Springer, D. Marshall, and I. Gellman, *The relation between process water quality characteristics and its reuse potential in combination board mills*, Technical Bulletin No. 282, National Council of the Paper Industry for Air and Stream Improvement, New York, 1975.

31 L. Rakosh, *Pulp and Paper Magazine of Canada*, 75 (3):63 (1974).

32 "Water and stock conservation on Fourdrinier machines," *Tappi*, 54 (10):1734 (1971).

33 W. Brecht, H. L. Dalpke, and F. Borner, *Wochenblatt für Papierfabrikation*, 101:113 (1973).

34 C. W. Honeycutt, *Reuse of white water on a bleached board Fourdrinier paper machine*, M.S. thesis, North Carolina State University, 1970.

35 S.K. Porwal, A. M. Springer, and A. Proctor, "Scale deposits on the Fourdrinier wire of a fine paper machine," *Tappi*, 63 (6):67–69 (June 1980).

36 M. U. Priester, *Tappi*, 47 (1):145A (1964).

37 D. Wenzl, "Closure of paper and board mill production systems and its effect on production conditions," paper presented at the TAPPI Annual Meeting, Chicago, March 2–5, 1981.

38 A. Springer, D. Marshall, and I. Gellman, *The relation between process water quality characteristics and its reuse potential in non-integrated manufacture of tissue and toweling*, Technical Bulletin No. 289, National Council of the Paper Industry for Air and Stream Improvement, New York, 1976.

39 S. D. Alexander, R. J. Dobbins, and R. W. Roscoe, "Water reuse experience in the manufacture of fine and specialty papers," paper presented at the National Council of the Paper Industry for Air and Stream Improvement Central-Lake States Regional Meeting, Chicago, September 1974.

40 P. Heller, W. E. Scott, and A. M. Springer, "Potential operational problems under conditions of complete water reuse," *Tappi*, 62 (12): 79–84 (December 1979).

41 W. J. McMahon, *An investigation into the effects of inorganic and organic contaminants on the performance of retention aids (closed whitewater systems)*, M.S. thesis, Miami University, 1983.

42 A. Springer, D. Marshall, and I. Gellman, *The relation between process water quality characteristics and its reuse potential in fine paper mills*, Technical Bulletin No. 287, National Council of the Paper Industry for Air and Stream Improvement, New York, 1976.

43 J. R. Smith, "In plant water conservation and effluent flow reduction at Brunswick Pulp and Paper Company," paper presented at the TAPPI Environmental Conference, Atlanta, 1983.

44 K. S. Razumalski, "History of a successful abatement program," paper presented at the TAPPI Environmental Conference, Atlanta, 1983.

45 W. L. Lapp, J. L. Parr, and K. C. Ayers, "Escanaba program to eliminate incremental water pollution from an expansion," paper presented at the TAPPI Environmental Conference, Atlanta, 1983.

46 A. Geller and L. Göttsching, "Closing water systems completely in the Federal Republic of Germany," *Tappi*, 65 (9):97–101 (September 1982).

8

Process Modifications In the Wood Yard

In the wood yard, effluent can arise from three sources—hydraulic conveyance of the logs, log washing, and wet debarking. The long-term industry trend is away from all three of these processes involving water. These processes all have in common the fact that the wood comes to the mill as logs. As the forest products industry continues its trend toward maximum use of the tree, it is likely that more wood will be delivered as chips. The hydraulic flumes for transporting logs are being phased out in favor of mechanical conveyers.

With a mechanical conveyer, it is necessary to wash the log before debarking to lessen equipment abrasion (1). In most installations, the water shower is activated by the log itself on the conveyer, so that a minimum of water is used. It is difficult to ascertain the precise amount of water used and its characteristics in this application because of the wide variation in log size. The flow is from 100 to 300 gal. per ton of product. The BOD_5 ranges from 1 to 8 lb/ton, and the TSS from 5 to 55 lb/ton, and the color is less than 50 units (2). The suspended solids are mainly silt, and their quantity fluctuates with the weather conditions. Little silt is present in dry weather and more during wet weather. This effluent is either disposed of on land together with grits and dregs from the pulp mill or combined with the barker effluent and sent to the treatment plant.

Bark consists of a physiologically active inner layer and a relatively inert outer layer. The dirt is held in the outer layer, but the soluble organics responsible for the BOD come from the inner layer. Bark comprises at least 10% by weight of the trunk of the tree, or about 5% of the total biological mass of the tree when the roots and leaves are included.

Debarkers can be categorized into six different types (3): drum barkers, bag (pocket) barkers, ring (rotary) barkers, cutterhead barkers, hydraulic barkers, and knife barkers.

Most pulpwood is small in diameter and is debarked in dry drum barkers. When round wood is either large in diameter or unusually long, wet barking may be employed, although this practice is declining. Wet barking is carried out in drum barkers, hydraulic barkers, or bag (pocket) barkers (4).

Bag (pocket) barkers are rapidly being phased out of use in the United States. These stationary machines abrade bark from timber by jostling and gradually rotating the confined wood stock against an endless chainbelt, on which pro-

Table 8-1 Analysis of Wet Drum Debarking Effluents

Mill	TSS (mg/l)	Nonsettleable Solids (mg/l)	Ash of SS (%)	BOD_5 (mg/l)	Color Units
1	2,017	69	—	480	20
2	3,171	57	21	605	50
3	2,875	80	18	987	50

Source: Copyright © 1981. TAPPI. Reprinted from *Introduction to Environmental Management for the Pulp and Paper Industry*, Study Guide, Volume 6, with permission.

jections, called "dogs," raise the wood pile and allow bark to pass between the chains. Water is sprayed through apertures in the side of the pocket at rates of 330–600 gpm for pockets handling 2.8–5.7 cords per hour (2).

Wet drum barkers consists of a slotted drum equipped with internal staves that knock the bark from the wood as the drum rotates in a pool of water. Bark is removed through the slots with the overflow water. Drum barkers handle 7–45 cords of wood daily. The water used in the drum is recycled within the barking unit itself, and is often spent process water. Drum barkers contribute from 15 to 20 lb BOD per ton of wood barked and from 50 to 100 lb suspended solids. The analysis of drum barker effluent from three mills is presented in Table 8-1. Modern drum debarkers often operate without water, and such a debarker is shown in Figure 8-1.

Hydraulic barkers employ high-pressure water jets to blast the bark from the log, which is either conveyed past the jet or rotated under a moving jet that traverses the log. Water use in 5,000–12,000 gal. per cord of wood barked. Usually only fresh process water is used because of the narrow clearances in the high-pressure pumping system. The effluent passes through coarse screens to remove pieces of bark and wood slivers, which are sent to disposal, and then passes through a fine screen to remove finer pieces; Then it is discharged. The total waste flow is about 5,000–7,000 gal. per cord; this flow carries 1–10 lb BOD and 6–55 lb suspended solids per ton of product.

Fine screen effluent from hydraulic barkers has been analyzed by several investigators (5–8), and their results are summarized in Table 8-2. These data indicate that the TSS content ranges from 521 to 2,350 mg/l with an ash content of 11–27%. BOD_5 values range from 52 to 250 mg/l. These values are low because the contact of water with the bark is short and no grinding action on the wood takes place. In drum and bag barkers, where grinding takes place and the bark is in longer contact with the water, higher BOD values result (5). The BOD values are greatly affected by the species of wood barked and the season in which the wood was cut, because the BOD is due to wood juices and water extractables. The suspended matter in the effluent contributes little to the BOD (6). Hydraulic barkers produce very fine suspended solids because they break the bark into small particles that do not settle well.

Figure 8-1 Dry drum debarker

Table 8-2 Analysis of Hydraulic Barking Effluents

Mill	TSS (mg/l)	Nonsettleable Solids (mg/l)	Ash of SS (%)	BOD_5 (mg/l)	Color Units
1	2,362	141	27	85	< 50
2	889	101	14	101	< 50
3	1,391	180	17	64	< 50
4	550	66	11	99	< 50
5	521	53	13	121	< 50
6	2,017	69	21	56	< 50
7	2,000	< 200	19	97	—
8	600	41	10	250	35

Source: Copyright © 1981. TAPPI Reprinted from *Introduction to Environmental Management for the Pulp and Paper Industry*, Study Guide, Volume 6, with permission.

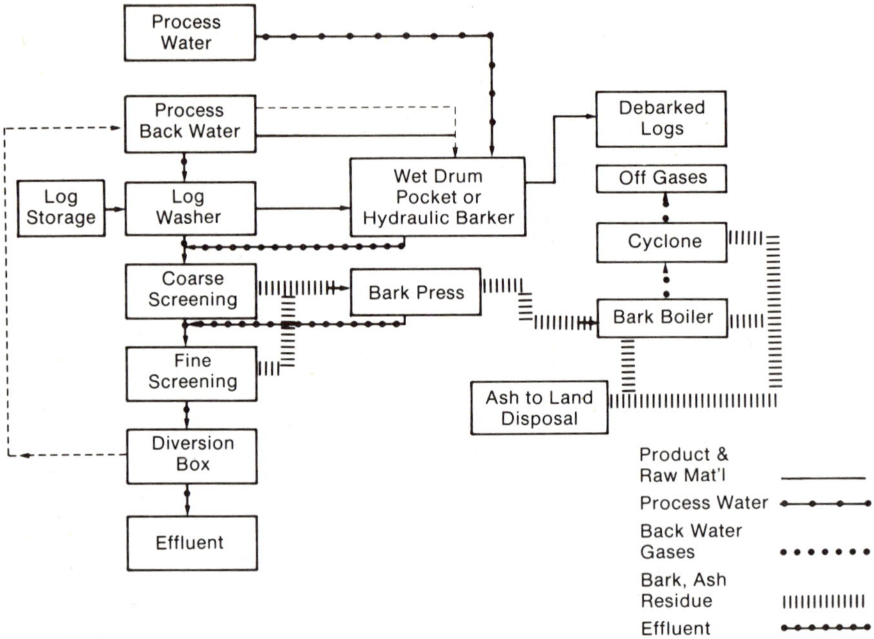

Figure 8-2 Wet barking process diagram

A block and line flow diagram illustrating a wet debarking system is shown in Figure 8-2. Note that a reduction in waste load can be achieved by recycling most of the debarking water. If recycling is to be used in hydraulic debarking, however, a high degree of solids removal must be achieved to prevent abrasion in pumps and barker nozzles.

The effluent from the process is normally sent to some form of separate sedimentation clarification system before being combined with the pulp mill effluent and sent to the mill's treatment system. The combined barker and log washing effluent contains a lot of silt and settles very rapidly. From 70 to 90% of the TSS are removed in 30–60 minutes settling, and additional time does not yield additional removal. It is common to employ a heavy-duty clarifier for this effluent, with an overflow rate of 1000–1,200 gal/ft^2/day and a retention time of about 2 hours. The clarifier must be equipped to handle the dense sludge that results from the silt present. The clarifier must also be equipped with a surface skimmer to handle floating material. It is common to dewater the underflow on a vacuum filter whose medium frequently consists of 120-mesh stainless steel wire cloth. Dry solids loading rates are about 10–12 lb/ft^2/hr. The solids content of the cake is about 30%.

The solids from debarking produce the largest solid waste problem from a mill. Formerly, the paper industry either burned the bark in open teepee burners or put the material into landfills. Since about 1970, the Clean Air Amendments have brought teepee burning to an end. The escalating costs of energy during

Table 8-3 Variations in the Caloric Value of Bark with Moisture Content

Moisture (%)	Pine and Spruce ($Mcal/m^3$)	Birch ($Mcal/m^3$)
30	500	590
40	480	570
50	460	550
60	420	520
70	360	460

the 1970s made bark a valuable fuel source for meeting mill energy needs. Some mills made garden mulches and seedling blocks from bark, but this business has been discontinued since bark attained value as an energy source. The practice of landfilling bark has also been largely discontinued, mainly because the leachate from bark has a high BOD content. As much as 20% of the bark material consists of water-soluble components (9).

Bark is now burned in special bark furnaces or in combination furnaces. The caloric value of bark depends largely on its moisture content, as illustrated in Table 8-3. The moisture content of bark can be decreased by practicing dry debarking or by pressing the wet bark. It may be economically advantageous, from the perspective of both capital investment and operating costs, to convert to dry debarking (10).

The efficiency of a steam boiler also decreases with increasing moisture in the fuel. If a furnace has a 90% efficiency when burning oil alone, the efficiency might drop to 75% when bark with a 60% moisture content is used as supplementary fuel. Therefore, not only the caloric value but also the resulting furnace efficiency must be considered when evaluating the benefits of using bark as fuel.

Burning bark produces 0.5–5.0% ash, largely grit resulting from the dragging and handling of the logs. In order to meet particulate emission requirements, some type of collection device is required. There are no particular odors or toxic emissions from the burning of bark.

The chipping operation produces no effluent, so in the next chapter we will turn our attention to pulping.

REFERENCES

1 P. J. Kleppe and C. N. Rogers, *Survey of water utilization and waste control practices in southern pulp and paper industry*, (Project No. A-036-NC), Water Resources Research Institute, University of North Carolina, Raleigh, 1970.

2 H. Gehm, *State of the art review of pulp and paper waste treatment*, Environmental Protection Series EPA-R2-73-184, U.S. Government Printing Office, Washington, D.C., 1973.

3 R. G. Macdonald and J. N. Franklin, eds., *Pulp and paper manufacture*, Vol. 1, 2nd ed., McGraw-Hill, New York, 1969.

4 H. Kronis and D. A. Holder, *Pulp & Paper Magazine of Canada*, 69:62 (February 2, 1968).

5 R. E. Draper and F. S. Mercier, "Hydraulic barker effluent clarifier at Wood Products Division, Weyerhauser Co.," in *Proceedings of the 11th Pacific Northwest Industrial Waste Conference*, 1962.

6 R. O. Blosser, *Practice in handling barker effluents in mills in the U.S.*, Technical Bulletin No. 194, National Council of the Paper Industry for Air and Stream Improvement, New York, 1966.

7 *Pollutional effects of pulp and papermill wastes in Puget Sound; report on studies conducted by the Washington State Enforcement Project*, National Council of the Paper Industry for Air and Stream Improvement, Northwest Regional Office, Portland, Ore., March 1967. (Prepared in cooperation with the Washington State Pollution Control Commission, Olympia, Wash.

8 W. Rudolfs, *Industrial wastes—their disposal and treatment*, American Chemical Society Monograph No. 118, Reinhold, New York, 1953.

9 *SSVL Environmental Care Project Technical Summary*, Stockholm, Sweden, 1974.

10 *Wastewater technology*, Environmental Protection Service EPS 3-WP-78-3, Beak Consultants, Ltd., Environment Canada, Toronto, Ontario, 1978.

9
Kraft Mill Process Modifications

The dominant pulping process in the world is the kraft process, largely because it can produce a strong pulp from a wide variety of species. Seventy-six percent of the pulp produced in the United States in 1982 was made by the Kraft process. The process also has an excellent chemical recovery system that, in itself, can be considered a pollution control device but was not developed for that reason. The high cost of chemicals led to the development of the recovery system.

PULP WASHING

In order for the pollutants and chemicals to reach the recovery system, they must be separated from the pulp in the pulp washer system. The pulp washer system thus has an important role as a pollution control device. The number of washers required depends on whether the pulping is carried out on a batch or continuous basis, because continuous digesters have incorporated diffusion washing as a final step. This final diffusion washer may be equivalent to one to three vacuum drum washers, the type commonly employed in most mills. Three to four stages of drum washing commonly follow the diffusion washer. For batch cooking, four stages of drum washers are common and five are not unusual in a modern mill.

In the continuous digester, the diffusion washing zone generally takes 1–4 hours (1) of retention time. Longer times are required in digester wash zones than in rotary drum washers because the pulp is still in chip form and considerable time is required to diffuse the liquor from the chip. Diffusion washers are one approach used to replace drum washers.

The efficiency of a washer is measured in terms of its displacement ratio (DR) (1):

$$DR = (S_v - S_e)/(S_v - S_w) \tag{9-1}$$

where S_v is the fraction of dissolved solids in the liquor entering the stage with the pulp; S_e is the fraction of dissolved solids in the liquor leaving the stage in the washed pulp; and S_w is the fraction of dissolved solids in the wash water supply.

Ideally, the displacement ratio should be unity, which is possible with very high dilution. High dilution, however, is undesirable because the recovered liquor must be sent to the multiple effect evaporator system, the size of which limits the total amount of wash water that can be used. It is obviously of little value to collect the liquor and then sewer it because of limited evaporator capacity. Since evaporation is very costly, it is desirable to keep washer dilution to a minimum. The dilution factor (*DF*) can be defined as the weight of wash water introduced into the black liquor per weight of pulp washed (2):

$$DF = L_w - L_e(1 - S_e)/1{,}000 \qquad (9\text{-}2)$$

where the dilution factor *DF* is expressed in kilograms of water per kilogram of air dry pulp; L_w is the wash water flow to the last stage, in kilograms per ton; L_e is the liquor leaving the last stage in the pulp, in kilograms per ton; and S_e is the fraction of dissolved solids in the liquor leaving the last stage with the pulp.

Recovery efficiency can be improved by increasing the dilution factor, but the response obtained depends on the number of washing stages used. The more stages used, the less gain additional dilution will yield. Thus, from the viewpoint of both pollution control and energy efficiency, more washing stages are desirable.

The operating conditions of a drum washer are important to its efficiency. The mat that is formed must be uniform. The showers must spray evenly across the surface, and the temperature of the water must be maintained at about 60–70°C. The operating temperature is important because it positively affects liquor viscosity, but if the temperature is too high, the resulting increased vapor pressure will interfere with the washer vacuum. Entrained air in the stock is also undesirable, because it forms a barrier to drainage.

Washing is obviously an interactive process, and a number of mathematical models have been developed to describe it (3–7). These models are quite complex and will not be discussed here. They can be useful to those who wish to improve washer efficiency through computer control.

One model that is useful in comparing different washing schemes is the Norden efficiency factor (3). This model computes the number of mixing stages that will give the same results as the washer or washing sequences under consideration. For uniformity of comparison, the factor has been redefined as mixing stages at 10% consistency and at the same dilution factor (4). Table 9-1 gives the modified Norden factors for some common washing equipment. The factors give a basis for comparing different types of equipment. This table clearly shows the advantages of washing within the digester. It also introduces some of the types of systems that are challenging conventional drum washing. The single-stage diffusion washer is clearly more efficient than the single-stage drum washer.

Diffusion washing is not new technology; it was practiced before drum washing became popular. What is new is the special mechanical configuration that Kamyr has developed to accomplish diffusion washing (8). The Kamyr configuration has several advantages over drum washing. The residence time in the diffusion

Table 9-1 Norden Efficiency Factors (modified)

Apparatus	Factor
Single-drum washer	2.5–4
Converted decker	2–2.5
Single-stage diffuser	3–5
Double-stage diffuser	6–10
Kamyr digester wash zone	
1 hour 30 minutes	4–6
3 hours	7–11
4 hours	9–14
Diffusion/extraction	1.5–2

Source: Reference 1.

zone is about 5–10 minutes, which is much longer than in drum washing, giving more time for the chemicals to diffuse out of the fibers. The washer is completely enclosed, so there is no exposure to air, and several diffusion washers can be put together in a tower, possibly eliminating the need for a building. The power requirement is reduced. There is no dilution between stages in diffusion washing, so less pumping energy is required.

Several commercial installations (9, 10) show that digester washing followed by a double-stage diffuser resulted in salt cake losses of less than 7.5 kg/ton (15 lb/ton). Formerly, when salt cake losses were considerably higher than this, a good direct correlation existed between salt losses and BOD losses. With the development of more efficient washing and the resulting low levels of salt loss, the two parameters are no longer linearly related.

Another washing alternative that is very conservative of space is the chemi-washer marketed by Black Clawson. This is a horizontal belt washer that bears some resemblance to a paper machine, except that it is completely enclosed under a pressure hood and operates under positive pressure. The pulp pad on the wire is also much thicker than the one on a paper machine. Results obtained with this device have been good (11), and it may see more application in the future.

A third device that seems to have promise as a replacement for drum washers is the Impco Compaction Baffle Filter (12). It is a slightly smaller unit than a drum washer and bears some similarity to a cylinder paper machine. The unit operates under pressure and eliminates the drop leg requirement of a drum washer. Because it is fully enclosed, it is free of odor and foam problems. It operates at high feed consistency, and the volumes of liquor handled are reduced to approximately 25% of those required by a drum washer. The displacement ratio at a given dilution factor appears to be superior to that of a drum washer. Its compaction zone gives it some of the properties of a press.

The merit of a press as the last stage of washing is that, in principle, more of the residual liquor entrained within the pulp can be recovered. However, for Kraft pulps, it has been found that the sodium and lignin contents in the liquid pressed out were lower than those in the liquid with which the pulps were initially in equilibrium (13). Thus the benefits might not be as large as a simple material balance would indicate, and pressing as a last washing stage has not found great acceptance.

KRAFT MILL EFFLUENT

The Kraft industry has an impressive historic record of reduction in the volume of effluent produced per ton of production. Figure 9-1 shows that effluent decreased from an average discharge volume of 20,000 gal./tonne in 1900 to about 5,000 gal./tonne in 1982 (14). This impressive record is due to a combination of process water reuse and conservation efforts and to an increase in machine productivity.

In a Kraft mill, effluent can be thought of as arising from three different sources—decker filtrate, evaporators, and intermittent uncontrolled losses.

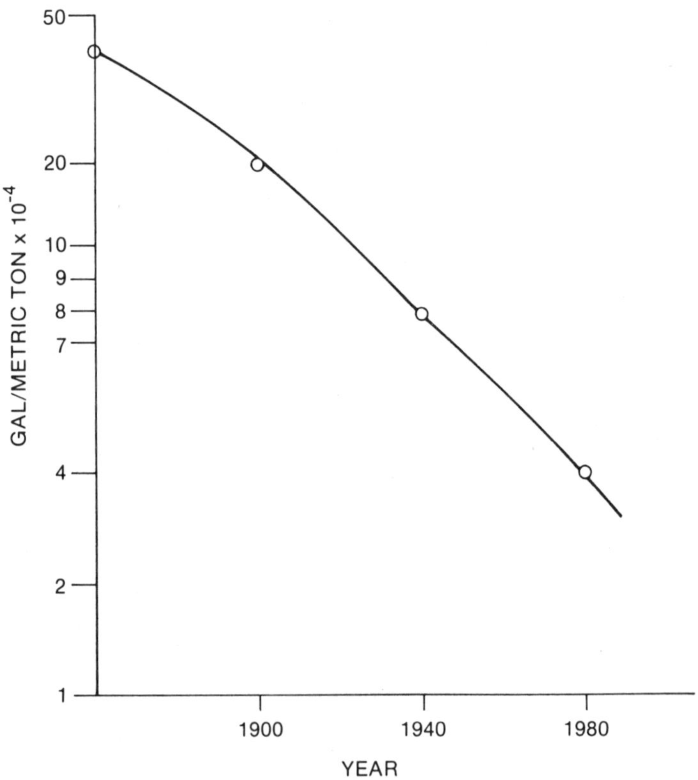

Figure 9-1 History of volumetric discharge, Kraft pulp and paper industry

Decker Filtrate

In older mills, decker filtrate accounted for a major portion of the sewer losses. It was common practice not to screen the pulp until after the brown stock washers. The pulp was diluted to 1–2% consistency and sent through a series of coarse and fine screens to remove slivers and knots. This dilution required 40–90 m^3 water per ton of pulp. In addition, depending on the design of the screens, another 10–50 m^3 could be required for additional dilution and as screen sealing water. In total, then, a poorly designed open-screen room could require as much as 150 m^3 water per ton. Just by practicing internal water recycling without changing the screen design, this amount could be reduced to 20–25 m^3/ton. After screening, the pulp was thickened on a decker for storage. Commonly, a large portion of the decker filtrate was sewered.

It has been demonstrated that the decker filtrate can be recycled to dilute the incoming pulp. It is possible to close the screen room totally with respect to water recycling. This has been accomplished both with pressure and with nonpressure screening systems, but it is easier with a pressurized screening system since there is less of a problem with foaming. One important consequence of closing the washing and screening system is that the temperature is increased to as much as 70–90°C, which makes special demands on screen design.

A second alternative for eliminating the decker filtrate as a source of pollution is to employ hot stock screening. After cooking, the pulp is passed through a fiberizer that fractionates the knots remaining with the pulp. Before washing, the pulp passes through a specially designed hot stock screen that effectively removes slivers. This sequence avoids the need to dilute the pulp for screening after washing and thereby reduces the losses from deckering.

In addition to hot stock screening, consideration is being given to eliminating screening entirely and replacing it with high-consistency refining, in which knots and slivers would be broken down while the stock is being refined.

Elimination of the decker filtrate as a source of pollution, by whatever method, should reduce the mill's BOD_5 load by about a third.

Evaporators

Once the material is captured by the washing system, the liquor must be concentrated in the evaporators before it can be burned in the recovery furnace. In the collection of condensates, it is always environmentally more advantageous to use surface condensers rather than direct-contact condensers. Digester and evaporator condensates together with decantation water normally account for about a third of the BOD load (15). Two major sources of pollution load in the evaporator area are boil-out solutions and evaporator condensates.

The normal procedure with hot water boil-outs is to send the first batch of solution back to the strong liquor storage tank, the second batch to the weak liquor storage tank, and the remainder to the sewer. Instead of sewering the last portion, a special tank should be constructed to store the boil-out solution, from which it should slowly be fed into the weak liquor system after the evaporator capacity has been restored by boil-out. Significant savings in BOD can be achieved by this method, which also eliminates a potential shock load to the treatment plant (16).

Contaminated condensates come from three main sources—continuous digesters, batch digesters, and multiple-effect evaporators. For the continuous digester, the principal source of contaminated condensates is the chip steaming vessel and the no. 2 flash tank. A condenser separates noncondensables for incineration from condensables. The principal source of condensables from the batch digester is relief gases during the cooking cycle and discharge from the blow tank during discharge of the pulp. The evaporators are another source of contaminated condensates. Evaporator exhaust vapors are collected in a hotwell and condensed using a barometric condenser. The cooling water from the condenser is thus the source of contaminated condensates.

Table 9-2 describes the main impurities involved, where they are involved, and their typical concentrations. Condensates are highly contaminated with low-molecular-weight organic matter. Methanol constitutes 80% of this organic matter and most of the BOD_5 (18). Other alcohols, ketones, small quantities of phenolic substances, sulfur compounds, and terpenes account for the remainder.

The logical way to solve the condensates problem is to reuse the condensates as much as possible. Digester and evaporator condensates can be used for brown stock washing, for makeup of caustics, in lime kiln stack scrubbers, and for dissolving tank makeup (19). Table 9-3 gives a more extensive list of possible uses for various water sources.

Extensive reuse of condensates has resulted in other internal problems in the mill. Increased wet strength resin usage has been linked to condensate usage in brown stock washing. Recycling also causes odor problems where none existed before, mainly at tank vents and vacuum pump exhausts. Many of the side effects of such recycling of condensates have yet to be defined.

The odor problems associated with condensate reuse can be solved by stripping the condensate before reuse. Two alternative techniques have been investigated—air stripping and steam stripping. Steam stripping seems to be preferable because of its effectiveness in removing BOD_5 as well as odorous compounds (18, 20–23). Air stripping is less costly and simpler to operate but does not have the flexibility of steam stripping (20, 24–27). [As of 1982, only one mill in Scandinavia was practicing waste gas (air) stripping of Kraft foul condensates (27).] The stripping column is either a separate unit operated on live steam or integrated into the evaporator plant and operated on liquor vapor. The latter alternative is preferred to keep operating costs reasonable. A separate column has a steam consumption of about 0.2 ton per ton of condensate.

The stripped-off gases contain substantial amounts of hydrogen sulfide and methyl mercaptan, as well as smaller amounts of dimethyl sulfide. The sulfur-containing materials are highly toxic, and the portion of them remaining in the condensate constitutes a significant fraction of the toxicity of the mill effluent (1). Several methods have been proposed for treating the stripped gases including

Table 9-2 Condensate Impurity Concentrations

Impurity	Batch Digester Vent Condensate	Batch Digester Blow Condensate	Continuous Digester Flash-Steam Condensate	Evaporator Combined Condensate	Evaporator Condenser Condensate
H_2S, ppm	30–270	1–230	210	1–90	1–240
CH_3SH, ppm	20–5,300	40–340	70	1–30	1–410
$(CH_3)_2S$, ppm	15–7,400	40–190		1–15	1–15
$(CH_3)_2S_2$, ppm	5–4,100	2–210		1–50	1–50
Methanol, ppm	1,300–12,000	250–9,100	670–8,900	180–700	180–1,200
Ethanol, ppm	90–3,200	20–900		1–190	1–130
Acetone, ppm	8–420	5–95		1–15	1–16
Methyl ethyl Ketone, ppm	27			1–3	2
Terpenes, ppm	0.1–5,500	0.1–1,100	100–25,000	0.1–150	0.1–620
Phenolics, ppm	12				3
Guaiacol, ppm				1–10	
Resin acids, ppm				28–230	
BOD_5, ppm	800–11,500	720–9,200	1,950–8,800	60–1,100	450–2,500
pH	9.5		9.2–9.6	6.0–11.1	6.7–8.2
Suspended Solids, ppm				30–70	
Color, APHA					280–5,500
Sodium, ppm				4–20	20–370

Source: Reference 17.

Table 9-3 **Reuse of Effluent from Different Unit Operations in a Kraft Mill**

Type of Effluent	Kiloliters/kkg (gal/ton)	Place of Reuse
Blow gas condensate, direct	5.7–22.4 (1,500–5,900) Average: 8.6 (2,000)	Brown stock washing Screen room or decker operation Hor water supply Mud washing Dissolving of additives
Blow gas condensate, indirect	1.33–1.52 (350–400)	None (sewered)
Cooling water for blow gas condenser, indirect	(1,500–5,900) Average: 8.6 (2,000)	Hot water supply Brown stock washing Bleached stock washing Screen room or decker operation
Turpentine decanter	0.038–0.63 (10–165) Average: 0.19 (50)	Showers on knotter Showers on brown stock washers
Cooling water for turpentine condenser	2.47–9.12 (675–2,440)	Hot water supply Screen room Boiler makeup water Direct blow heat condenser
Evaporator condensate	2.56–106 (675–2,800) Average: about 5.7–7.6	Brown stock washing Lime kiln scrubber Cooking liquor preparation Mud washing or dreg washing Woodyard, wash-ups, sewer Boiler makeup water
Evaporator barometric effluent	30.4–57.0 (8,000–15,000)	Transport of bark boiler fly ash Recycled through cooling tower
Cooling water for evaporator surface condensers		Hot water supply Machine showers in paper mill
Evaporator seal pit, discharge from surface condensers	1.52–5.89 (400–1,550)	Brown stock washing

absorption in white liquor, scrubbing and oxidation in the black liquor oxidation tower, scrubbing with chlorine, and incineration. Chlorine converts the sulfur compounds into relatively nonvolatile derivatives. The most widely used method of disposal is to burn the stripped gases, converting the sulfur content to SO_2. Incineration can take place in the lime kiln (28), a recovery furnace (29), or a separate incinerator (25). The lime kiln is the most popular alternative because of its less explosive hazard and because it is less costly than a separate furnace.

Intermittent Losses

Spills, overflows, and wash-ups together form the third major source of raw waste load from a Kraft mill. Approximately one-third to one-half of the BOD and TSS are attributable to this source. Such losses are due to a variety of factors, including breakdown of equipment, routine maintenance, planned shutdowns and startups, power failures, and grade changes. These losses can be minimized by the following techniques:

1 Periodic boil-out of evaporators to remove scale. The boil-out solution is saved in a special tank, as described previously, and returned to the process. As of the late 1970s, approximately 65% of mills employed this practice
2 Storage facilities for weak black liquor, strong black liquor, and recovery plant chemicals should be sized to avoid overflow in at least 90% of process upsets. As of the late 1970s, special spill collection systems were in use in about 50% of mills
3 Production curtailments should be used to prevent overflows if they cannot be avoided by other means
4 Continuous monitoring of mill sewers can be used to show when spills occur. Conductivity probe systems are especially useful in this regard
5 Personnel should be trained to avoid spills whenever possible and to take immediate corrective action when they occur
6 Surge lagoons before biological treatment should be provided
7 Installation of modern process control equipment can both aid in running the plant more efficiently and help to avoid upsets

To minimize sewer losses and spills, it is necessary for the mill to develop a segregated sewer system and to treat each sewer as a separate entity. Sewer systems are commonly classified into four categories: low suspended solids, high suspended solids, strong wastes, and sanitary sewers (23). Low suspended solids wastes are streams such as cooling water, bleaching effluents, and evaporator condensates that do not require clarification. High suspended solids wastes are streams such as decker filtrate, paper machine effluent, and wood yard effluent. Strong wastes sewers catch floor drains, overflows, spills, and strong condensates. These strong wastes should be contained in these areas and reused if possible.

The sanitary sewer contains human waste and is usually discharged separately to a municipal system or treated in a packaged, activated sludge system.

CLOSED KRAFT MILL CONCEPT

Both the techniques just discussed and some additional technology are part of the closed-cycle bleached Kraft pulp mill concepts developed by Dr. W. H. Rapson of the University of Toronto (30). Figure 9-2 shows the overall scheme employed. The system is completely closed, and water (pulp machine white water or other process water) is added only to the bleached pulp decker or to the last dioxide stage washer of the bleach plant. The bleach plant is completely countercurrent, and a major portion of the filtrate from this plant is recycled to the brown stock washers, after which it flows to the black liquor evaporators and then to the recovery furnace. The evaporator condensate is stream stripped and is then used as a major water source at various points in the pulp mill.

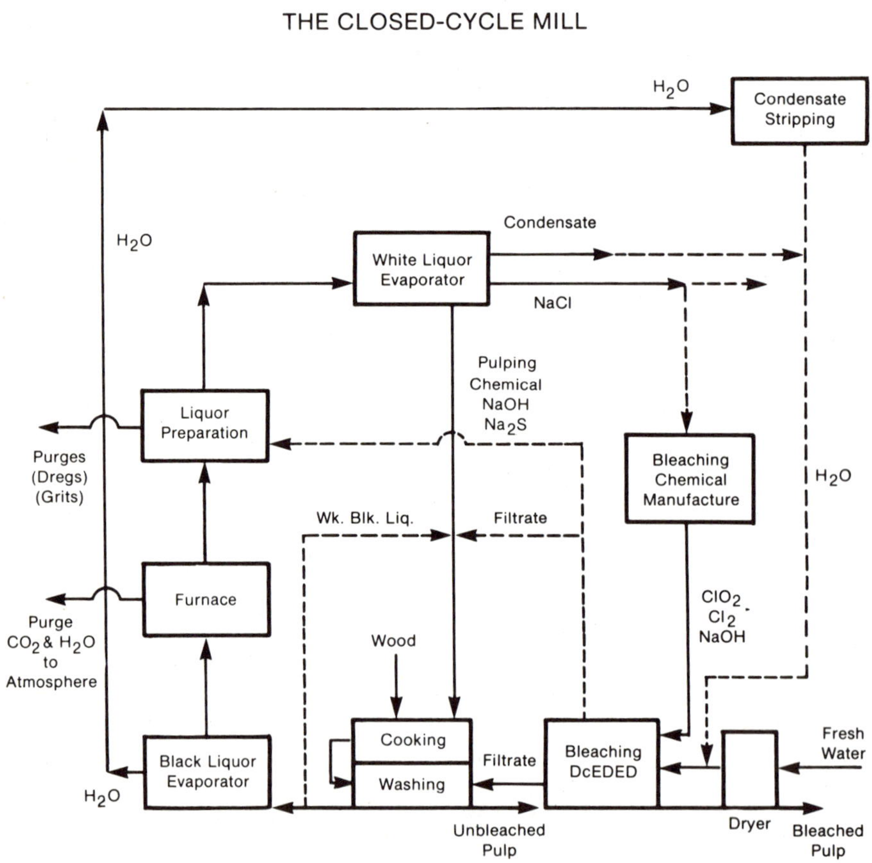

Figure 9-2 Closed-cycle mill

The malodorous sulfur compounds and the methanol removed by stripping are burned in the lime kiln. The smelt from the recovery furnace flows to the normal liquor preparation sequence. It contains a large amount of NaCl because the bleach plant liquors are recycled to the recovery furnace; this NaCl is removed by an evaporation crystallization technique. The while liquor is then used to do the cooking. The NaCl recovered is used to produce bleaching chemicals.

The main features of the closed-cycle bleached Kraft mill are:

1 Use of contaminated condensates in pulp processing after steam stripping
2 A closed screen room
3 A spill collection system
4 Countercurrent washing in the bleach plant
5 A salt control system

The spill collection system calls for separating sewers with contaminated flows from those with clean flows. Conductivity probe-activated gates divert spills to four different spill tanks, one for each separate area:

1 Fiber spills (typical size 120,000 gal.)
2 Acid bleach filtrate (typical size 180,000 gal.)
3 Alkaline bleach filtrate and black liquor spills (typical size 230,000 gal.)
4 Causticizing spills (typical size 270,000 gal.)

In order to accomplish countercurrent washing in the bleach plant, the first stage of bleaching becomes a DC stage rather than a straight chlorination. With countercurrent washing, all stages operate at a temperature of about 60°C, and a straight chlorination stage could degrade pulp strength (31). The countercurrent procedure, shown in Figure 9-3, reduces the DCEDED bleach plant recycle volume to about 3,000 gal/ton (22), which is required to maintain the water balance in the closed-cycle mill (32). By contrast, in 1980 the lowest effluent volume discharged from a North American CEDED bleach plant was 6,900 gal/air-dried ton.

Countercurrent washing in the bleach plant requires 317 extra low carbon (ELC) stainless steel washers due to the presence of chlorine dioxide. The recycled chlorination filtrate is expected to have chlorine concentrations somewhat above the recommended limits for 317 ELC stainless steel. The chlorination/dioxide stage washer should be of titanium or of a high-molybdenum stainless steel, such as Sandvik Alloy 2RK65.

Salt (NaCl) enters the system from a variety of sources, as listed in Table 9-4, but mainly from recycling of the bleaching filtrate. Different five-stage bleaching sequences impose different NaCl loads on the system.The NaCl imposes a dead load on the recovery system and leads to potential corrosion difficulties. The salt control process is a fractional crystallization procedure (33, 34).

When white liquor is increased in concentration, the solubility of sodium

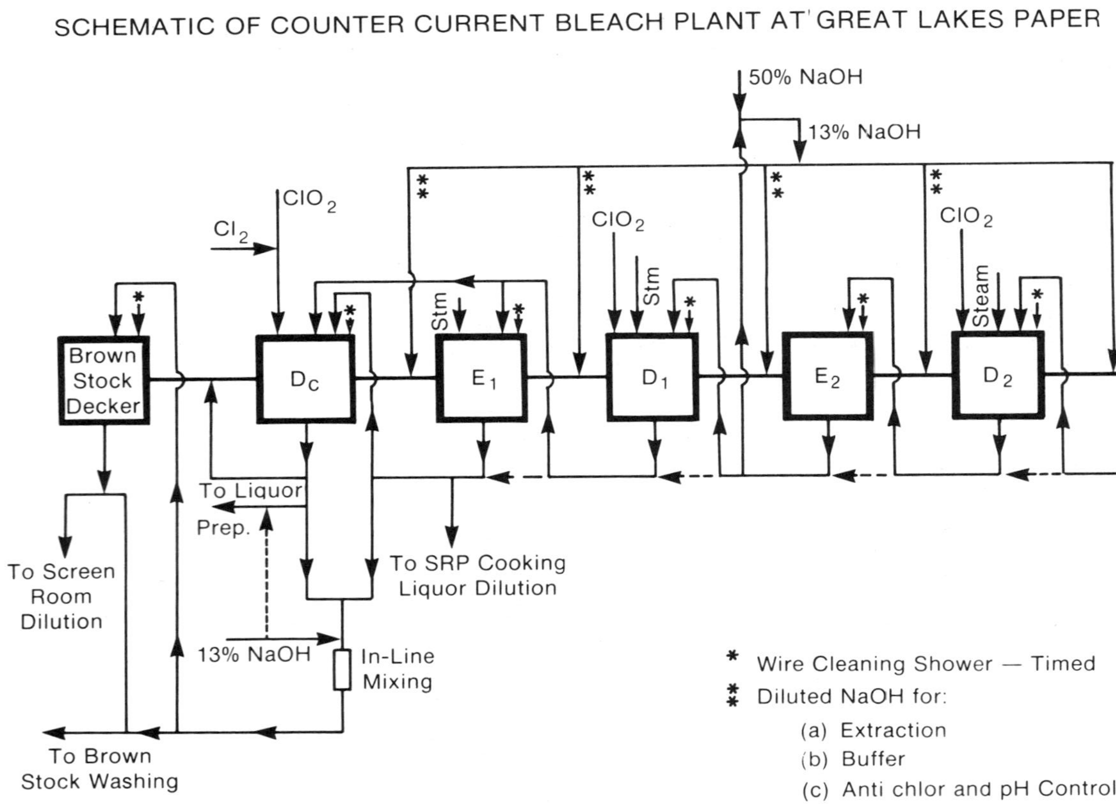

Figure 9-3 Closed-cycle Kraft mill; schematic of a countercurrent bleach plant

Table 9-4 Major Sources of Sodium Chloride Contamination

Sources	Range (lb/ton)
Saltwater-'borne' logs	2–55
'Brakish' process water	5–50
Salt cake	0.1–1.5
Other makeup	0.1–15
NaOH filtrate reuse	0.2–20
Dioxide 'spent liquor'	0.1–0.5
Bleachery filtrate recycling	30–350

Source: Copyright © 1981. TAPPI Reprinted from *Introduction to Environmental Management for the Pulp and Paper Industry*. Study Guide, Volume 6, with permission.

chloride decreases and it crystallizes. The salt recovery process (SRP) system uses a two-stage process, as illustrated in Figures 9-4 and 9-5. In the triple-effect first stage of evaporation, sodium carbonate plus sodium sulfate, in the form of Burkeite (a double salt containing 1 mole sodium carbonate per 2 moles sodium sulfate), is crystallized. About 90% of the total evaporation occurs in the first stage. The crystals are filtered and recycled, partly to the green liquor to recausticize the carbonate and partly to the black liquor ahead of the furnace to reduce the sulfate to sulfide. Cyclones are used instead of thickeners in the SRP process in order to eliminate a variable settleability problem with the salt (35).

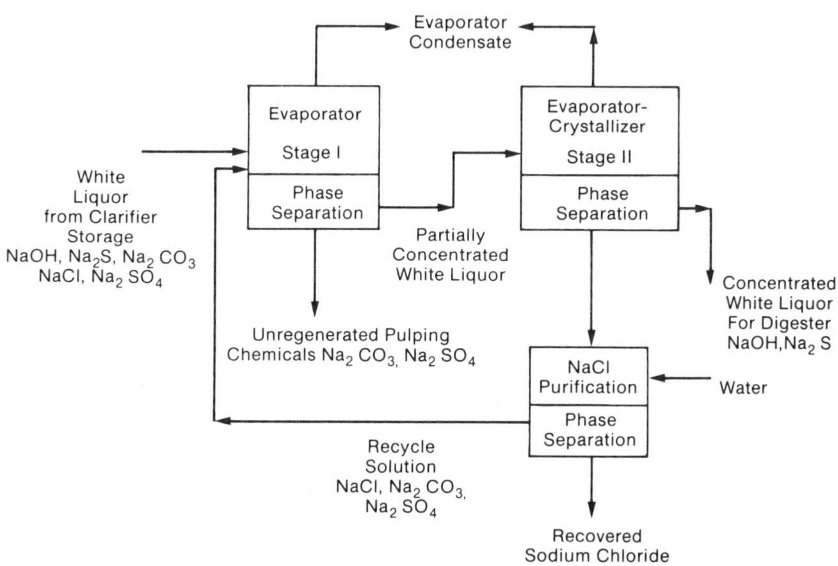

Figure 9-4 Salt control process. Copyright © 1981. TAPPI. Reprinted from *Introduction to Environmental Management for the pulp and Paper Industry*, Study Guide, Volume 6, with permission

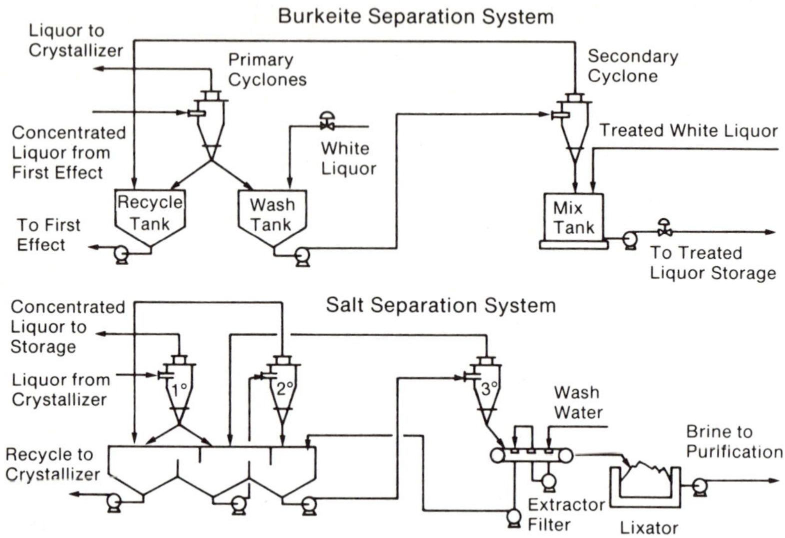

Figure 9-5 Salt cyclone separation system. (From D. W. Reeve et al., "Chemical recovery in the closed cycle mill, Part I: Superheater corrosion," *Pulp & Paper Canada*, 84(1): T25–29, Jan. 1983. Presented at the 1981 International Conference on Recovery of Pulping Chemicals, Technical Section, CPPA and TAPPI.)

In the second-stage, single-effect evaporator-crystallizer, sodium chloride is crystallized along with a small amount of sodium carbonate and sodium sulfate. The concentrated white liquor is diluted and sent to the digester for pulping. The controlled NaCl concentraton in the white liquor is about 20–30 g/l. The impure sodium chloride is leached with water to dissolve the sodium carbonate and sodium sulfate; this solution is returned to the first stage of evaporation. The purified NaCl is availabe for electrolysis to produce sodium hydroxide and chlorine, which are used, in turn, to make sodium chlorate or as feed chemical for the manufacture of chlorine dioxide.

In a mill in which water and all chemicals are recycled, the sodium chloride removed must balance stoichiometrically the sodium hydroxide, chlorine, and chlorine dioxide used in the process. If it does not, the chemical inventory will increase or decrease. It also follows that the amount of sodium chloride removed is exactly that required to produce the sodium hydroxide, chlorine, hydrochloric acid, and sodium chlorate used as bleaching chemicals.

High levels of sodium chloride in the recovery system can lead to two potential problems—increased corrosion and increased risk of recovery boiler explosions. In alkaline solutions above pH 11, the presence of chlorides has no significant effect on corrosion. Consequently, in the digester, the evaporator, and the green and white liquor systems, there should be no intensification of corrosion (36).

An increased chloride ion concentration does have a substantial effect on the corrosivity of acid solutions toward steels and stainless steels. However, only in the chlorination stage is the acidity high enough to allow the chloride to have a substantial effect; this means, as previously mentioned, that more resistant metals must be used in this stage. The brown stock washer system may also have to be made of more corrosion-resistant materials. The actual severity of the corrosion problem can only be determined from actual mill experience.

Laboratory experiments have shown that the sensitivity of smelt to explosions increases as the concentration of sodium chloride rises (36). However, smelt is already very sensitive to smelt water explosions due to sensitization by sodium sulfide, a major component. A survey by special teams investigating all Kraft mill recovery boiler explosions concluded that in no case was the explosion due to the presence of chloride (33).

The evidence seems to indicate that there will be no severe problems due to chloride. Many mills have operated for more than 25 years with higher concentrations of sodium chloride in their liquor systems than will be experienced in the closed-cycle mill. More than a dozen have more than 15 gNaCl/l in their white liquor, due mainly to the transportation of logs to the mill in sea water. The Powell River Mill of MacMillan Bloedel Ltd., in British Columbia, keeps the NaCl in its white liquor at 60–70 g/l, and has done so for 4 years with no apparent problems (32).

The closed-cycle bleached Kraft mill concept can be modified easily to incorporate oxygen bleaching or dynamic bleaching instead of the bleaching system described (37).

The closed-mill concept has been incorporated in the new 250,000 ton/year bleached Kraft pulp mill of the Great Lakes Paper Company at Thunder Bay, Ontario. This world's first closed-cycle mill put all systems into operation during 1977. Operating experience will show how successful the concept will be in practice. So far, bleaching chemical consumption appears to be higher than in conventional plants, and there have been corrosion problems in the superheater tubes of the recovery boiler. Some of the tubes were replaced with Incoloy 800, since the original ASA 209-T22 tubes could not withstand the conditions of sulfidation and oxidation present. The uniformity of the temperature distribution across the superheater tubes also had to be improved (38).

For the Great Lakes Paper Mill, it was estimated that the capital costs for the closed-cycle mill would be approximately equal to those required for conventional secondary biological treatment. In terms of operating costs, however, biological treatment represents an ongoing expense, whereas the closed cycle approach claims to yield savings due to

1. Heat savings from decreased steam consumption and from the additional steam produced from reclaimed organics
2. Fiber savings
3. Yield increases
4. Chemical savings
5. Water savings

Thus the closed-cycle approach appears to be economically advantageous provided the operating problems do not preclude it. From the standpoint of energy conservation and the environment, the closed-cycle mill is certainly advantageous because there is no contaminated effluent to be dealt with.

PRACTICAL MINIMUM EFFLUENT KRAFT MILL

The closed-cycle concept as implemented by Great Lakes Paper is one way to approach water reduction. Another is to examine practical effluent minimums that might be widely achievable. At the 1984 TAPPI Environmental Conference, in a workshop session on in-plant modifications, a survey conducted to determine the present state of the art for effluent from bleached Kraft market pulp mills, what effluent level was possible through conservative application of minimum effluent technology, and what absolutely minimum level of effluent appeared to be potentially possible was reported on. These three situations are outlined in Tables 9-5, 9-6, and 9-7; the numbers in parentheses in Tables 9-6 and 9-7 are the amounts of effluent reduction obtained by the represented technology in comparison with the current state of the art shown in Table 9-5 (39). The survey also included visits to the mills in North America and Scandinavia with the best (i.e., lowest) effluent volumes and loads, from which were compiled a list of the technologies being used and their operating experience (Table 9-8). We have previously discussed these technologies in this chapter except for the bleaching technologies (Chapter 13). The molecular adsorption resin system mentioned in Table 9-6 is discussed in Chapter 14.

From Table 9-6 it is apparent that significant gains can be made over current practice, and all Kraft mills should be able to achieve these levels of effluent loadings with minimal difficulties. If they can be achieved, the potentially possible

Table 9-5 Current State-of-the-Art (SOA) Kraft Mill Effluent Loading (Untreated)

	Untreated Effluent Loadings			
Unit Operation	Flow (m^3/ton)	BOD (kg/ton)	TSS (kg/ton)	Color (kg/ton)
Wood preparation	2	1	1	5
Pulping (including brown stock washing)	6	2	2	20
Bleaching (CEDED)	30	15	2	100
Recausticizing	1	—	1	—
Recovery/power	3	3	—	45
Machine (including stock preparation)	6	2	6	—
Filter plant	2	2	1	—
Total	50	25	13	170

Table 9-6 Practical Minimum Effluent Kraft Mill Technology Contributions[a]

	Untreated Effluent Loadings			
Technology	Flow (m^3/ton)	BOD (kg/ton)	TSS (kg/ton)	Color (kg/ton)
SOA Kraft mill	50	25	13	170
Bleach sequence (COD/improved washing)	(13)	—	—	(20)
Spill containment	(8)	(5)	(3)	(65)
Extensive white water filtration/reuse	(3)	(2)	(3)	—
Treatment of residual effluents (molecular adsorption)	(2)	(10)	(2)	(60)
Total	24	8	5	25

[a]Numbers in parentheses represent reductions achieved by each technology in comparison with the current SOA Kraft mill.

Table 9-7 Potentially Possible Minimum Effluent Kraft Mill Technology Contributions[a]

	Untreated Effluent Loadings			
Technology	Flow (m^3/ton)	BOD (kg/ton)	TSS (kg/ton)	Color (kg/ton)
SOA Kraft mill	50	25	13	170
Bleach sequence (nonchlorine)	(30)	(15)	(2)	(100)
Spill containment	(8)	(5)	(3)	(65)
Extensive white water filtration/reuse	(3)	(2)	(3)	—
Treatment of residual effluents		(None required)		
Total	9	3	5	5

[a]Numbers in parentheses represent reductions achieved by each technology in comparison with the current SOA Kraft mill.

minimum effluent levels described in Table 9-7 would totally eliminate the need for effluent treatment and would result in the discharge of a minimal pollution load from the mill. As of 1984 no mill had achieved this minimum effluent state, and technical issues such as corrosion, higher chemical consumption, buildup of thermal energy, product quality considerations, other interactions within the system that might appear, and effects on air emissions might make these minimum effluent levels impossible to achieve.

Table 9-8 In-Plant Technology Assessement

Technology/concept	Assessment of 19 Mills (North America/Scandinavia)[a]																		
Sewer segregation/spill collection		+			+	×	+	+	+		+	+			+		+/×	+	+
Closed brown stock screen room	×/+	+	n	+	+	+	+	+	+		n	+	+	+	+	+	+	+	+
Condensate stripping/reuse		+	n		n	+		+	+	+	n	+	+			+	+		+
Freshwater reduction devices																			
(mechanical seals, etc.)			×	+	+				+		+	+			+	+	+	+	
White water reuse	+	+	+	+	+			+	+		+	+			+		+	+	+
Counterflow bleach washing	+	+	n	+	n	+		+	+	+	n	n		+			+	+	+
Brown stock oxygen			n	+	n		+		+		n	n	+	+	+				
Operating/control strategy		+		+	+				+		+				+		+	+	+
Management commitment		+	+	+	+	+			+	+	+		+		+	+		+	

[a] +, technology operating successfully; ×, technology not working; n, not applicable to the mill process.

Nevertheless, these minimums provide a goal toward which the industry should be working, since such in-plant changes provide a more cost-effective means to reduce pollution while still producing a quality product. This problem is the process engineering challenge of the future. Such technology, in combination with the concept of millwide computer control, should lead to a more environmentally responsible as well as cost-effective Kraft process.

REFERENCES

1 J. R. G. Bryce, "Alkaline pulping," J. P. Casey, ed., *Pulp and Paper: Chemistry and Chemical Technology*, Vol. 1, Wiley, New York, 1980, pp. 443–452.

2 H. K. Waters and R. E. Bergstrom, *TAPPI*, 38(3):169 (1955).

3 H. V. Norden, V. J. Pohjola, and R. Seppänen, *Pulp & Paper Magazine of Canada*, 74(10):T-329 (1973).

4 J. R. Phillip and J. Nelson, *Pulp & Paper Magazine of Canada*, 78(6):T-123 (1977).

5 F. Kommanen, *Paperi ja puu*, 50(6):347 (1968).

6 M. Perron and B. Lebeau, *Transactions of the Technical Society of the Canadean Pulp and Paper Association*, 3(1):TR-1 (1977).

7 A. Tomiak, *Pulp & Paper Magazine of Canada*, 75(9):T-331 (1974).

8 J. Richter, *Tappi*, 49(6):48A (1966).

9 T. E. Jenkin, *Tappi*, 59(6):138 (1976).

10 J. P. Morgan, *Pulp & Paper Canada*, 76(2):57 (1975).

11 E. O. Ericssen and D. M. Moody, "Operating experience with a new horizontal brown stock washer," *Proceedings of the TAPPI Pulping Conference*, Toronto, Ontario, 1982, pp. 57–61.

12 B. H. Sanjana and L. D. Shockford, "Westvaco starts up new brown stock washer," *Proceedings of the TAPPI Pulping Conference*, Toronto, Ontario, 1982, pp. 39–44.

13 *SSVL Environmental Care Project Technical Summary*, Stockholm, Sweden, 1974.

14 S. W. Bush, "The closed mill concept," *Tappi*, 61(10): 54–56 (1978).

15 G. N. McDermott, *Sources of wastes from Kraft pulping and theoretical possibilities of reuse of condensates*, Technical Bulletin No. 72, National Council of the Paper Industry for Air and Stream Improvement, New York, 1954.

16 R. R. Perrault, *Tappi*, 65 (8):77–80 (1982).

17 "Review of Kraft foul condensates," *Tappi*, 62(10):33 (1979).

18 B. F. Hrutfiord et al., *Steam stripping odorous substances from Kraft effluent streams*, EPA-R2-73-196, U.S. Environmental Protection Agency, Washington, D.C., April 1973.

19 P. J. Kleppe and C. N. Rogers, *Survey of water utilization and waste control practices in southern pulp and paper industry* (Project No. A-036-NC), Water Resources Research Institute, University of North Carolina, Raleigh, 1970.

20 M. J. Matteson, L. N. Johanson, and J. L. McCarthy, *Tappi*, 50(2):86 (1967).

21 H. G. Maahs, L. N. Johanson, and J. L. McCarthy, *Tappi*, 50(6):270 (1967).

22 S. Bengkvist and E. Foss, "Treatment of contaminated condensates in Kraft pulp mills," International Congress on Industrial Waste Water, Stockholm, Sweden, November 2–6, 1970.

23 W. G. Timpe, E. Long, and R. L. Miller, *Kraft pulping effluent treatment and reuse—state of the art*, Environmental Protection Technology Series EPA-R2-73-164, U.S. Environmental Protection Agency, Washington, D.C., February 1973.

24 J. P. Morgan and F. E. Murray, *Pulp & Paper Magazine of Canada*, 73(5):T-124 (1972).

25 R. S. Rowbotton and J. G. Wheller, *Pulp & Paper Magazine of Canada*, 76(2):T-44 (1975).

26 E. Hebbon and H. Mannisto, "Evaluation of condensate stripping in Kraft mills," *Proceedings of the TAPPI Environmental Conference*, Minneapolis, 1982, pp. 149–159.

27 B. R. Blackwell, C. Chalk, and J. Betts, "Waste gas stripping of Kraft foul condensates," *Proceedings of the 68th CPPA Annual Meeting*, January 1982, pp. A207–A213.

28 A. A. Coleman, *Tappi*, 41(10):166A (1958).

29 S. Lindberg, *Svensk Papperstidning*, 69(15):484 (1966).

30 W. H. Rapson and D. W. Reeve, *Tappi*, 56(9):112 (1973).

31 W. H. Rapson, C. B. Anderson, and D. W. Reeve, "The effluent-free bleached Kraft pulp mill (VI): Substantial substitution of chlorine dioxide for chlorine in the first stage of bleaching," *1975 Alkaline Pulping Conference*, Technical Association of the Pulp and Paper Industry, Atlanta, 1975, p. 177.

32 D. W. Reeve, G. Rowlandson, and W. H. Rapson, "The effluent-free bleached Kraft mill (VIII): Bleach plant renovation and design," *1976 International Pulp Bleahhing Conference*, Technical Association of the Pulp and Paper Industry, Atlanta, 1976, p. 117.

33 D. W. Reeve and W. H. Rapson, *Pulp & Paper Magazine of Canada*, 74(1):87 (1973).

34 D. W. Reeve, J. A. Lukes, K. A. French, and W. H. Rapson, *Pulp & Paper Magazine of Canada*, 75(8):67 (1974).

35 D. W. Reeve et al., *Pulp & Paper Canada*, 84(2):59–62 (1983).

36 D. W. Reeve, *Pulp & Paper Canada*, 77(8):35 (1976).

37 D. W. Reeve, G. Rowlandson, J. Kramer, and W. H. Rapson, *Tappi*, 62(8):51–54 (1979).

38 D. W. Reeve et al., *Pulp & Paper Canada*, 84(1):58–62 (1983).

39 Personal communication, D. C. Haavisto, Special Projects Manager, Energy and Environmental Technology, Weyerhauser Company, Tacoma, Washington, 1984.

10

Process Modifications in Neutral Sulfite Semichemical Pulping

The neutral sulfite semichemical (NSSC) pulp produced in the United States is used for coarser-grade products such as corrugated board, the manufacture of which consumes 75% of this pulp. NSSC pulping is a two-stage process in which wood chips are softened by brief cooking with either a neutral sodium or an ammonium sulfite solution and then defibrated in a refiner.

From the perspective of process water reuse, sodium- and ammonium-based systems are similar, except that the ammonia produces additional potential pollution problems in the system using it.

It has been estimated (1) that the capital investment required to institute a water recycling program in an NSSC mill would be 15% of the cost of a new mill.

STOCK PREPARATION

Stock preparation can be handled in a manner similar to that discussed in Chapter 7. The key element in the recycling system is the use of saveall water free of long fiber and with a TSS below 240 ppm (2 lb/1,000 gal) in all paper machine showers, consistency regulators, and wash-up hoses (1, 2). Data on NSSC mill saveall water quality are available (3). In critical applications, a slotted screen is used to provide protection from plugging with long fiber. Some mills (4–7) have chosen to follow their saveall with an in-line straining device to remove additional large suspended solids. A review of industry experience with such devices has been published (8). The Sweco Microstrainer, Bauer Hydrosieve, and Dorr Oliver DSM Screen have all found application in this use. Some mills contend (4) that edge or cut squirts, breast roll showers, and seal water showers on suction box rolls must remain on fresh water, but other mills (5, 6) have used saveall clarified water in these applications.

Since pulp washing is usually not very efficient, the white water systems are highly contaminated. This rich white water is used as is in as many applications as possible, such as

1 Wastepaper repulping
2 Machine broke repulping
3 All stock dilution
4 Cooking liquor makeup
5 In the heat recovery scrubber of the liquor combustion plant

CASE HISTORY

The Owens-Illinois mill at Tomahawk, Wisconsin (5, 6), was able to add an entire paper machine without expanding its effluent treatment facilities by following these principles. Elements of their program included:

1 Use of cloudy saveall white water for pulp dilution
2 Upgrading of clear saveall white water with Sweco microstrainers to produce water usable on all paper machine showers
3 Routing of excess vacuum pump separator water to the white waste system
4 Segregation of noncontact waters for separate discharge or for recirculation to the incoming mill water during the cold months of the year
5 Installation of a vacuum pump seal water recirculation system with freshwater makeup for temperature control
6 Use of excess seal water from the vacuum system for ash flowing and mixing in the boiler area
7 Replacement of fresh water with evaporator condensates for pulp washing
8 Addition of an extra 400,000 gallons (400 gal./ton/day) of white water storage capacity to prevent intermittent losses

SPILL PREVENTION

In the liquor preparation, pulping, and recovery portions of an NSSC mill, nonequilibrium losses account for a major fraction of the effluent load—two-thirds of the total losses in one reported instance (9). Reference 10 discusses how best to control intermittent losses. Control of nonequilibrium losses should include the following:

1 Prevention of spills through process control modifications
2 Redirection of overflow pipes to trenches leading to "contaminated" surge vessels for ultimate reuse
3 Individual modifications in level controls and storage tanks to minimize overflows and spills

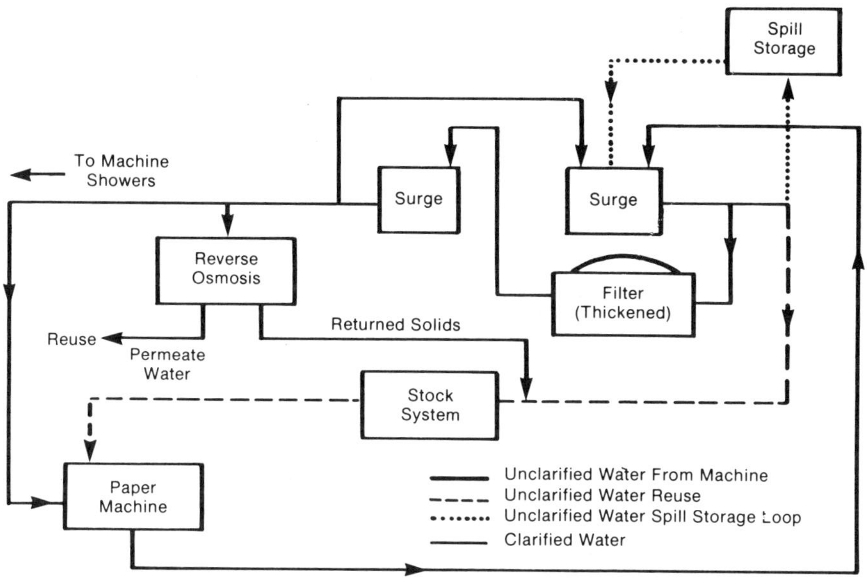

Figure 10-1 Water reuse schematic, NSSC mill

4 Redundant installation of key pumps and other equipment to avoid losses due to equipment failure and routine maintenance

5 Monitoring systems to alert operating personnel of potential and actual spills so that corrective action can be initiated promptly

6 Possible provision of storage lagoons before biological treatment to accept longer-term shock loads

7 Training of personnel to avoid spills whenever possible and to take immediate corrective action when they occur

Many valuable losses can be prevented by instituting sewer separation. An NSSC mill should have three distinct sewer systems—a sanitary sewer, a clear-water sewer receiving clean streams such as cooling water, and an upset sewer that is capable of returning upset losses and spills to the process system. A surge basin is used to collect spills. One mill producing 300 tons/day installed a 200,000-gal. surge basin (2). Figure 10-1 illustrates the location of the spill storage basin in one mill; this basin returns material to the system as the total system can absorb it. Figure 10-1 also shows a reverse osmosis unit as part of the system, the purpose of which is to maintain the system's volume within the limitations of the fixed surge storage that is available. The reverse osmosis unit becomes especially important if the liquor-burning unit is down for any reason (11, 12).

EFFLUENT REDUCTION

NSSC mills have been able to achieve fairly high degrees of recycling so that effluent levels on the order of 1,000–2,000 gal./ton of production have been achieved. In so doing, they reduce the BOD and suspended solids loads appreciably. One mill (1) went from a BOD of 97 to 11 lb/ton, and with more stable performance thought it could reach 4.8 lb/ton. This same mill went from suspended solids of 10.5 to 1.7 lb/ton and believed that 1 lb/ton could be achieved. A second mill (2) with higher liquor losses moved from a BOD of 100 lb/ton at 40,000 gal. of effluent per ton to 40 lb/ton at 11,000 gal./ton. A survey (8) of exemplary NSSC mills showed discharge volumes of 268–8,980 gal./ton, BOD of 2.1–37.7 lb/ton, and color of 17–450 lb/ton.

TECHNICAL PROBLEMS

With such high levels of recycling, the dissolved solids in the white water reach very high levels and the system temperature rises to as high as 71°C (160°F). The first mill cited above (1) experienced 6% dissolved solids in its white water without serious operating difficulties, but the second mill (2) experienced serious difficulties at a dissolved solids level of 1.5% and an effluent discharge level of 1,500 gal/ton. The problems observed were:

1. Variable paper quality due to wet streaks in wet felts
2. Decreased wet felt life due to plugging from fines
3. Increased slime deposits
4. Higher maintenance costs due to more frequent cleaning of machine elements
5. Higher corrosion rates
6. Increased calcium scaling
7. Greater chemical demands for sized and wet-strength grades
8. Buildup of contaminants from waste paper used in the furnish
9. Increased incidence of sheet breaks, particularly at the presses, due to "stickiness" caused by solubles buildup and by apparent reduction in wet web strength

To minimize these problems, freshwater usage was increased to 2,000 gal/ton (8,346 l/kkg).

The disparity of experience between the two mills may be attributable to equipment differences. A third mill followed the same approach as did the first and did not have these problems (13). Judicious application of technical knowledge might solve or alleviate the nine difficulties listed. A totally closed NSSC mill might be possible at some time in the future.

Some of the sheet strength problems can be overcome by increasing the quantity of waste paper in the furnish. The dissolved solids problems may have an effect on sheet strength. By increasing the quantity of old corrugated boxes in the furnish, the strength and runability can be improved. The use of waste corrugated to replace part of the furnish for corrugating medium is growing. The impact of this move on NSSC mill raw waste loads is unclear. The EPA Development Documents for Phase III (14) attempted to define this impact. The EPA concluded that raw waste flow tended to decrease with increased waste paper usage, but the exact functional relationship was uncertain. The EPA also concluded that in semichemical mills using 10–67% waste paper, raw waste BOD loads were not related to waste paper usage.

IMPACT OF REFINING

One factor that does seem to affect BOD production is the amount of refining (15, 16). Berg and co-workers (15, 16) showed that birch NSSC and spruce bisulfite washed to 95% efficiency (sodium losses), defibrated in a laboratory Sprout-Waldron refiner, and refined in a PFI mill produced a BOD of 40 lb/ton for birch and 30 lb/ton for spruce. Although there was significant variation in the results, refining was thought to contribute about 20 lb/ton. This result indicates that if it is at all practical, the refining of NSSC pulp should be done before pulp washing. It must be remembered, however, that refining will probably affect washer performance adversely.

In general, the loss of organics to wastewater should be related to one or more of the following factors:

1 Generation rate of soluble organics in pulping and refining
2 Ability to route pulping organics to the recovery system
3 The amount of waste paper used
4 The amount of organics retained in the sheet

Historically, NSSC pulp has not been very well washed. Some mills do not even use washers, but merely put the pulp through two stages of dilution and pressing (4). Some of the newer mills, however, have switched to five stages of pressing or three stages of drum washing, and these methods should have a beneficial impact on the effluent BOD load.

CORROSION

As with other types of mills, closure of the white water system of an NSSC mill seems to bring about an intensification of mill corrosion problems. The NSSC mill environment seems to be an exceptionally corrosive one, which may be due partly to the possible presence of sodium hydrosulfite. Sodium hydrosulfite has a major influence on the initiation of corrosion on 304 stainless steel cupons (17). Sodium hydrosulfite might act in a manner similar to chlorides and sulfate to destabilize stainless steel and allow the initiation of corrosion. In the NSSC system, pitting corrosion can result from the destabilizing effect of one or more of these ions, or it can result from other causes. Deposits of sulfate-reducing bacteria can form large crust-covered areas under which pitting corrosion can occur. In some instances, *Desulphoribrio desulphuricans* has been identified as the responsible species (18, 19). These two references also discuss the mechanism by which bacteria attack stainless steel.

The usual approach taken to solve the corrosion problem is to replace the affected material with a more corrosion-resistant substance. The beneficial effects of molybdenum on the corrosion resistance of stainless steel are well recognized. For the environment of a closed white water system, a stainless steel with at least 2% molybdenum content will be needed. The 316, 316L, and 444 stainless steels all fulfill this criterion. The new 444 ferritic stainless steel is finding application in the paper industry where stress corrosion cracking may be a problem, and it is less expensive than 316L (20).

The effects of system closure on corrosivity are being studied to find the solution to the problem (19, 21, 22).

REFERENCES

1 W. R. Nelson, G. O. Walraven, and D. C. Morris, *Tappi*, 56 (7):54 (1973).

2 K. E. Lowe, "Control of effluent at a NSSC mill by reuse of white water," *7th TAPPI Water and Air Conference*, Technical Association of the Pulp and Paper Industry, Atlanta, 1970, p. 211.

3 S. Nortox and R. Miner, *Factors impacting raw waste loads from semi-chemical corrugating medium mills*, Technical Bulletin No. 385, National Council of the Paper Industry for Air and Stream Improvement, New York, November 1982.

4 G. O. Walraven, W. R. Nelson, P. E. DeRossi, and R. L. Wesneski, *Closed process water loop in NSSC corrugating medium manufacture*, Environmental Protection Technology Series, EPA/600-77-241, U.S. Environmental Protection Agency, Washington, D.C., December 1977.

5 C. W. Stahr, S. F. Galeano, and J. R. Retzke, "How Owens-Illinois expanded effluent control capacity using internal rather than external controls," *Tappi*, 63 (8):39–42 (August 1980).

6 K. R. Boike, "Tomahawk project of Owens-Illinois wins API/NFPA management award," *Paper Trade Journal*, 163 (18):40–42, 44, 47 (September 1979).

7 Y. M. Mehta, "Effects of process water recycle on pollution control and energy conservation," *Pulp & Paper Canada*, 80 (9):T-270–T-275 (September 1979).

8 R. Miner, *A review of industrial experience with selected internal process solids separation devices*, Technical Bulletin No. 314, National Council of the Paper Industry for Air and Stream Improvement, New York, July 1978.

9 D. W. Reeve and W. H. Rapson, *Pulp & Paper Magazine of Canada*, 74 (1):87 (1973).

10 J. McKeown, *Development of a management strategy for intermittent losses*, Technical Bulletin No. 341, National Council of the Paper Industry for Air and Stream Improvement, New York, February 1981.

11 W. R. Nelson and G. O. Walraven, "A role for reverse osmosis in neutral sulfite semichemical

pulp and paperboard mill," *Proceedings of the XXIII Industrial Waste Conference*, Part 1, Purdue University, May 7–9, 1968. *Engineering Extension Series* No. 134, p. 540.

12 D. C. Morris, W. R. Nelson, and G. O. Walraven, *Recycle of papermill waste waters and application to reverse osmosis*, ORM, EPA Program #12040 FUB, Green Bay Packaging Inc., Mill Division, Green Bay, Wisco., January 1972.

13 H. C. Hentzel, "Pollution abatement by whitewater system closure at Consolidated Packaging Corp.," *Proceedings of the TAPPI Environmental Conference*, Savannah, Ja., April 9–11, 1984, pp. 123–130.

14 *Proposed development document for effluent limitations guidelines and standards for the pulp, paper and paperboard and builders' paper and board mills point source categories*, EPA 440/1-80/025-b, U.S. Environmental Protection Agency, Washington, D.C., December 1980.

15 L. Berg, N. Hartler, and H. Norrstrom, "Removal of organic substances during defibration and refining of semichemical pulps. Part 1: Birch NSSC pulp," *Svensk Papperstidning*, (9):291–297 (1978).

16 L. Berg and H. Norrstrom, "Removal of organic substance during defibration and refining of semichemical pulp. Part 2: Spruce bisulfite pulp," *Svensk Papperstidning*, (11):375–378 (1978).

17 C. C. Robinson, "White water corrosion of stainless steel in a paper mill," *Appita*, 35 (4):303–306 (1981).

18 W. A. Mueller and J. M. Muhanen, *Tappi*, 55 (4):589 (1972).

19 M. Lapointe et al., *Pulp & Paper Canada*, 80 (8):T-251 (1979).

20 R. Jackson and R. Fraser, "A new stress corrosion resistant stainless steel for pulp storage tanks," *Appita*, 36 (2):144–146 (1981).

21 D. F. Bowers, "Changes in water properties and corrosivity with closure," *Proceedings of the TAPPI Environmental Conference*, Atlanta, April 18–20, 1982, pp. 95–97.

22 D. F. Bowers, *Tappi*, 61 (3):57–61 (1978).

11
Groundwood, RMP, and TMP Process Modifications

GROUNDWOOD

Groundwood pulp is used mainly in the manufacture of newsprint, toweling, tissue, wallpaper, and coated specialty papers (1). The pulp, as its name implies, is essentially ground up wood, so the yield generally exceeds 85%. A wide range of processes are used to produce mechanical pulp—stone groundwood, chemi-groundwood, refiner groundwood, thermomechanical, and chemimechanical techniques. Specific details of these processes are given elsewhere in this book and in other references (1–4). This section will discuss the effluents involved in the processes and those modifications that can minimize or eliminate them.

The processes can be divided into two types, depending on whether the pulp is produced from round wood or chips. In older processes, round wood is pressed against large rotating stone grinders by hydraulic rams while water is sprayed on the stones (2). The ground pulp is then screened to remove wood slivers and coarse debris and is thickened on a decker. The pulp is sent to a stock chest for use in the paper mill. The decker filtrate is clarified by sedimentation, filtration, or flotation and is returned to the process. A flow sheet for this process is shown in Figure 11-1. The newest groundwood mills are considering pressurizing the grinding operation in an effort to improve pulp quality.

Effluent arises from gland seals, spills, screening of rejects, and screening of decker filtrate. Gland seal water should be kept to a minimum by using limiting orifices in the seal lines. Most mills do a successful job of keeping spills to a minimum. The rejects from the screening operation should be passed to a reject refiner and returned to the process just ahead of the fine screen, as indicated in Figure 11-1. Thus there should be no effluent from this source. However, as of 1980, some mills were still sewering screen rejects. The decker filtrate is largely returned to the process, but it is common practice to sewer a portion of it in order to prevent the buildup of solubles and the slime problems that might result. These problems can be controlled by good housekeeping and the addition of proper chemicals. In some modern newsprint mills, the groundwood operation is completely closed, and the only process effluent is from the paper machine (5). Chapter 7 discusses in depth the closure of the paper machine white water system.

A well-operating stone groundwood mill will have a discharge of about

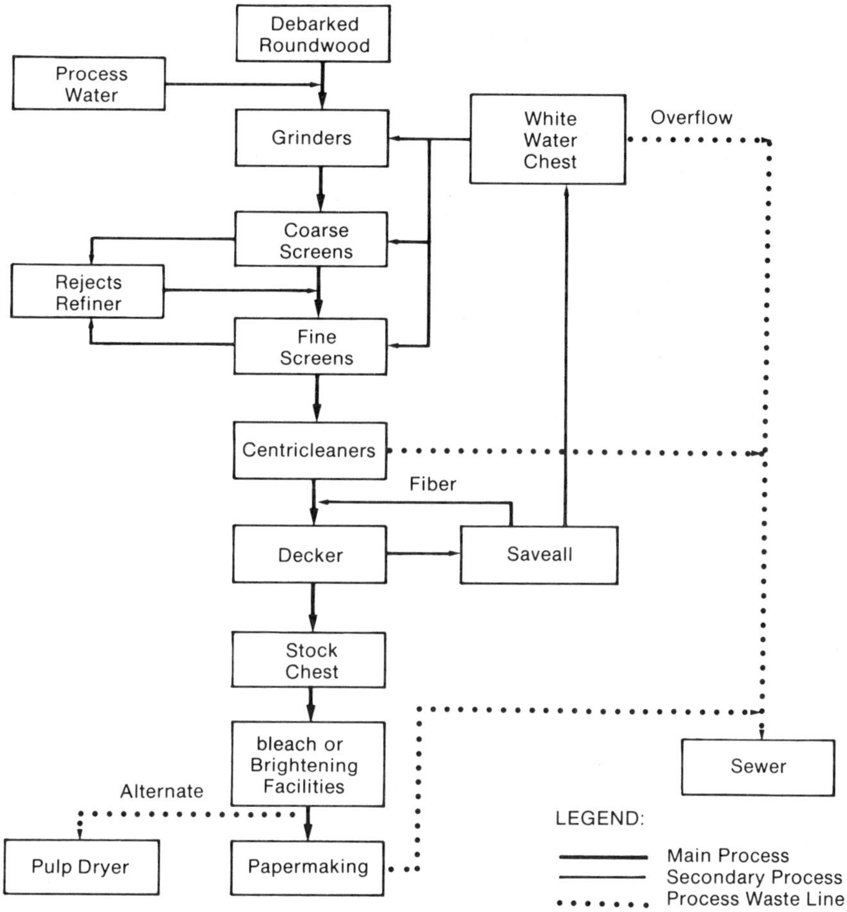

Figure 11-1 Stone groundwood pulp mill process flow diagram

2,000 gal/ton. The BOD_5 of stone groundwood mills ranges from 4.5 to 9 kg/ton (10–20 lb/ton) and the TSS from 3.6 to 23 kg/ton (8–50 lb/ton). The flow ranges from 6 to 100 m^3/ton, with an average of 25 m^3/ton (6).

A totally closed system was tried at two chemigroundwood mills in which the logs were penetrated with a sodium sulfite cook before grinding. The system was abandoned due to severe equipment and piping corrosion. To be successful, the mill's water system would have to be totally rebuilt, using expansive, corrosion-resistant materials throughout.

REFINED MECHANICAL PULP (RMP)

In recent years, a considerable amount of groundwood has been produced as RMP by passing wood chips through disc refiners. This process requires less power than does stone groundwood and is amenable to the use of saw mill waste. Two stages of refining are normally employed. The refiners contain fixed and rotating discs between which the chips pass together with a stream of water. The pulp is discharged as a thick slurry and is screened and thickened in a manner similar to that of stone groundwood pulp. Figure 11-2 is a block and line flow diagram for such a process. Effluent arises from the same sources as those used with stone groundwood, and similar comments apply. Generally, less water is

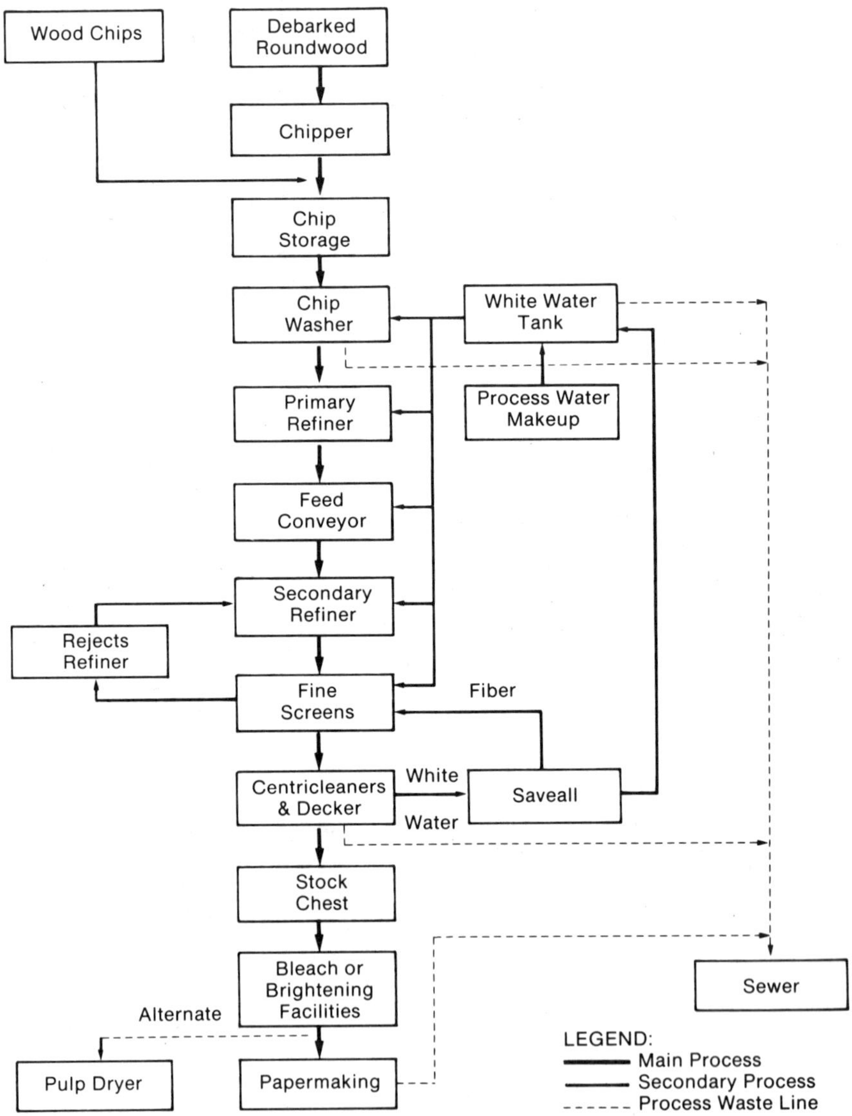

Figure 11-2 Refiner groundwood pulp mill process flow diagram

required with refiner groundwood. BOD_5 ranges from 7 to 55 kg/ton (15–120 lb/ton) and TSS from 13 to 50 kg/ton (30–110 lb/ton).

THERMOMECHANICAL PULPING (TMP)

TMP, which is increasing in populatiry in the United States, is an outgrowth of the refiner process that uses a presteaming or reaction vessel before the primary refiner, as do chemi-mechanical pulping processes. As the use of hardwoods, which require chemical pretreatment in order to produce a suitable groundwood pulp, increases in chemimechanical pulping systems, corrosion problems of high-water recycling in such chemical systems may become more important.

The pollution load created by TMP is greater because higher temperatures and pressures are used. In the presteaming stage, soluble components are removed from the wood. Figure 11-3 shows a generalized relationship between the pollution load and the various mechanical pulping processes. These pollution loads are lower than those from chemical pulping processes. The suspended solids discharges for mechanical pulping, however, are usually greater than those of chemical pulping because of the large quantity of fines produced.

A primary factor affecting the amount of pollution from a mechanical pulp mill is the wood species used. Most northern softwoods such as spruce, western red cedar, and hemlock are low in resin and fatty acids and produce less pollution than do the pines. Mechanical pulping effluents before biological treatment are toxic (7). Most of the toxicity comes from the resin and unsaturated fatty acids, but neutral compounds such as pimarol, juvabione, and diterpene alcohols also

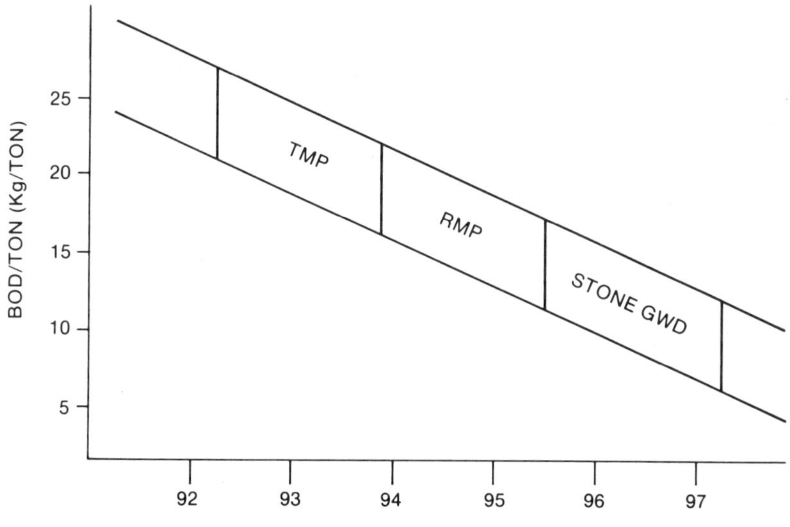

Figure 11-3 Generalized variation of a BOD load with a mechanical pulping process

contribute. This toxicity is lowered to a tolerable level by dilution and biological treatment.

The TMP process can recycle water in the pulping system in the manner shown in Figure 11-2. If the mill is integrated, the principles for recycling water in the paper machine area and stock preparation area as outlined in Chapter 7 apply. A set of guidelines that have been developed for mechanical printing paper mills may be helpful (8). The term "short circuit" applies to the tray water returning directly to the fan pump, while all other water is in the "long circuit."

Guidelines for Mechanical Printing Papers (8)

1 White water with the highest fiber content must be retained in the short circuit
2 All discontinuous white water use shall be from the long circuit. The long-circuit white water surge tank must have a volume corresponding to the need for discontinuous white water (broke, pulpers, showers)
3 Fresh water makeup should be at one point only if possible
4 The white water overflow should be as constant as possible and should come from only one source
5 The level in all of the tanks should be maintained with white water having a consistency and temperature similar to those in the corresponding tank
6 The shower water should be taken from the white water whenever possible
7 New sealing systems that preclude water leakage into the circuit should be used whenever possible

These rules were developed from experience at the Anjala Paper Mill in Finland.

Reference 9 discusses ways of reducing environmental loads from a TMP mill. It advocates making each operation as water efficient as possible and then recycling water from the paper machine back to the pulp mill. This procedure would lead to a higher concentration of dissolved materials in the pulp mill than in the paper mill, but the pulp mill can better tolerate these higher concentrations.

In Chapter 5 as demonstrated it is desirable to operate a mill at the minimum cost combination of external and internal means of pollution abatement. For the Anjala Paper Mill, the combination of costs and savings for internal measures was so low relative to the cost of external treatment that the optimum solution for this mill was to maximize water system closure. If any external treatment is required, the BOD strength should then be high enough to make a combination of anaerobic treatment followed by aerobic treatment the most feasible solution (8).

For the wood room and chipping operation, the principles discussed in Chapter 8 are applicable.

One of the unique problems of mechanical pulps is brightness stability. Stability is affected by the amount of iron and manganese left in the pulp. These ions build up in a closed white water system and may be a limiting factor in the decrease of effluent discharge from mechanical pulping and papermaking systems.

A major drawback of TMP has been the large energy input required and the large amount of contaminated waste produced. The energy requirements have been greatly reduced by the development of a TMP heat recovery system. One mill reports 71% clean steam recovered from the electric power input to the refiners (10). Steam is recovered from both the first- and second-stage refiners of the TMP process. The first-stage steam is generally used to preheat the chips and to supply the hot water needed for the paper machine. The second-stage steam goes to the heat recovery unit, which consists of a scrubber followed by a heat exchanger. In one mill, a venturi scrubber was used (10) to clean up the TMP steam.

The dirty steam generated by the TMP process contains fiber and resinous compounds that would foul the recovery heat exchanger. This heat exchanger can be either a tube-type unit or a lamellar unit. One of the first heat recovery systems installed chose the lamellar exchanger (10), and it performed satisfactorily. The unit was designed to produce clean steam with no more than a 6°C difference from the dirty steam and has in fact produced steam within 4° of the incoming dirty steam.

The steam produced by the heat recovery system is usually not enough to provide all of the energy required by the paper mill. It is not uncommon for the TMP heat recovery system to provide about half of the steam required by the paper machine. The other half must come from another source such as the bark boiler.

The TMP process is increasing in populatiry, and some newsprint is being manufactured from 100% TMP. The problem is always to get rid of the last small amount of chemical pulp in the furnish that seems to be necessary for machine runability. The OPCO-CTMP process appears to offer this capability (11). The OPCO process will be less polluting than the normal practice of producing newsprint from a combination of groundwood and sulfite pulp.

SULFITE PULPING PROCESS MODIFICATIONS

The small amount of chemical pulp used in conjunction with groundwood is often sulfite pulp. Because of the existence of such a wide variety of sulfite systems, which are gradually diminishing in importance as processes for pulp production in the United States, this category will be discussed only briefly.

The first requirement of any environmentally sound pulping system is that it contain a means of recovering the pulping chemicals. Therefore the calcium-based sulfite systems can be ruled out and only systems based on soluble components such as sodium, magnesium, and ammonia can be considered. References 2, 3, and 12 can be consulted for discussions of the wide variety of recovery technologies available for sulfite and their advantages and disadvantages.

As with other pulping systems, the main source of BOD is spent liquor, which must be washed from the pulp and efficiently contained to keep it from the white water system. Closed-screen rooms that do not spill screening water must also be designed into sulfite mills. A recently-modernized sulfite mill (13) incorporated the following features in order to minimize losses:

1 A spill collection system for the pulp washing and screening area
2 High-level alarms on the chemical and stock tanks
3 Flow recorders and continuous samplers on main sewers
4 Overflow from heavy to weak liquor tanks
5 Extra equipment capacity to handle spills and upset conditions
6 Ability to return heavy liquor from the evaporator to the weak liquor tanks instead of sewering it

If liquor is recovered and spills are contained successfully, the evaporator condensates contain the process BOD_5. One mill reported (13) that 60% of the BOD load was contained in the condensates. Two ways to deal with this situation exist—neutralization of weak liquor before evaporation or reuse of condensates. Neutralization decreases the volatility of acetic acid, which is a major BOD source in condensates. A pH of 5.0–5.5 causes 75–80% of the acetic acid to be retained in the heavy liquor and destroyed in the recovery furnace. The Swedish Steam Users Association has become proficient in designing systems for handling condensate, and reference 14 provided additional information on available techniques.

It is possible to design an efficient, modern sulfite pulping system that will contain most of the pollution load within the system. Very few sulfite plants are now being built in the United States, however, because the Kraft process has greater flexibility and produces higher pulp yields. The typical U.S. sulfite mill is diagrammed in Figure 11-4 (15). It incorporates far fewer recycle loops than does a normal Kraft mill. In particular, the sulfite mill uses three major loops:

1 Wastewaters from liquor preparation are used for pulp washing and screening
2 White water from the paper machine is used for final bleach stage washing and pulp washing. White water is also used for all stock preparation dilution and consistency regulation
3 Press effluents are used in the machine white water system. If felt hairs are a problem, press effluents are used for stock dilution before the cleaners.

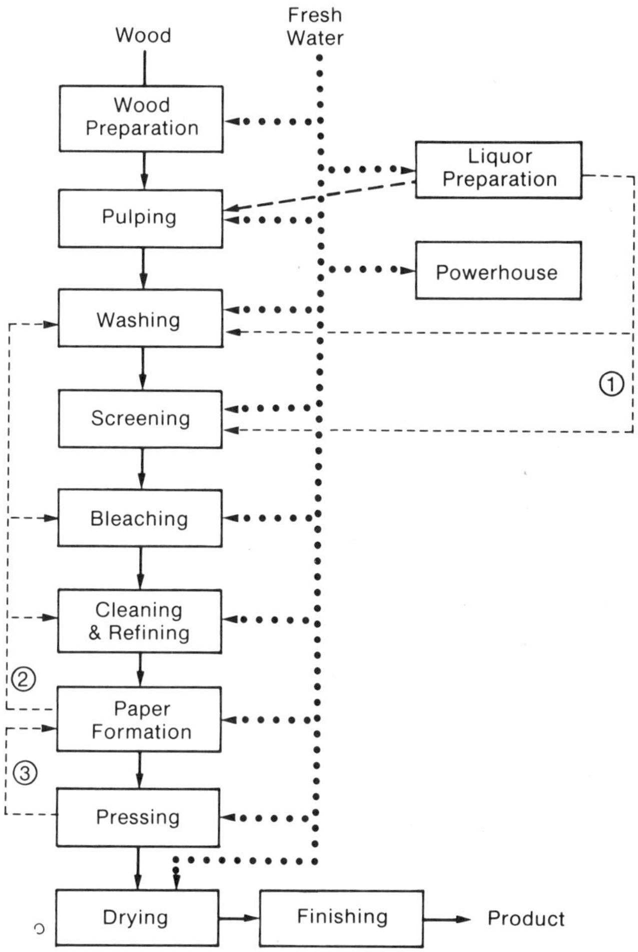

11-4 Sulfite mill process flow diagram

REFERENCES

1 R. G. Macdonald and J. N. Franklin, eds., *Pulp and Paper Manufacture*, Vol. 1, 2nd ed., McGraw-Hill, New York, 1969.

2 S. Rydholm, *Pulping Processes*, Interscience, New York, 1967.

3 K. W. Britt, *Handbook of pulp and paper technology*, 2d ed., Van Nostrand Reinhold, New York, 1970.

4 J. P. Casey, ed., *Pulp and Paper: Chemistry and Chemical Technology*, vol. 1, 3d ed., Wiley, New York, 1980, pp. 169–196 (stone groundwood pulping); 197–252 (refiner mechanical and thermomechanical pulping).

5 H. Gehm, *State of the art review of pulp and paper waste treatment*, Environmental Protection Series EPA-R2-73-184, U.S. Government Printing Office, Washington, D.C., 1973.

6 J. Orivuori, T. Nystrom, and W. Aho, "Closure of water systems in newsprint and mechanical

printing paper mills," *Proceedings of the TAPPI Environmental Conference*, Minneapolis, April 1982, pp. 161–169.

7 J. M. Leach and A. N. Thakore, *Tappi*, 59 (2):129 (1976).

8 J. Orivuori, "Closure of water systems in newsprint and mechanical printing paper mills," *Proceedings of the TAPPI Environmental Conference*, Minneapolis, April 1982, pp. 161–173.

9 A. Järvine, M. Vahtila, B. Mannström, and J. Sundholm, "Reduce environmental loads in TMP," *Pulp & Paper Magazine of Canada*, 81 (3):39–42 (1980).

10 H. J. Ahonen, "TMP heat recovery from Jylha Tandem TMP line," *Proceedings of the TAPPI Pulping Conference*, Toronto, Ontario, October 25–27, 1982, pp. 147–149.

11 B. Schobas, "Pulping process reduces water pollution," *Chemical Engineering*, 89 (22):38–39 (November 1982).

12 J. P. Casey, ed., *Pulp and paper; Chemistry and Chemical Technology*, Vol. 1, 3d ed., Wiley, New York, 1980.

13 P. W. Schriver, *Pulp & Paper Canada*, 80 (3):T-62–64 (1979).

14 *SSVL Environmental Care Project, Technical Summary*, Swedish Steam Users Association (ed.), Stockholm, Sweden, 1974.

15 J. C. Wyvell, J. C. Adams, G. E. Valentine, and T. Shelnutt, *An assessment of the potential for water reuse in the pulp and paper industry*, Report No. RU-84-1, Bureau of Reclamation, U.S. Dept. of the Interior, Washington, D.C., March 1984.

12
The Potential of New Pupling Technologies for Pollution Abatement

In addition to the processes discussed already, a wide variety of other pulping processes that might offer significant potential for reduction of pollution loads have been researched. They include hydrotropic pulping (1,2), alchohol pulping (3), holopulping (4,5), nitric acid pulping (6), phenol pulping (7), soda anthroquinone pulping (8, 9), and oxygen pulping (10–13).

MEDIUM-CONSISTENCY PULPING

One prerequisite for success that all new pulping processes should have in common is operation at medium to high consistency rather than low consistency. This trend is already seen in modern Kraft mills with continuous digesters, displacement washers, and hot stock screening. The pulp mill of the future likely will operate at medium consistencies of 12–14% from the digester to the paper machine (14). Medium-consistency mixers, pumps, and screens have already been developed and have been tested in full-scale mill trials. Medium-consistency cleaners and headboxes are still at the laboratory stage of development, but work is progressing. The medium-consistency headbox appears to be a major problem, since the production of a sheet with good formation has so far been difficult.

In general, working at medium consistency produces problems when the flow is stopped, because high torques are required to start the slurry moving again. The power required to overcome the shear strength of the fiber network is high. If this problem can be overcome, less power overall will be required to operate at medium than at high consistency because a smaller mass of material will be moved (14).

OXYGEN PULPING

As of 1984, the one emerging process that seems closest to full-scale development is oxygen pulping. A number of investigators have amply demonstrated that interesting pulps can be produced in the laboratory by various combinations of oxygen alkali processing steps (15–21). Both single-stage and multistage processes have been tried. Oxygen pulping offers simplified pollution control because its chemicals (oxygen- and sodium-based alkali) contain no source for generating odorous compounds and because chemical recovery should be relatively simple. In reality, it may be difficult to keep an oxygen pulping system totally free of sulfur, because sulfur compounds occur as minor constituents of wood, other chemicals, and fossil fuels. Other benefits of the process are that greater selectivity in degrading lignin, even at low lignin content, has been demonstrated and that substantially less energy is needed to refine oxygen pulp.

The oxygen pulping methods that are being considered can be divided into two classes: two-stage soda-oxygen and single-stage oxygen pulping. In the two-stage process, the chips are cooked first in caustic soda to a point at which they can easily be mechanically separated into fiber (50–60% yield), and then these fibers of high lignin content are delignified with oxygen to give a yield slightly better than that obtained with Kraft pulping (22). This process uses existing mill technology with the addition of an oxygen stage.

This is not the situation with the single-stage process. The key to this process is getting the oxygen to penetrate the wood chips. To accomplish this, an alkaline solution is used to swell the chips and provide a transport medium, but the oxygen solubility is low. To increase the solubility, the pressure is increased to 2,000–2,755 kPa (280–400 psi) in a high-pressure reaction vessel (23). As an additional measure to increase the penetration of oxygen, very thin chips, usually less than 1.5 mm in thickness, are used. The cooking is carried out at a pH of 7–9 (24) at 140–150°C to minimize cellulose degradation. Since considerable heat is generated by this exothermic reaction, low consistency is preferred so that the water can act as a heat sink.

Oxygen pulping has two principle disadvantages: it produces pulp that is not very strong, and the pulping process is nonuniform when normal-sized chips are used due to slow transport of the oxygen into the wood. The strength of the pulp is improved if potassium iodide is used as a protective reagent, but the cost is prohibitive unless the potassium iodide is recovered. Oxygen pulping must also overcome the following economic and engineering problems:

1 There is no mechanism for the recovery of oxygen, so its consumption is a major operating expense
2 The recovery of tall oil and turpentine from resinous softwoods is reduced due to the oxidation of these by-products
3 Prolonged exposure to oxidative conditions causes the spent liquor to have a lower fuel value
4 Corrosion rates increase, especially where metal surfaces are repeatedly wetted and dried in the presence of oxygen
5 There is a potential fire and explosion hazard from increased combustibility of fines and terpenes in the presence of oxygen

6 There are toxicity and explosion hazards from the carbon monoxide produced during the pulping, and frequent venting of the oxygen reactor is necessary to minimize these hazards

7 Surplus heat from the exothermic reaction of oxygen pulping could be difficult to dissipate

The successful commercialization of oxygen pulping will depend on how well these economic and engineering difficulties can be overcome. A 300-ton/day pilot plant was constructed in the United States to further evaluate the process, but a commercial venture did not result from this effort.

Because there is a wide range of oxygen pulping alternatives rather than just a single process, it is difficult to define the environmental benefits in absolute terms. With oxygen pulping, the effluent volume at which the process operates affects the amount of pollution discharged. Worster and Pudek (25) demonstrated in the laboratory that in a comparison of Kraft and single-stage soda-oxygen pulping of hemlock, the Kraft process produced a TOC of 519 kg/ton (1,145 lb/ton), while soda to a Kappa number of 140 produced 283 kg/ton (624 lb/ton) and oxygen to a permanganate number of 23 produced a TOC of 111 kg/ton (244 lb/ton), a total for the soda-oxygen system of 394 kg/ton (868 lb/ton). Thus there was a 24% reduction in TOC for the oxygen system at an equivalent state of discharge. These investigators also demonstrated that oxygen pulp mill white waters had about one-tenth the toxicity of comparable Kraft mill liquors. The simulated effluents also had lower color and less foam.

Another potential environmental benefit of all of the oxygen approaches is related to the effect of oxygen treatment on bleach requirements and bleach response. If oxygen pulping results in shorter bleaching sequences, the benefits derived would be both a reduction in bleaching costs and a reduction in the volume and load of bleach plant effluents (26).

ANTHROQUINONE (AQ)-ASSISTED PULPING

Because the sulfide present in Kraft pulping produces odorous compounds, a replacement chemical has been sought. Hydrazine, hydroxylamine, and sodium borohydride have been tried successfully, but all are too expensive for use on a commercial basis. Recently, however, AQ and similar compounds have been shown to yield the desired beneficial effects at very low concentrations (8, 9). Both AQ and the sodium salt of anthroquinone-2-sulfonic acid have been found to be effective. Amounts as low as 0.05–0.1% on wood have been demonstrated to produce delignification rates, pulp yields, and strength properties similar to those obtained by Kraft pulping. AQ has also been used commercially in the Kraft process to give improvements in yield and pulp properties (27–32). It is possible to get about a 5% improvement in yield compared to that of straight Kraft cooking and to obtain slightly better properties. The same results can be obtained with soda alone, but larger amounts of AQ are required (0.25% on wood rather than 0.13%).

The AQ seems to function somewhat as a catalyst and is not completely consumed in the cooking process. It appears to give higher yields by stopping the carbohydrate peeling reaction through oxidation of the reducing end group to aldonic acid (31). The mechanism of action of AQ in enhancing the rate of delignification has not yet been determined.

If AQ behaves as favorably under commercial conditions as it has in laboratory studies, it could solve the odor problem of Kraft mills. It could also lead to a significant reduction in color in Kraft mill effluents.

REFERENCES

1 R. H. McKee, U.S. Pat. 2,308,564 (1943); *Industrial and Engineering Chemistry*, 38:382 (1946); Pulp & Paper Magazine of Canada, 55 (2):64 (1954).

2 T. A. Robert, *Tappi*, 38 (9):149A (1955).

3 T. N. Kleinert, *Tappi*, 57 (8):99 (1974).

4 R. P. Whitney, N. S. Thompson, G. A. Nicholls, and S. T. Han, *Pulp & Paper Canada*, 43:68 (July 1969).

5 H. Asaoka, T. Nagasawa, and M. Morita, "Pilot plant test for pollution-free pulping," *1977 TAPPI Environmental Conference*, Technical Association of the Pulp and Paper Industry, Atlanta, 1977, p. 137.

6 D. L. Brink, *Tappi*, 44 (4):256 (1961).

7 R. Hartmuth, German Pats. 326, 705 and 328, 783 (1919).

8 H. H. Holton, *Pulp & Paper Canada*, 78 (10):T-218 (1977).

9 B. Bach and G. Fiehn, *Zellst. Papier*, 21 (1):3 (1972).

10 P. J. Kleppe, A. Backlund, and Y. Schidt, "Oxygen/alkali delignification at Kamyr digest blow time consistency—Slotus Report," *Proceedings of the International Pulp Bleaching Conference*, Technical Association of the Pulp and Paper Industry, Atlanta, 1976, p. 141.

11 B. S. Kirk, M. J. Kirschner, and J. K. Tsend, "Law consistency oxygen delignification in a Pipline Reactor–Pilot Study," *Proceedings of the TAPPI Alkaline Pulping and Secondary Fibers Conference*, Technical Association of the Pulp and Paper Industry, Atlanta, 1977, p. 147.

12 L. D. Markham, C. E. Courchene, E. F. Elton, and V. L. Magnatta, *Tappi*, 63 (11):79 (1980).

13 L. Nasman and G. Annergren, *Tappi*, 63 (4):105 (1980).

14 J. M. Macleod, *Pulp and Paper*, 56 (4):194–96 (April 1982).

15 J. L. Minor and N. Sanyer, *Tappi*, 57 (5):120 (1974).

16 H-M. Chang, W. T. McKean, and S. Seay, *Tappi*, 56 (5):105 (1973).

17 M. Saukkonen and I. Palenius, *Tappi*, 58 (7):117 (1975).

18 R. Marton, A. Brown, and S. Granzow, *Tappi*, 58 (2):64 (1975).

19 A. G. Jamieson, O. Samuelson, and L. Å. Smedman, *Tappi*, 58 (2):68 (1975).

20 T. Nagano, S. Miyao, and K. Takeda, "Hopes oxygen pulping process—its basic concept and some aspects for the reaction of oxygen pulping," *1974 Non-Sulfur Pulping Symposium*, Technical Association of the Pulp and Paper Industry, Atlanta, 1974, p. 53.

21 R. H. Hart (Beloit Corp.), U.S. Pat. 3,298,903 (1967).

22 J. E. Carles, R. Angelier, E. Muratore, and P. Manzie, *Revue ATIP*, 27 (2):147 (1973).

23 A. Braden and O. Samuelson, *Paperi ja Puu*, 57 (9):607 (1975).

24 J. J. Renard, D. M. Mackie, H. G. Jones, H. I. Balker, and D. W. Clayton, *Trans. Tech. Sect. CPPA*, 1 (1):(1975).

25 H. E. Worster and M. F. Pudek, *Pulp & Paper Magazine of Canada*, 74 (3):95 (1973).

26 W. T. McKean, "Potential pollution loads from oxygen pulping," *1976 TAPPI Environmental Conference*, Technical Association of the Pulp and Paper Industry, Atlanta, 1976, p. 17.

27 H. H. Holton and F. L. Chapman, *Tappi*, 60 (11):121 (1977).

28 K. Goel, A. M. Ayroud, and B. Branch, "Anthroquinone in Kraft pulping," *Proceedings of the TAPPI Alkaline Pulping Conference*, Seattle, 1979, pp. 213–220.

29 P. J. Kleppe, *Proceedings of the EUCEPA Conference*, Helsinki, 1980.

30 J. S. Shaw and J. J. Renard, "Pulping additives," *Proceedings of the TAPPI Pulping Conference*, Technical Association of the Pulp and Paper Industry, Atlanta, 1980, p. 472.

31 L. Löwendahl and O. Samuelson, *Svensk Papperstidning*, 80 (17):549 (1977).

32 J. Renord, R. Phillips, H. Jamul, and A. Rudie, *Tappi*, 64 (8):51–54 (1981).

13
Process Modifications in Bleaching

CONVENTIONAL BLEACHING PROCESS MODIFICATIONS

The chemicals most commonly employed for bleaching chemical pulps are chlorine, sodium or calcium hypochlorite, chlorine dioxide, peroxide, and oxygen. These chemicals are used on a once-through basis, resulting in a large pollution load. Typically, this load may be 40% of the BOD, 25% of the suspended solids, and 70% of the color of the pollution load of an entire integrated mill complex. The exact numbers will depend on many factors, including the bleaching sequence used, bleaching conditions, wood species, type of pulping process, and so on. A variety of conditions and sequences exists in bleaching, as illustrated by Table 13-1, which lists the common sequences used to bleach Kraft pulp to various degrees of brightness. For highly bleached Kraft pulp, the CEDED sequence appears to be the industry standard. A 1971 survey (1) showed that water use in bleacheries ranged from 12,000 to 57,000 U.S. gal./ton, with average use being 20,000–30,000 gal./ton.

Bleaching ordinarily is performed in a number of stages to preserve the strength of the pulp by avoiding excessively strong chemical treatments. All stages of bleaching are usually carried out at greater than 10% consistency with the exception of the chlorination stage, which is performed at 3% consistency. Thus, if stock comes to the chlorination stage at 10% consistency and leaves at 10% consistency, considerable process water must be both added and extracted during chlorination—21 m^3/ton (5,600 gal./ton). This amount of process water could be equal to that used in the rest of the bleach plant. The chlorination stage is therefore a focal point for pollution control considerations. Four approaches have been suggested:

1. Replace the chlorination stage with an oxygen or ozone stage
2. Recycle the chlorination stage process water
3. Use water from the countercurrent washing system in the chlorination stage
4. Perform high-consistency gas phase chlorination

Table 13-1 Common Sequences Used to Bleach Kraft Pulp to Various Degrees of Brightness

Range of GE Brightness	Sequence
70–80	CEH
	CEHH
80–85	CHEH
	CEHEH
	CC/HEHH
	CED
85–92	CEHD
	CHED
	CEHDD
	CC/H ED/H
	CEDED
	CEDHED

Source: Copyright © 1981. TAPPI. Reprinted from *Introduction to Environmental Management for the Pulp and Paper Industry*, Study Guide, Volume 6, with permission

A combination of the second and third approaches seems to be the most popular. Reduction in effluent flows can be achieved by countercurrent recycling of shower and seal box waters, which is possible in most bleach plant configurations and which eliminates flows to sewers from downstream stages. The three main types of countercurrent washing used in bleacheries—split flow, jump stage, and direct (1)—are illustrated in Figures 13-1, 13-2, and 13-3. Some use of countercurrent sequences is practiced by most mills.

Direct countercurrent washing produces the least effluent. The use of a direct countercurrent flow is restricted, however, because serious corrosion problems are encountered when washers or piping in contact with chlorine dioxide filtrates are not made of a material at least as corrosion resistant as 317 stainless steel. Partial or jump stage washing can be used with some water savings when washer and pipeline materials prohibit direct countercurrent washing. The use of direct countercurrent washing requires that the construction materials of at least the chlorination and first caustic extraction stage washers be changed. Serious corrosion difficulties would be expected with a chloride content above 1,700 ppm. Either titanium, a combination of titanium and fiber-reinforced plastic, or one of the ferritic–high molybdenum content stainless steels could be used. One of these materials would be essential for the chlorination stage washer and would be desirable for the first extraction stage washer. It might be possible to use 317 ELC stainless steel for the first extraction stage washer. The high cost of construction materials for the countercurrent sequence is partially offset by the resulting steam savings (2). Steam savings are the largest benefit of water recycling in bleaching.

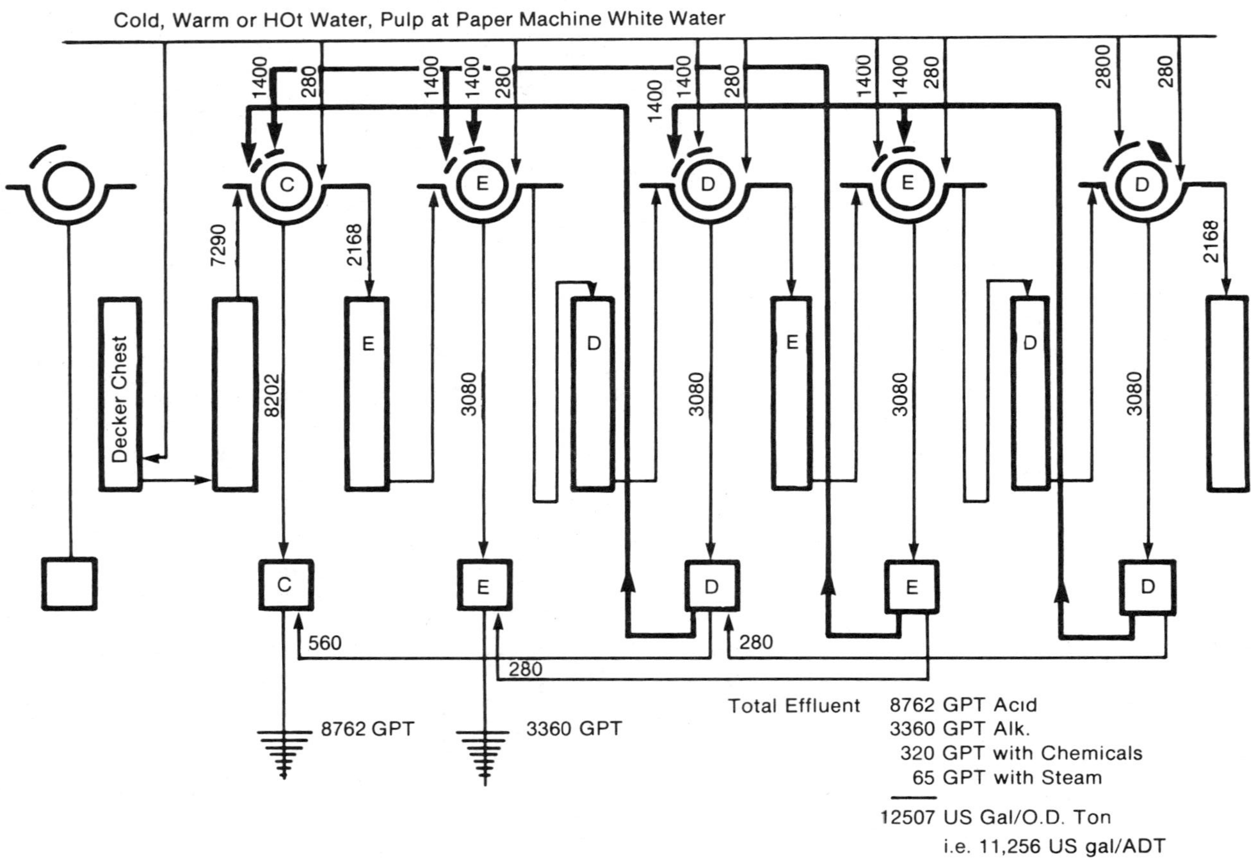

Figure 13-1 Split-flow countercurrent

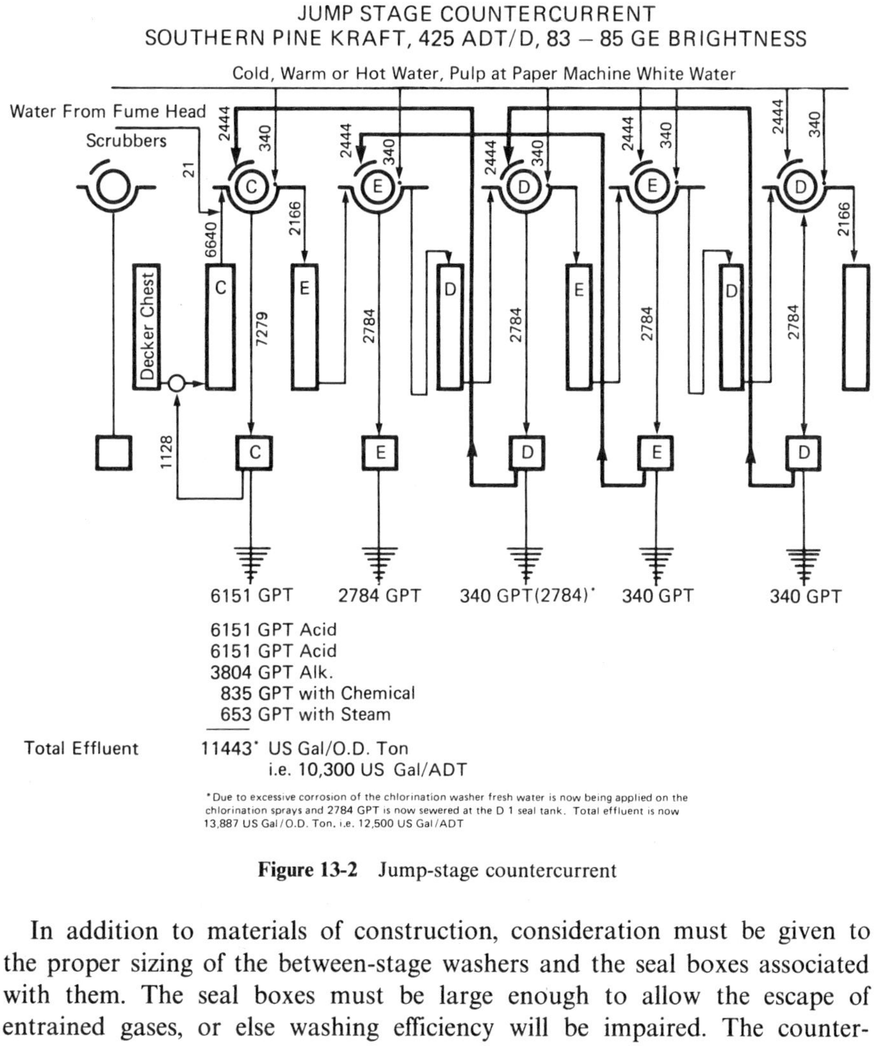

Figure 13-2 Jump-stage countercurrent

In addition to materials of construction, consideration must be given to the proper sizing of the between-stage washers and the seal boxes associated with them. The seal boxes must be large enough to allow the escape of entrained gases, or else washing efficiency will be impaired. The countercurrent system makes the stages interactive, which increases the foaming tendency, making it more crucial to have adequate seal tank capacity.

Countercurrent recycling brings thermal energy to the chlorination stage, rising the temperature to as high as 60°C (140°F). Excessive degradation of viscosity can be avoided if 40–70% chlorine dioxide is substituted for chlorine. Exactly how much chlorine dioxide is optimal is still a subject of discussion (3–5). In many instances, high substitution has been shown to be no better than low substitution, as measured by pulp yield, shive and dirt removal, physical strength, viscosity, and the effluent parameters of total organic carbon content and chlorophenolics. There may be marginal gains for high chlorine dioxide substitution in the parameters of pulp delignification, chemical consumption, and the effluent parameters of COD and toxicity (3). In hardwood Kraft pulps, high

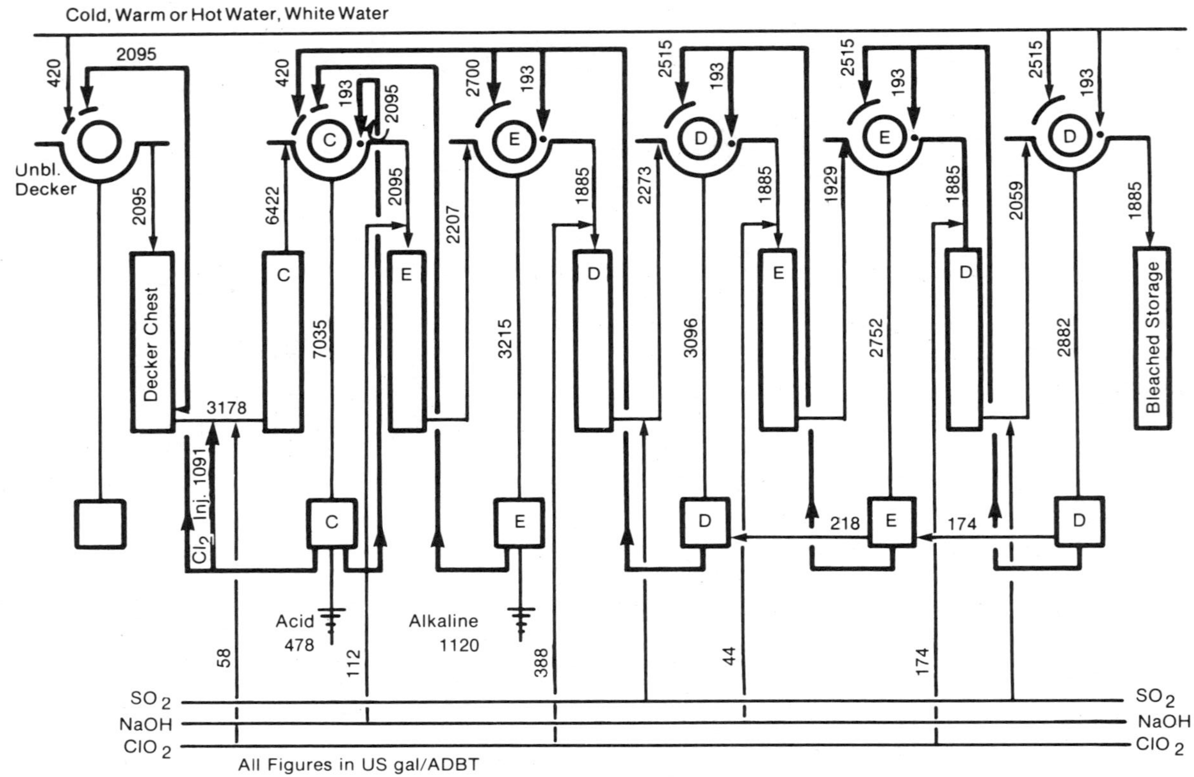

MINIMUM EFFLUENT COUNTER CURRENT WASHING SYSTEM

Direct Countercurrent Washing, Maximum Chlorination Filtrate Recycle. Filtrate Used for Face Cleaning Showers
Total Bleachery Effluent 1598 gal/ADBT

Figure 13-3 Direct countercurrent

chlorine dioxide levels improve resin removal, but the resin ends up in the effluent and may increase toxicity. High substitution does seem to provide a clear advantage in the areas of effluent color, chlorine content, and mutagenicity. There is conflicting evidence in the literature on the effect of chlorine dioxide in the chlorination stage on BOD (3). The level chosen may often be based on the cost of chlorine dioxide at a particular mill.

Alternatively, the high temperature in the chlorination stage can be compensated for by decreasing the reaction time to about 5 minutes, but precise control of the time would be needed to prevent degradation. A single additional minute of reaction time can cause substantial degradation of carbohydrate.

It should be mentioned that the displacement ratio on the washers must be less than 1:1, or countercurrent washing is impossible. It is also critical to have large enough seal tanks to avoid spillage from upsets.

Direct countercurrent bleaching can reduce the effluent level for the bleach plant to 3,100 gal./AD ton, as illustrated in the effluent-free Kraft mill concept. It has been demonstrated in the laboratory that with the CEDED sequence it is possible to reduce the effluent to 1,600 gal./AD ton (1). Figure 13-3 illustrates the flow pattern used to achieve this level of discharge. Any difficulties that result from using the direct countercurrent bleaching sequence are apparently not overwhelming, since Great Lakes Paper implemented the concept in 1979.

Other techniques for reducing freshwater use in the bleach plant include the use of intermittent washer showers and paper machine white water. The latter is common practice.

While water recycling in the bleach plant has dramatically reduced the effluent volume, the quantity of pollutants discharged per ton of pulp has not been much influenced. Reductions in BOD_5 on the order of 25% may be possible (6), but the main effect of water reuse is to concentrate the effluent stream. The BOD_5 load of a modern bleaching operation is at least 25 lb/ton. Pulp bleached to a given brightness level must have an equivalent amount of colored organic matter removed, which must appear as a pollutant in the effluent stream. Dramatic reductions in pollutant loads will be achieved only by returning the bleach plant

Table 13-2 Representative Bleach Plant Effluent Characteristics for High-Brightness Pulp

Sequence	Pulp Type	Color (PCU)[a]	BOD_5 (lb/ton)	COD (lb/ton)	Chlorides (lb/ton)	Total solids (lb/ton)
CEHDED	SWK	370	35	110	224	570
CEH	HWK	84	21	84	168	—
CEDED	HWK	304	28	—	205	—

Source: Copyright © 1981. TAPPI. Reprinted from *Introduction to Environmental Management for the Pulp and Paper Industry*, Study Guide, Volume 6, with permission.

[a] PCU, platinum cobalt units.

Table 13-3 Effluent Characteristics, by Stage, for High-Brightness Hardwood Kraft Pulp

	Stage				
Characteristic	C	E	D	E	D
Color (PCU)	23	69	4	3	1
Chlorides (lb/ton)	66	11	14	3	6
COD (lb/ton)	35	36	12	8	9
Total solids (lb/ton)	40	32	14	5	9

Source: Copyright © 1981. TAPPI. Reprinted from *Introduction to Environmental Management for the Pulp and Paper Industry*, Study Guide, Volume 6, with permission.

effluent to the recovery system, with or without replacements of the chlorination stage with oxygen. Recycling bleach plant effluent to the recovery system results in a net loss of energy, because it takes more energy to concentrate the effluent than is gained by burning the organics present in the effluent stream.

Representative effluent characteristics from established conventional bleaching sequences are given in Tables 13-2 and 13-3. Table 13-3 clearly illustrates that most of the pollution load arises from the first two stages of bleaching. An important aspect of this pollution load not included in the tables is the toxicity of the chlorinated aromatic compounds, resin acids, and fatty acids produced during bleaching or extracted from the pulp. The question of toxicity will receive more emphasis in the future. Polymer resin systems for absorption of color and toxic materials have been developed (7). Some mills have substituted a buffered hypochlorite stage for the normal hot caustic extraction stage to reduce pollution, especially color load (8).

MODERN BLEACHING TECHNOLOGIES

Dynamic Bleaching (Displacement Bleaching)

The ultimate in water recycling and compactness of bleaching sequences is dynamic bleaching, the basic concepts of which have been well documented in laboratory studies for a considerable period of time (9–13). In the 1960s, the Kamyr Corporation developed the diffusion washing system (14), which has evolved into the multistage system illustrated in Figure 13-4. This system made possible the engineering realization of the dynamic bleaching system. The key to its operation is the concentric baffles that move with the pulp as it flows through the tower and are returned to zero position hydraulically at the end of a wash cycle. The diffusion assembly is a series of concentric double-sided baffles. Each baffle is made of two perforated plates spaced and joined approximately 2 in.

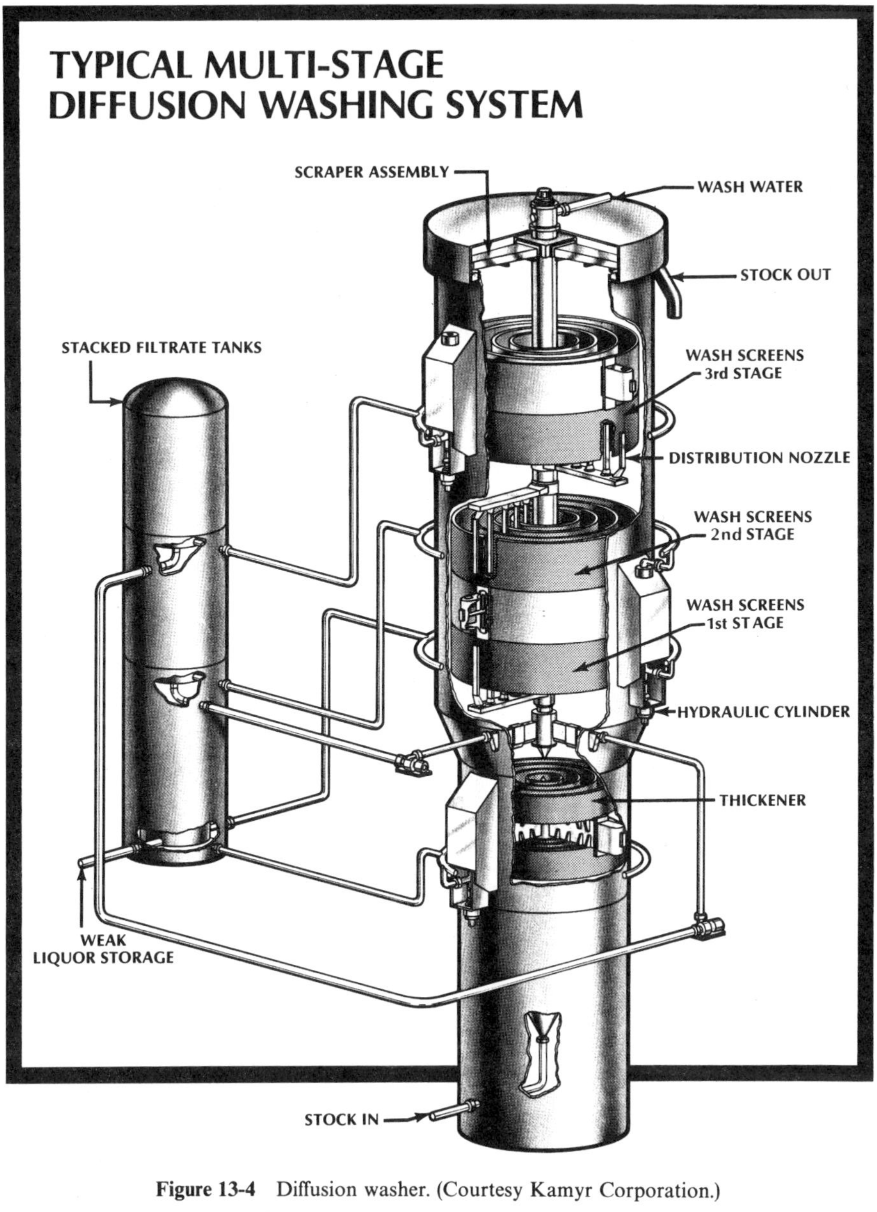

Figure 13-4 Diffusion washer. (Courtesy Kamyr Corporation.)

apart. The baffles are spaced approximately 20 in. apart. The baffle plates are welded together at the top and bottom to form a hollow shell. The bottom of the shell contains a drainage channel. The baffle assembly, consisting of several baffles, is simultaneously connected with and supported by radial drainage arms.

The washing solution is introduced into the pulp through rotating distribution nozzles. The nozzles rotate at approximately 7.5 rpm, passing between the baffles.

The washing solution rapidly displaces liquor toward the nearest baffle. Washing solution is forced hydraulically through the pulp across to the baffles, thus effectively displacing the liquor. This is a very efficient washing system, providing at least 50 times as long a wash cycle as a drum washer. The pulp is kept at high consistency during the wash cycle and is not exposed to air, resulting in no odor problems and no entrained air effects on washer efficiency. The use of defoamers is practically eliminated.

When this diffusion washing system is used with conventional bleaching simply to replace drum washing, it is commonly termed "diffusion bleaching," which is a separate and distinct process from dynamic (displacement) bleaching.

In dynamic (displacement) bleaching, the chlorination stage in carried out in a conventional manner, and the E_1, D_1, E_2, and D_2 stages are performed in a single tower in four Kamyr displacement diffusion washers stacked on top of one another. Figure 13-5 shows a schematic diagram of the process. The reaction time is 5–10 minutes per stage, and the consistency is greater than 10%. Bleaching chemicals are displaced through a pulp mat rather than being mixed in the

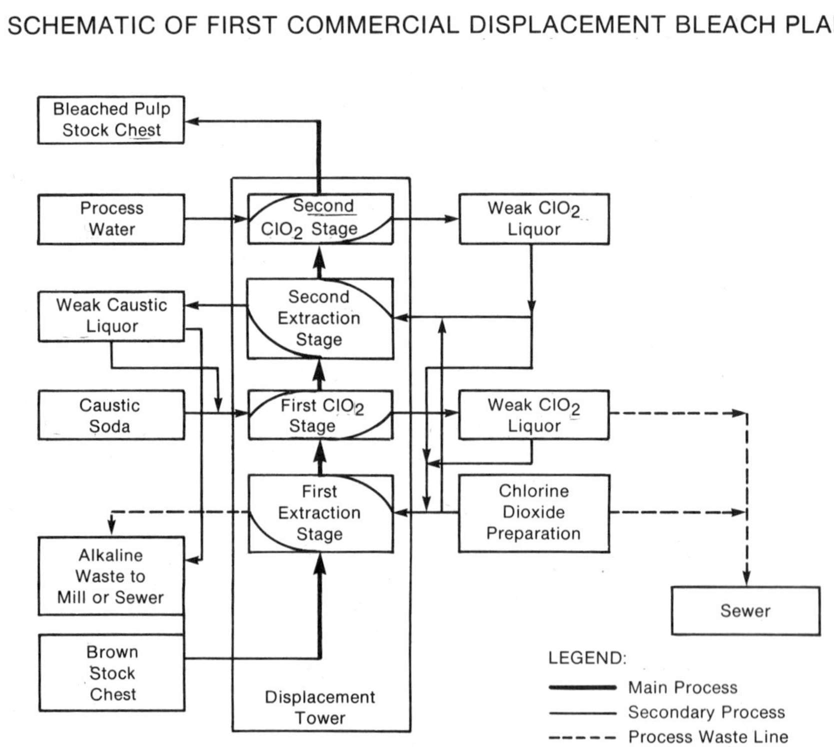

Figure 13-5 Displacement bleaching

conventional manner. Very rapid bleaching can be accomplished due to the high reaction rate. Filtrate withdrawn at each stage is fortified with makeup chemicals and reused (15). Unbleached pulp moves sequentially up the tower, white liquor moves countercurrently.

The pilot plant data were developed in Finland (16), and Table 13-4 summarizes some of the pilot test results on BOD. The system has several distinct advantages over conventional bleaching technology for a new mill. The installed capital cost is less, and the operating cost in 1983 dollars has been estimated to be \$6–10 less per AD ton. The capital cost is less, in spite of the fact that the system must be made of titanium. The operating cost savings result from lower steam, lower electrical energy, and possibly lower chemical requirements. The system is very compact and so requires minimum space. Since it is completely sealed, there are minimum atmospheric emissions. It is operated completely at high consistency with no dilution between stages, so there is minimum water use and effluent volume. The system would integrate well into the "closed-mill" concept discussed in Chapter 9. Table 13-5 gives an economic comparison between displacement and conventional bleaching.

Some variations on the traditional displacement bleaching scheme of a conventional chlorination stage followed by the EDED stages in a single diffusion washing tower are as follows. For certain pulps to achieve the high

Table 13-4 BOD Comparison of Conventional and Displacement Bleaching

	BOD_5	
Bleaching Stage	Displacement	Conventional
E_1	7.50	7.2
H	0.50	2.4
E_2	0.30	1.0
D	0.08	1.7
Total	8.38	12.3

Table 13-5 Economic Comparison D/C EDED Bleaching Sequence

	Displacement	Conventional
Relative space	0.15	1.00
Effluent volume, m^3/BDT[a]	10 –13	30–40
Heat consumption, MJ/BDT	800	2000
Electric power required, kWh/BDT	40	140

[a]BDT, bone dry ton.

brightness desired, it is necessary to allow additional time in the last dioxide stage. To accommodate this need, it may be advisable to split the stages into two displacement towers, with the last tower devoted to the final dioxide stage and a final wash. Thus a potential scheme would be conventional chlorination followed by EDE in the first displacement tower and D wash in the second displacement tower.

The reason the chlorination stage is usually done at 3% consistency is the low solubility of chlorine gas and the need for uniform dispersion to assure uniform bleaching. Kamyr has developed an alternative solution, however, with its high-efficiency mixer, termed the twister mixer, which may allow chlorination to be carried out at 10% consistency. In this case, chlorination could be incorporated into the displacement tower system. The twister mixer operates on the principle that if pulp is accelerated to a relatively high velocity, it will become fluidized. In the fluidized state, an introduced gas will uniformly and rapidly mix with the pulp (17). At 10% consistency there is not enough liquid available to dissolve the gas, but after reaction there is. The residence time in the mixer plus the first diffusion washing stage is sufficient for the reaction to occur. To assure uniform mixing, the pulp is actually sent through two stages of twister mixers before entering the tower.

As of 1984, there were 10 full-scale displacement bleaching installations in existence—three in the United States, three in Scandinavia, and one each in Canada, Japan, Brazil, and Russia. The two full-scale plants in the southern United States with significant operating experience have operated well but have not been completely trouble free. These mills have used both hardwood and softwood furnishes with a final brightness of 86 to 88 GE. There have been difficulties with corrosion of components not made of titanium, and some scaling problems from deposition of calcium carbonate and calcium oxalate (18). It has also been discovered that hot chlorination is more demanding in terms of gas

Table 13-6 Commercial Experience with Dynamic Bleaching

	Softwood	Hardwood	Sawdust Softwood	Hardwood
Bleaching sequence	D/C EDED	D/C EDED	D/C EDED	D/C EDED[a]
Kappa number	17	12	19	12
G. E. brightness	86–88	88	88	86–88
Total available chlorine, %	8	6.7	9.6	6.3
Total NaOH, %	4.8	2.7	4.4	2.2
Steam consumption, MJ/ADT	1000	1250	1250	600
Electric power, kWh/ADT	60	60	60	40
Effluent volume, m^3/ADT	15	15–20	15–20	14
Reference	19	20	20	21

[a] Uses medium-consistency chlorination.

dispersion, and requires excellent mixing and precise control. The entire system is more sensitive to operate and requires a higher degree of control than do conventional bleach plants. The Polarox instrument is being used in most oxidative stages of displacement bleaching for mannual or online computer control of chemical addition. Despite these problems, the displacement bleaching system appears to be quite promising and will undoubtedly see further application.

Table 13-6 summarizes some of the commercial experience with displacement bleaching. It evidently represents a practical alternative that can give equivalent pulp quality.

A more detailed treatment of dynamic bleaching can be found in reference 22.

Oxygen Bleaching

For a number of years, people tried to develop a commercially viable means of bleaching with oxygen (23, 24). The first applications were in dissolving pulps where the strength degradation encountered was not a problem. It was not until Robert and associates discovered that certain inorganic chemicals (inhibitors) such as magnesium carbonate could aid in controlling the degradation problem that oxygen bleaching became a technically viable alternative for other types of pulp (25, 26). Oxygen bleaching can be thought of as two competing series of reactions, one desirable—lignin degradation—and the other undesirable—carbohydrate degradation. One of the purposes of developing oxygen bleaching was to reduce the pollution load from the bleach plant. Since the chlorination stage is a major contributor to the pollution load, it is not surprising that one of the first commercial applications of oxygen bleaching was to replace this stage. This installation was at South African Pulp and Paper Inudstries, Ltd., and was a joint venture of L'Air Liquid of France and Kamyr of Sweden (27).

A parallel process was developed that employed oxygen as a stage in conventional bleaching after chlorination, which reduced the chlorine requirement. The motivating forces here were the low cost of oxygen and the improvement in pulp quality. This procedure was a joint development of Modocell (Sweden), CIL (Canada), and Sunds Akliebolog (Sweden) (28).

Oxygen bleaching has evolved considerably from these early beginnings, and today it is generally accepted that about half of the lignin left in brown stock can be removed by an oxygen stage without harming the pulp strength properties. In order to get environmental benefits from oxygen bleaching, the oxygen stage effluent must be sent to the pulp mill recovery cycle, Thus the discovery that white liquor could be used instead of sodium hydroxide in the oxygen stage was important in maintaining the proper chemical balance in the mill.

Using oxygen to reduce bleach plant pollution is a reality, and in the near future it may be possible to eliminate chlorine entirely from the bleaching system by substituting a combination of oxygen, ozone, and peroxide stages (29). This procedure would make the entire bleach plant effluent recyclable to the recovery system, eliminating the bleach plant as a source of pollution. Recycling the bleach plant effluent to the recovery system requires additional evaporator and causticizing capacity and a larger recovery furnace, however, so it is not an inexpensive means of pollution abatement. The best estimates to date seem to indicate that adding an oxygen stage to a softwood Kraft mill would require increasing the evaporator capacity 5–6%, assuming an oxygen stage discharge of 0.3–0.5 ton/AD ton. The recovery furnace would need 3–5% additional capacity and the causticizing plant lime kiln 7%, assuming white liquor is used in the oxygen stage (30).

The chief factor retarding further implementation of oxygen bleaching is the high capital cost of the equipment required for the process. The development of medium-consistency oxygen bleaching is an attempt to reduce these capital costs.

High-Consistency Oxygen Bleaching. "Consistency" refers to the consistency of the pulp in the oxygen reactor, and 20–30% is considered high. At this consistency level, the water present is contained within the pulp fibers, so the pulp responds like a fluff rather than a fluid. All of the early oxygen bleaching installations were high-consistency processes.

Figure 13-6 shows a flow diagram of the Sapoxal process, representing the SAPPI–L'Air Liquid–Kamyr development (31). The pulp must be dewatered before bleaching. It may also be necessary to treat it with acid if it contains a high level of metal ions such as manganese, iron, or copper. This treatment can be carried out in the filter washer before thickening. It may be desirable to add an inhibitor or carbohydrate protection agent. The alkali required in the bleaching must be evenly distributed throughout the pulp. In Figure 13-6 it is added to the pulp stream flowing into the steam mixers. Other alternatives are adding it before the thickener and recirculating the liquid from the thickener or waiting until the pulp is a fluff and spraying it on.

Thickening can be carried out either on a vacuum filter, as shown, or in a press. Fluffing of the pulp takes place in the top of the reactor because the pulp must flow in a plug to this point to allow sealing of the reactor. The reactor itself is a downflow tower. The decks and segments of the Kamyr-designed oxygen reactor

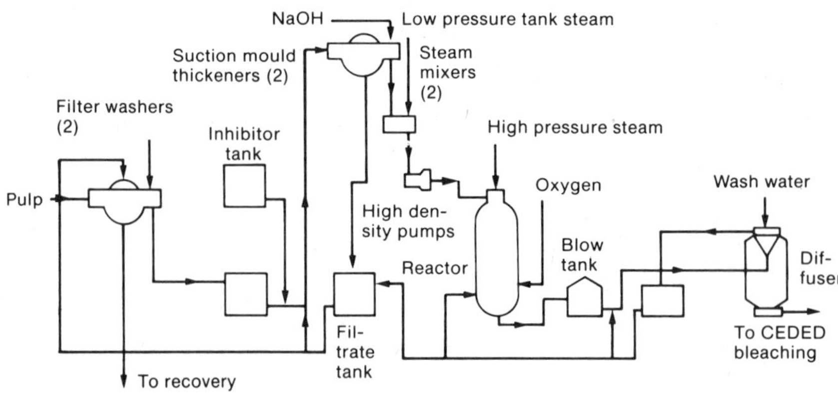

Figure 13-6 SAPOXAL process. Copyright © 1979, reprinted from *The Bleaching of Pulp*, TAPPI PRESS, No. BO47, pp. 187 and 196, with permission

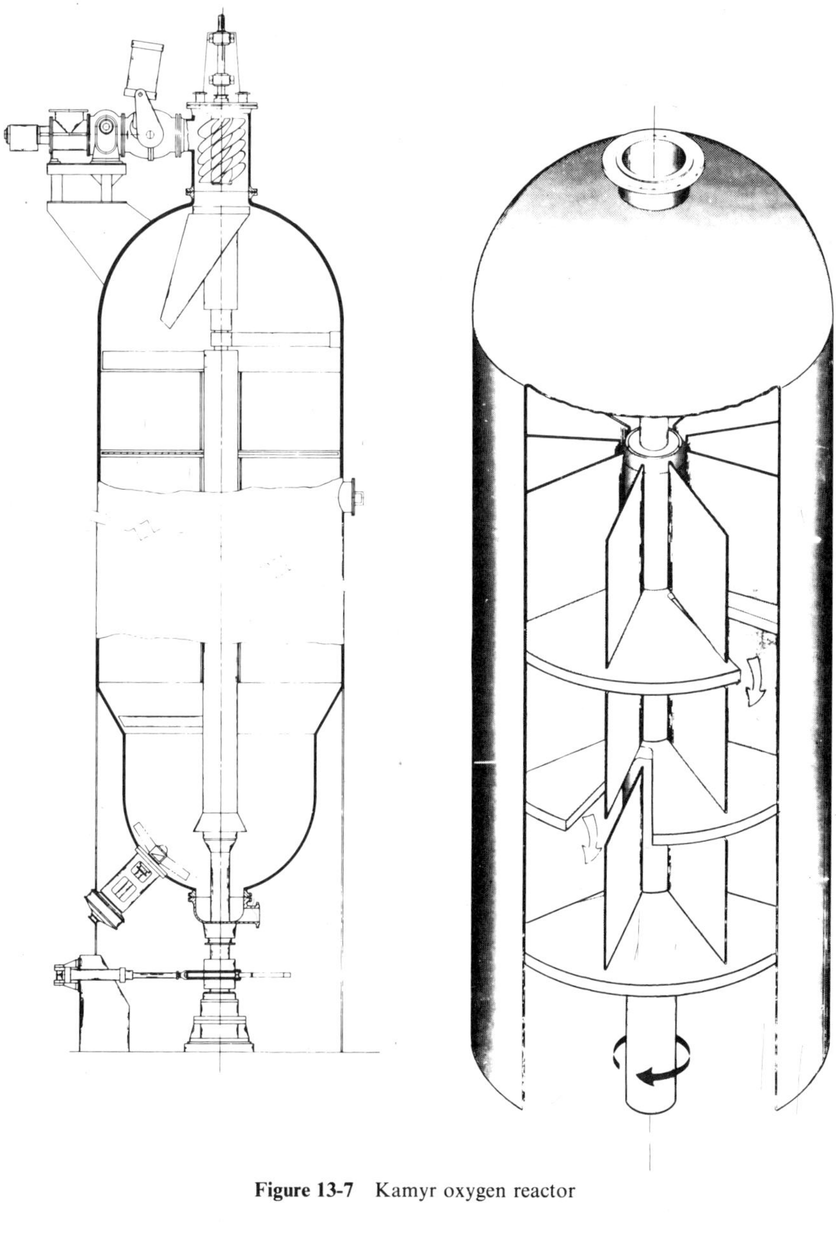

Figure 13-7 Kamyr oxygen reactor

of Figure 13-6 are shown in detail in Figure 13-7. The pulp enters one deck, is rotated to the gap between decks, falls to the second deck, makes another circuit, and falls to the next deck (32). Pressure on the pulp is low, and consistency can be as low as 22%.

An alternative reactor system developed by Sunds does not contain this baffle arrangement, and the consistency must be kept above 25% to avoid excessive compression and allow enough gas volume to pass through the pulp (33). Oxygen is fed to the reactor at a pressure of 750–950 kPa and reacts rapidly with the pulp. The temperature is kept at 115–125°C. At the bottom of the reactor, the pulp is diluted to about 3% consistency before discharge to the flow tank.

Fires or explosions are a potential source of trouble in an oxygen reactor. The gases given off are carbon dioxide, carbon monoxide, and turpentines. The carbon monoxide and turpentines can form an explosive mixture with oxygen, so the reactor must be purged if the concentrations get too high. The purged gases can be catalytically oxidized and returned to the reactor (34), so that the purged oxygen is not wasted.

An even greater source of potential trouble is the possibility that pulp can become stuck at various points within the digester. Pulp of 30% consistency is not easily ignited, but pulp of 38% or higher consistency can be ignited with an electrical heating element and would, in turn, ignite lower-consistency pulp (35). The resulting pressure buildup could cause an explosion. It is important that pulp not cling to the inside of the reactor and dry out.

Another potential problem is contamination of the pulp with oil or grease. Special lubricants must be used with this equipment.

These reactors are normally equipped with several independent safety systems that are triggered by either temperature or pressure. The lowest-level response is a water shower at the top of the reactor, and the last resort is a rupture disk to provide instant pressure release in case of a dangerous buildup. A pressure release valve can handle minor upsets. Oxygen feed is cut off whenever any of these safety devices goes into operation.

The reactor for high-consistency oxygen bleaching must be made of 309 L stainless steel because corrosion occurs in the top zone of the reactor, where pulp tends to stick to the walls and become acidic. These pockets of acidity under the pulp can cause pitting, which must be avoided because metal ions can affect the bleaching. All auxiliary equipment must be made of corrosion-resistant material for the same reason.

High-consistency oxygen bleaching is a commercial reality. If the oxygen stage effluent is returned to the recovery system, significant environmental benefits are possible, especially if the oxygen stage is used partially to replace the chlorination stage. Table 13-7 summarizes some pollution load data for the two popular types of high-consistency oxygen bleaching. Several facts from this table are worth noting. First, the OCEDED sequence definitely produces less pollution load than the CODED sequence; therefore OCEDED is more advantageous environmentally. Second, both sequences produce more BOD and total solids than do either the CEDED or CEHDED sequences if the oxygen stage effluent is not recycled. Third, if the oxygen stage effluent from the OCEDED sequence is recycled, 85% color reduction, 31% BOD reduction, 61% COD reduction, 40% chloride reduction, and 32% total solids reduction can be achieved in comparison to the CEHDED sequence, and slightly smaller reductions can be achieved in comparison to the CEDED sequence. Thus it can be seen that significant gains can be made by incorporating even a single oxygen stage.

Table 13-7 Laboratory Softwood Kraft Total Bleach Plant Loadings

	Pollution Loads (lb/ton of pulp)				
Sequence	Color	BOD	COD	Chlorides	Total Solids
CEDED	545	28	—	205	—
CEHDED	370	35	110	224	570
OCEDED	240	48	164	142	814
OCEDED with oxygen-stage effluent recycling	56	24	43	134	387
CODED	273	47	167	143	727
CODED with oxygen-stage effluent recycling	233	25	108	134	580

Source: Modified from reference 36.

These significant environmental benefits, together with the satisfactory pulp quality, prompted investigators to try to reduce the capital cost of the equipment used for oxygen bleaching by designing systems to operate at lower consistency. The alkaline oxygen medium itself is not very corrosive even to carbon steel.

Medium-Consistency Oxygen Bleaching. A considerable effort has been made to develop a commercially viable medium-consistency oxygen bleaching sequence (37, 38). It appears that the following sequence may be possible:

1 Stop a Kraft cook at the point of 60–65 Kappa number.
2 Use the EOD sequence to reduce the Kappa number to 30.
3 Complete the bleaching with CEHD to achieve the desired Kappa number and brightness.

This process would be comparable to a Kraft cook followed by a CEHDED bleaching sequence.

For the oxygen stage, the reaction conditions needed to obtain a decrease in Kappa number of 25–35 are 10–15% pulp consistency, 15-minute reaction time, temperature of 100–115°C, 100 psig total pressure, and 3–4% sodium hydroxide dose. This procedure appears to give an increased yield on softwood pulps, and the yield advantage carries through the bleaching sequence. Figure 13-8 compares the medium-consistency oxygen procedure with a conventional sequence. The oxygen procedure appears to use about 3% less sodium hydroxide, which may reduce the load on the recovery system. There may also be a slight improvement in strength properties with the medium-consistency oxygen procedure (37).

This concept can easily be retrofit in a mill to increase digester throughput and to reduce the pollution load. The oxygen pulp is easier to bleach, so there

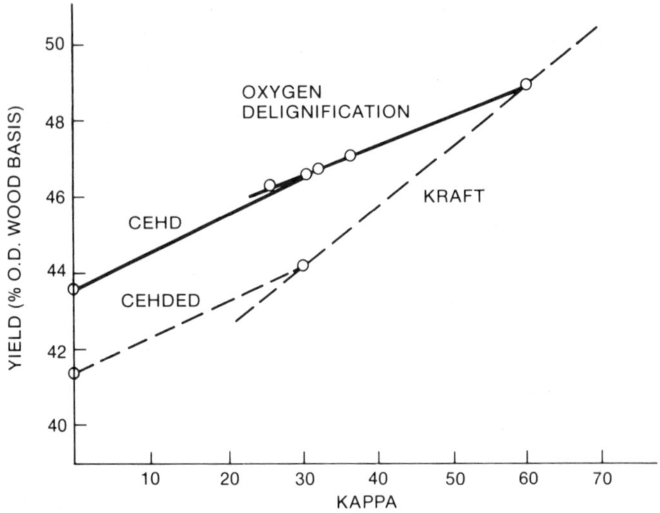

Figure 13-8 Medium-consistency oxygen compared to conventional Kraft. (From F. White, "Oxygen delignification: State of art report on advances," *Paper Trade Journal*, Feb. 28, 1982.)

is an overall reduction in bleaching chemical costs of about \$4–8/ton (37). The predicted decreases in pollution load have not yet been reported, but they should be appreciable if the oxygen stage effluent is sent to the recovery system.

A medium-consistency oxygen bleaching pilot plant is being operated by M. Peterson and Son in Norway (39). Its results show that medium consistency produces minimum chemical consumption and maximum pulp viscosity.

An even more interesting development than medium-consistency oxygen bleaching was the discovery that oxygen could be introduced into the alkaline extraction stage. The process is simple, requiring only a mixing device for upflow towers or a mixing device plus an upflow retention tube for downflow towers. This technique has been used in Sweden since 1980. It introduces the possibility of significantly reducing the number of bleaching stages (40, 41).

Short-Sequence Oxygen Bleaching. Oxygen is the least expensive chemical among the oxidizing agents used in the commercial bleaching of pulp, so significant savings can result if oxygen is used for a majority of the oxidation. Laboratory experience with sequences incorporating E_o seems to indicate that, if an 84–86 ISO brightness is acceptable, it can be obtained with the sequence C/DE$_o$D. This sequence offers low capital investment and lower power use but does not reduce chemical costs or have many environmental benefits (35). An existing C/DEDED plant can be converted to two parallel C/DE$_o$D plants with only a small capital investment, giving a tremendous gain in capacity.

If the E$_o$ is used in conjunction with a first oxygen stage, significant gains can be made. In a study (42) comparing an OD_cE_oD sequence with the C_DEDED sequence, the sequence with E_o saved 57% of chemical costs for softwood and 64.1% for hardwood. The two sequenes gave equivalent strength properties for hardwood, and the E_o sequence gave about 2–4% less for softwood. The E_o sequence also gave somewhat lower steam demand, water use, and decreased effluent volume and load—a 50% decrease in BOD and a 65% decrease in color. The E_o pulps seemed to be cleaner than conventional pulps.

If 90 ISO brightness is required, it can be achieved with the sequence OC/DE_oD (35). This sequence saves substantial chemical costs compared to C/DEDED but is comparable in steam or power requirements. It is applicable to both hardwoods and softwoods. It has been reported (42) that the OC/DE_oD sequence can save about \$8–12/ton for hardwoods and \$17–21/ton for softwoods due to savings in chemicals and, to some extent, in energy. The reductions in BOD. color, and total organic carbon can save an additional \$2/ton in the recovery-plant. These cost advantages make these short sequences very attractive, but their implementation has been retarded by some difficulties in reliably and consistently producing pulp of adequate brightnesss and good strength. These difficulties will probably be overcome as mills gain experience with the new procedures.

The E_o pulps seem to wash easier than normal extraction stage pulps (42), and the effluent can be recycled to the recovery systems, thus giving some environmental gain over conventional bleaching. References 35 and 41 can be consulted for a more detailed discussion of short-sequence bleaching conditions, chemical use, chemical costs, strength properties, and energy consumption.

As of April 1984 there were 34 known E_o installations: 9 in Finland, 6 in the United States, 6 in Canada, and the remainder in various other countries (43). It appears to be a viable technology.

Ozone Bleaching

Some experimental work has been done with ozone as a bleaching agent (18, 44–48), which has advantages similar to those of oxygen. It is hoped that ozone may even surpass oxygen as a bleaching agent and eventually lead to the elimination of chlorine. If this result occurs, it would be relatively easy to recycle completely all bleaching effluent to the recovery system and to eliminate the bleach plant as a source of pollution (47). Ozone is a very powerful and nonselective oxidizing agent, so control of such a bleaching process would always be difficult. Laboratory and pilot plant trials have demonstrated that it is possible to conduct ozone bleaching of hardwoods (46). For pulps of 82–85 brightness, an ozone-caustic extraction-peroxide (ZEP) sequence can replace the more conventional CEHD or CEHED. For pulps of 86–88 brightness, a ZED or ZEDD sequence can replace CEDED or CEHED. Very bright pulps (88–90 brightness) require a ZEZP or ZEDED sequence for bleaching. In each case, the sequence is free of chlorine dioxide, making it possible to recycle the effluent completely.

The ozone stage is carried out at 35–45% consistency, a temperature of 27–58°C (80–137°F), a pH of 1–2, and an ozone dose of 0.9–1.2% of the pulp, for a period of less than 6 minutes (46). The ozone concentration in gas is only 2.0–3.5%, so good mixing is required.

For softwood pulps, it has been demonstrated in the laboratory that it is possible to perform a soda-AQ cook followed by oxygen followed by ZEZEZP bleaching to produce 88 brightness pulp with satisfactory strength properties, except for a minor problem with tear (49). Peroxide is used as the final stage to promote brightnesss stability.

These ideas, although scientifically exciting, had not moved past the pilot plant stage as of 1984. Commercial feasibility is yet to be demonstrated. One major problem is the cost of generating ozone (48).

Gas-Phase Bleaching

Conventional bleaching can also be done with reactants in the gas phase applied to pulp at 20–40% consistency with no free draining water. This procedure results in a significant reduction in the amount of chlorine and chlorine dioxide needed to bleach to a comparable brightness level. It is also claimed that a significant reduction in shrinkage is experienced. Fresh water input would be 1400 gallons per ton oven dry (OD) pulp (50). The equipment needed to accomplish gas phase bleaching commercially would probably be very expensive.

REFERENCES

1 J. A. Histed and F. A. Nicolle, *Pulp & Paper Magazine of Canada*, 74(12):80 (1973).

2 G. G. Nelson, Jr., A. A. Yankowski, G. H. Trostel, and M. P. Clark, *Tappi*, 55(6):933 (1972).

3 W. H. Rapson, *The Bleaching of Pulp*, TAPPI Press, Atlanta, 1979, pp. 413–421.

4 G. D. Quesenberry, Jr., "Mill experience with hot chlorination system," *Proceedings of the TAPPI-CPPA International Pulp Bleaching Conference*, Portland, Oregon, 1982, pp. 179–184.

5 G. W. Kutney, H. H. Holton, D. H. Andrews, and G. P. Dornnini, "C_D vs. D_C: Part I—pulp quality; Part II—effluent properties," *Proceedings of the TAPPI-CPPA International Pulp Bleaching Conference*, Portland, Oregon, 1982, pp. 195–222.

6 D. W. Reeve, G. Rowlandson, J. Kramer, and W. H. Rapson, *Tappi*, 62(8):51–54 (1979).

7 S. Lindberg and L. B. Lund, *Tappi*, 63(3):65 (1980).

8 R. J. Gall and F. H. Thompson, *Tappi*, 56(11):72 (1973).

9 W. H. Rapson, Canadian Patent No. 783, 483; Swedish Patent No. 306, 817.

10 W. H. Rapson and C. B. Anderson, *Tappi*, 49(8):329 (1966).

11 H. Makkonen, M. Pelkanen, and T. Laxen, *Tappi*, 57(2):113–116 (1974).

12 T. Laxen and U. Zellstoff, *Das Papier*, 23(6):169–173 (1974).

13 F. Herman and J. Wenzl, "Dynamic bleaching," *Tappi*, 60(5):3–4 (1977).

14 J. Richter, *Tappi*, 49(6):48A (1966).

15 S. H. Watkins, *Paper Trade Journal*, 158(17):28 (1974).

16 J. Gullichsen, *Tappi*, 57(10):111 (1974).

17 J. Gullichsen, *Tappi*, 59(11):106–109 (1976).

18 J.. Gullichsen, *Tappi*, 62(12):31–34 (1979).

19 P. L. Ashcraft, "Displacement bleaching at Temple-Eastex," *Proceedings of the TAPPI Pulping Conference*, New Orleans, 1978, pp. 67–68.

20 D. Ferguson, "Displacement bleaching at Weyerhauser Plymouth," *Proceedings of the TAPPI Pulping Conference*, New Orleans, 1978, pp. 79–80.

21 K. Yonekura, "Experiences with medium consistency chlorination and displacement bleaching," *Proceedings of the TAPPI Pulping Conference*, New Orleans, 1978, pp. 81–86.

22 J. Gullichsen, "Displacement bleaching," R. P. Singh, ed., *The bleaching of pulp*, TAPPI Press, Atlanta, 1979, pp. 275–295.

23 V. M. Nikiten and G. L. Akin, *Trudy. Leningr. Lesolekhn. Akad.*, 75:144–145 (1956); 80(2):77–90 (1958); 85:3–11 (1960).

24 V. M. Nikiten and G. L. Akin, *Bumazhnaya Promyshlemnast*, 35:5 (1960).

25 A. Robert, P. Rerolle, A. Viallet, and C. Martin-Bonet, *Revue ATIP*, 18(4):151–176 (1964).

26 A. Robert, A. Viallet, P. Rerolle, and J. P. Andreolety, *Revue ATIP*, 20(5):207–217 (1966); *Paper Trade Journal*, 152(32):49 (August 5, 1968).

27 *Development and exploitation of the Georges Claude processes*, Finnish Patent No. 1734/68.

28 A. Jameison, S. Noreus, and B. Pettersson, *Tappi*, 54:1903 (1971).

29 R. P. Singh, Canadian Pat. No. 966, 604 (April 29, 1975).

30 *SSVL Environmental Care Project Technical Summary*, Stockholm, Sweden, 1974.

31 J. Kalish, *Pulp and Paper International*, 17(3):58 (1975).

32 B. Coetzie, *Pulp & Paper Magazine of Canada*, 75(6):T-223 (1974).

33 H. E. Engström, L. Ivnös, and B. Pellesson, *Tappi*, 54(11):1899 (1971).

34 R. Hellström, B. Lendqvest, L. Smedman, and A. Jameison, *International Environmental Improvement Conference*, Montreal, 1976, p. 61.

35 R. W. Schleinkofer, "Short sequence bleaching with oxygen, Part II," *Proceedings of the TAPPI Pulping Conference*, Toronto 1982, pp. 303–308.

36 W. Carpenter, W. McKeown, H. Berger, and I. Gellman, *Tappi*, 59(11):81 (1976).

37 L. Markham, C. Courchene, E. Elton, and V. Magnotta, "Medium consistency oxygen delignification of Kraft pulp," *Proceedings of the TAPPI Pulping Conference*, Chicago, Illinois, 1981, pp. 165–174.

38 E. F. Elton, V. L. Magnotta, L. D. Markham, and C. E. Courchene, *Tappi*, 63(11):79–82 (1980).

39 L. E. Nasman, "A system using medium consistency oxyen bleaching works in Sweden," *Pulp and Paper*, 55(10):137 (1981).

40 G. Carre and G. Annergren, "Oxygen alkali extraction: A versatile tool towards a simplified bleaching technique," *TAPPI International Pulp Bleaching Conference*, 1982.

41 L. Nasman and G. Annergren, *Tappi*, 63(4):105–109 (April 1980).

42 R. W. Schleinkofer, "Short sequence bleaching with oxygen," *Pulp & Paper Canada*, 83(11):21–24 (T-299–T-302) (1982).

43 D. Reeve, "E_o installation survey," *Tappi*, 67(4):143 (1984).

44 N. Soteland, *Pulp & Paper Magazine of Canada*, 75(4):91 (1974).

45 S. Rothenberg, D. H. Robinson, and D. K. Johnsonbaugh, *Tappi*, 58(8):182 (1975).

46 R. P. Singh, "Ozone replaces chlorine in first bleaching stage. Advances in ozone bleaching, part I," *Tappi*, 65(2):45–48 (February 1982).

47 J. K. Perkins and R. W. Schleinkofer, "Advances in ozone bleaching, part III: pilot plant insallations and proposed commercial installation," *Proceedings of the TAPPI Annual Meeting*, Chicago, Illinois, 1981, pp. 263–266.

48 F. C. Tritschler and D. J. Shelton, "Commercial manufacture and industrial use of ozone as an oxidant," *Proceedings of the TAPPI Annual Meeting*, Chicago, Illinois, 1981, pp. 259–262.

49 S. Rothenberg, R. V. Jelinek, S. J. Aiken, and R. Aravamuthan, "Ozone bleaching of oxygen pulp," *Proceedings of the TAPPI Pulping Conference*, Toronto, Ontario, 1982, pp. 341–346.

50 N. Liebergott, H. I. Bolker, and D. W. Clayton, "Gas-phase bleaching of softwood and hardwood Kraft and Kraft-oxygen pulps," *Proceedings of the 1976 International Pulp Bleaching Conference*, Technical Association of the Pulp and Paper Industry, Atlanta, 1976, p. 167.

14
Overview of Color and Toxicity Removal Techniques

A number of physicochemical treatment processes have been developed to remove a variety of materials from pulp and paper effluents. These treatment processes include resin separation and ion exchange, aluminum oxide, adsorption on wood, MYCOR processing, lime precipitation, enzyme pretreatment, metal ions, membrane separation, irradiation, activated carbon, amine processes, and ozone. Some of them are capable of removing nearly everything from the effluent to produce water of almost drinking water purity. Both reverse osmosis techniques and activated carbon systems could probably achieve this purity. Although technically feasible, such systems can be costly. Physicochemical treatment processes must compete economically with secondary biological treatment and, in most instances, are not justified.

However, physicochemical treatment schemes can be useful when biological treatment cannot accomplish the needed purification. One such instance is the removal of color from pulp and paper mill effluents. Color removal technologies cost about \$10–\$20 per ton of production, equivalent to as much as 10% of the value of the final pulp produced. Thus it is easy to understand why mills have been reluctant to implement such systems.

Proper biological treatment is effective in removing toxic materials from pulp and paper effluents. However, in upset conditions, physico chemical methods may be useful as a safeguard. They may also be used to treat selected troublesome streams within the process.

In this chapter, we will discuss specific physico chemical treatment processes, but it is important to remember that process modifications can also have a considerable effect on color and toxicity. Changing to oxygen pulping or using the soda-AQ process instead of Kraft pulping would have a major impact, as would incorporating oxygen, chlorine dioxide, hypochlorite, peroxide, or ozone into the bleaching sequence. Either displacement bleaching or gas phase bleaching would change the color and toxicity loads. Of these various technologies, oxygen bleaching has the most potential for decreasing the color and toxicity produced.

These various internal process changes are covered elsewhere in this book. This chapter will discuss specific external treatment technologies.

COLOR REMOVAL

The color in pulp and paper industry wastewater results from three operations: chemical pulping, pulp bleaching, and colored paper production. The majority of the colored wastewater arises from the pulping and bleaching operations. The bleaching operation produces more color load than pulping, and the dominant color stream in the bleach plant is the caustic extraction stage. For a representative bleached Kraft mill, 300 lb of color per ton might be generated. The pulp mill would contribute 65 lb/ton, and the bleach plant 200 lb/ton, of which the caustic extraction stage would contribute 145 lb/ton. The remaining color, 35 lb/ton, originates in the wood yard, in recovery and causticizing, and from the paper mill. The bleach plant can be responsible for more than two-thirds of the total color load while contributing less than half of the effluent volume. Table 14-1 shows the range of color values for each stream (1); as expected, the alkaline extraction stage predominates.

Unfortunately, secondary biological treatment plants provide, at best, a 30% reduction in the color load (2, 3), and some biological treatment systems actually create color. The difficulty is that much of the color is associated with lignin compounds that are not easily biodegradable.

Since the early 1950s, a search has been underway for an economical way to remove color. A wide variety of coagulants have been tested for their effectiveness, including alum, ferric sulfate, lime, sulfuric acid, char, clay, activated carbon, activated silica, ferric chloride, phosphoric acid, waste pickle liquor, and a barium alumina silicate compound (4). Good color reduction has been obtained with several of these materials—at, however, a prohibitive cost.

Lime Precipitation or Coagulation

From the perspective of cost, the best precipitation agent is lime. Lime is also attractive because of the major role it plays in the Kraft pulping chemical recovery system. The Kraft recovery system also provides dual opportunity for:

Table 14-1 Range of Color Concentrations for Various Waste Streams (Bleached Kraft Mill)

Stream	Color (PCU)[a]
Pulp mill effluent	700–3,700
Bleach plant	
Chlorination effluent	600–2,300
Alkaline extraction	4,400–23,000
Total mill effluent	1,200–3,700

[a]PCU, platinum cobalt units.

1 Recovery of lime used for decolorization
2 Destruction of removed color by conventional thermal oxidation

Considerable research and development has gone into this process, and its history has been traced by several authors (5–7). The sludge of lime and color bodies that is produced is difficult to dewater.

To overcome this problem, the first approach was to use a massive dosage of lime so that the color body portion was only a small fraction of the material to be dewatered. In this "massive lime" process, the mill's total lime supply is slaked and reacted with a small-volume, highly colored stream, usually the bleach plant caustic effluent, at a lime concentration of 20,000 mg/1. The lime is settled, dewatered, and used for causticizing green liquor. The color bodies dissolve in the white liquor and eventually find their way into the recovery furnace. The effulent stream being decolored passes to a carbonation clarifier, where additional lime is removed by precipitation. This sludge flows to the mud washer and eventually to the lime kiln. The massive lime process, as tried under an EPA demonstration grant, achieved 94–95% color removal on the stream treated (8). This result was equivalent to a 74% color reduction for the overall mill, considering all effluent streams. The inability to treat the total mill effluent is a shortcoming of the process. Another disadvantage is that the cooking liquor is diluted about 15%, which necessitates an increase in the capacities of the recovery and chemical preparation equipment.

To overcome these shortcomings, three other lime treatment processes have been developed independently, each of which employs lime dosages of about 1,000 mg/l (9–11). All three processes have been tested in mill trials, and the one that has had the longest operating history is shown in Figure 14-1. Effluent from the mill, after grit and trash removal, is pumped to a clarifier, where lime at approximately 1,000 mg/l is added to the influent stream. It was found that the presence of fiber fines in the paper mill effluent enhanced color removal. About 60% of the lime, most of the color bodies, and virtually all of the settleable solids settle to the bottom of the clarifier. The materials settle well, since the clarifier is designed to run at 1,440 gal/day/ft^2, which is about double the traditional design capacity. The primary clarifier sludge is removed, thickened by centrifugation, and burned in the mill's lime kiln. The overflow from the clarifier is treated with carbon dioxide from lime kiln stack gas, which converts the soluble lime to insoluble carbonate. The insoluble carbonate is pumped to the causticizing process "mud" storage tank for dewatering on a lime mud filter and then is burned in the lime kiln. The kiln is 25% oversized to accommodate the color removal sludge, which does not adversely affect kiln operation. The process is capable of achieving 85–93% color removal on the total mill effluent (12). The particular mill involved is a cross-recovery NSSC mill; when NSSC effluent was present in the total mill effluent, a decrease of 15% in color removal efficiency was observed.

The other minimum lime processes that exist have elements in common with the one just described. The literature contains details on their differences (7, 9,

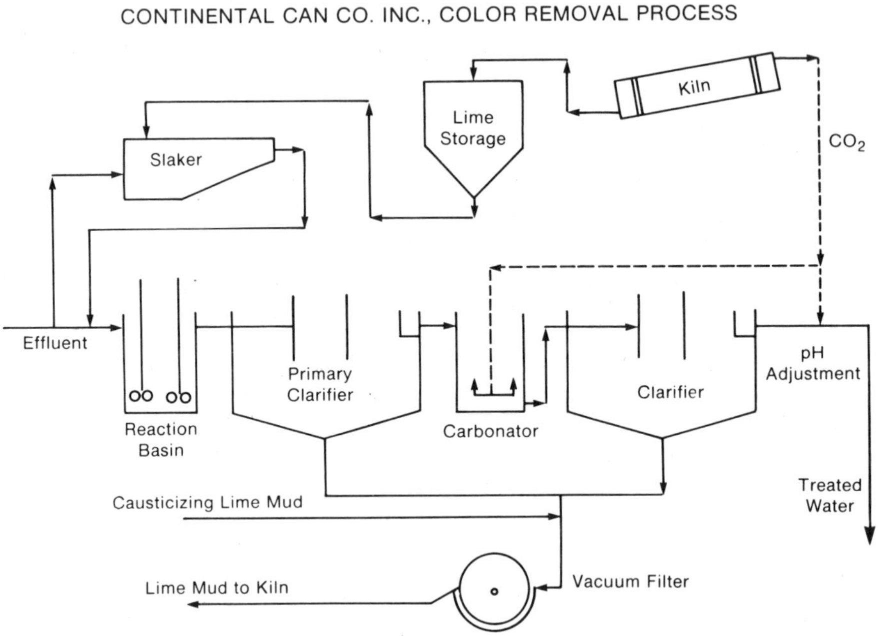

Figure 14-1 Color removal process

11, 13). In some of these minimum lime processes, a lime mud treatment scheme has been evaluated in which a minimum quantity of slaked lime and the total mill lime mud produced are mixed together and reacted with the effluent. The lime organic sludge is dewatered and burned in the lime kiln. This scheme produces an extremely voluminous, nonconsolidated sludge.

Enzyme Pretreatment

The lime color removal process is very effective on high molecular weight ($>5{,}000$) color bodies but not on those of low molecular weight. It was hypothesized that pretreatment with the enzyme peroxidase could polymerize the phenolic color bodies to higher molecular weight compounds that would be more easily removed by lime precipitation. It was found that pretreatment with 2.5 mg/l horseradish peroxidase plus 2.5 mg/l hydrogen peroxide allowed 94% color removal by lime at dosages that gave 10% color removal without the enzyme pretreatment (14). Thus color removal efficiency could be increased and the lime dosage decreased. Enzyme pretreatment may not be economically feasible, however, and hydrogen peroxide is difficult to use.

Magnesium and Lime

A second, and perhaps more economical, approach used to improve the efficiency of color removal by lime is a combination of lime and magnesium sulfate (15).

Experimental data have shown that equivalent color removal (90%) can be obtained by any of the following three treatments:

1. 2,000 mg/l lime (or more)
2. 800 mg/l lime plus 30 mg/l Mg^{2+}
3. 500 mg/l lime plus 60 mg/Mg^{2+}

Both magnesium and calcium can be recovered through recarbonation and recalcination. Only laboratory evaluations on paper mill effluents have so far been performed for this process. It appears to have considerable merit, providing the magnesium can be recovered at high efficiency.

Alum Coagulation and Precipitation

Alum treatment was suggested in early studies (16–18) because it had been shown capable of decolorizing potable water supplies. Laboratory studies have shown it capable of achieving 89% color removal at dosages of 30 mg/l for softwood Kraft effluents and 150 mg/l for hardwood wastes (16).

On a commercial scale in the U.S.S.R., the Baikol pulp mill treats 76 MGD of waste from the production of 110,250 ton/year of alpha cellulose and 110,250 ton/year of viscose cellulose. Treatment consists of neutralization, nutrient addition, equalization, activated sludge treatment, alum coagulation, sedimentation, sand filtration, polishing lagoon treatment, reaeration, and discharge to Lake Baikol (19). In the coagulation stage, alum is used at a rate of 30 mg/l as Al_2O_3, and polyacrylamide flocculent at a rate of 1 mg/l is also added. Detention time in the clarifier is 6 hours. The color of the raw wastewater averages 1,000 units and that of the treated water 101 units, yielding 90% color reduction.

In the United States, Gulf States Paper at Tuscaloosa, Alabama, used to treat 12 MGD by an alum process. Due to economic difficulty, this plant has been closed. Treatment consisted of primary clarification. bio-oxidation using high-purity oxygen, secondary sedimentation, and alum coagulation and sedimentation. Primary clarifier sludge, excess biological solids, and the alum sludge were blended, thickened, pressure filtered, and burned in a multiple-hearth furnace. The furnace ash, high in Al_2O_3, was reacted with sulfuric acid to recover aluminum as aluminum silicate. The recovery of alum was 94% (20). Alum mud from a nearby plant served as the initial alum source. The final effluent color was 50 American Public Health Association (APHA) units from initial color ranging from 800 to 1,200 APHA units, or removal exceeding 94%.

Ozone

Oxidation with a powerful oxidizing agent such as ozone or peroxide can destroy both color bodies and toxic compounds. The cost of the oxidizing agent is a significant issue. Experience with ozone on pulp and paper effluents is summarized in references 21–24.

To make these processes more economical, it has been proposed to use them as pretreatments before biological treatment. Pretreatment should partially degrade the compounds that otherwise resist biological treatment into forms that are biodegradable, thus eliminating color or toxicity. Studies have found that pretreatment with ozone or peroxide does increase the biodegradability of Kraft mill caustic extraction stage effluents (24). Ozone was found to be more effective than peroxide. Ozone at 20 mg/l performed better than peroxide at 20,000 mg/l. Ozone at 20 mg/l applied for 90 minutes achieved better than 90% color reduction, with 85% reduction achieved in the first 15 minutes of contact time (24). Free radical reactions are not involved, and the major oxidizing agent is ozone itself. Some investigators have found beneficial effects from using ozone in combination with ultraviolet (UV) light. UV light activates some organic compounds, making them more susceptible to ozonation. The effect is compound specific, however, and UV light was not found to be beneficial for the treatment of tannin, lignins, and caustic extraction stage effluent.

There is little question of the effectiveness of ozone; the problem is generation of ozone at low cost. Current processes for ozone generation lose about 90% of the required energy as light, sound, and heat.

Furhter work should be done to discover the minimum amount of ozone that will allow subsequent biological treatment to decolorize pulp and paper mill effluents. Studies to date have dealt mainly with the ozonation process itself, and not the combined sequence of ozonation followed by biological treatment. It is possible that the right combination sequence might be economically viable.

MYCOR Process

The MYCOR process is another attempt to accomplish color removal through bilogical treatment. MYCOR is an acronym for a process of mycelial color removal that uses the white rot fungus, *Phanerochaete chrysosporium burds*, to metabolize lignin color bodies and thus remove color (25, 26). The fungus is grown on a rotary biological contactor, and the effluent is decolorized as it passes through the contactor. The rate of color removal is dependent on color concentrations of up to 6,000 PCU/l (PCU are platinum cobalt units, a measure of color) and then appears to be concentration independent. The system can achieve high degrees of color removal at a rate of approximately 2,000 PCU/l/day (26), and is also very good at removing toxic material, since the effluent is free of chlorinated phenols. The system might be very economical if it could be made self-sustaining, but in experiments done so far, the fungus organism would not sustain itself and had to be reseeded constantly.

Polymeric Adsorbents and Ion Exchange Resins

While conventional ion exchange resins have not proven technically successful in removing pulping or bleaching color after regeneration with process liquors

(27), several specialized resins have offered encouraging results. Pilot plant work has been performed on the effectiveness of synthetic polymeric adsorbents in decolorizing Kraft bleach plant effluents (28, 29). Since resin systems must be used on streams that are very low in suspended solids and free from contaminants that will prevent the resin from being regenerated, they are usually applied to small-volume streams. Since 85% of the bleach plant color load is contained in the chlorination and first-stage caustic extraction effluents, these have received major attention for resin treatment. The performance of the resins is usually pH dependent. The systems seem to be capable of color removal efficiency of more than 90%.

Among the more promising resin treatments are the Rohm and Haas process and the Uddeholm–Kamyr process. The Rohm and Haas process is based on the use of a special polymeric adsorbent, Amberlite XAD8, a highly crosslinked 20–60 mesh resin possessing no ion exchange properties (30). This process functions best at a pH of 2.0–2.5, so mixing of acid and alkaline bleaching effluents proves favorable. Higher decolorization efficiency is achieved for hardwood (85–92%) than for softwood (75–85%) effluents. BOD removal is about 31–55% and COD removal 31–57% (31). A 5-year resin life is claimed, and adequate protection against resin fouling is provided by hydrosieve filtration. Resin regeneration is effected using weak wash, which transfers the color components to the green liquor in the smelt dissolving tank. Alternatively, white liquor has been proposed for regeneration.

A similar system using phenolic resins has also been developed (32). In the testing of this system, the chlorination and extraction stage effluents were blended to obtain pH 2. This blending caused some of the lignin to precipitate, and the precipitated lignin was then settled out of the effluent before it was fed to the phenolic column. Polymer was sometimes used to assist the settling. The precipitated lignin is dissolved in caustic and sent to the recovery system. The settled effluent is fed to the phenolic column, where the color bodies are adsorbed from it. More than 99% color reduction is claimed. The spent resin is washed with sodium sulfate to remove any chlorides, then eluted with an alkali stream from within the mill.

Neither this system nor the Rohm and Haas System has found full-scale application in the pulp and paper industry.

The Uddeholm–Kamyr process is a Swedish development. It combines absorption and ion exchange mechanisms (33–35), possibly based on a weakly anionic resin. The resin is regenerated using 1–2*N* caustic. It then requires an acidic reactivation, using either sulfurous acid or spent sulfuric acid from ClO_2 production. The regenerants are recoverable, and the eluted colored material is ultimately burned in the recovery furnace. Chloride separation is accomplished using sodium sulfate displacement, so that chlorides do not appear to present problems. There are full-scale installations in Sweden, Japan, and the United States.

At the U.S. installations, the first-stage caustic extraction effluent from a 300-ton/day bleached Kraft mill using a CEHDED bleaching sequence is treated. This waste stream flow averages 100 m^2/hr at 14,000 PCU of color. A 90%

color reduction is achieved. After prefiltration to remove fibers, this waste passes through ion exchange columns at a rate of five bed volumes per hour. Sixteen bed volumes of waste can be treated to an average of 90% color removal before regeneration of the resin is necessary.

The resin is used to treat both the chlorination stage effluent and the extraction stage effluent before being regenerated. Rather than switching columns, the resin itself is physically transported from one column to another. This mechanical transport results in a tolerable amount of mechanical damage to the resin. Moving the resin keeps it from becoming plugged with fibrous debris and minimizes channeling because of the periodic repacking.

The concentrated waste resulting from regeneration of the column is sent to the evaporators and then burned in the recovery boiler. At this level of treatment, the total bleach plant effluent color reduction is 86%.

Resin treatments have also been considered as a posttreatment after massive lime processing (28).

Aluminum Oxide

Another color removal system that has been used full scale is one that employs aluminum oxide as the resin. A West German bisulfite mill uses it on effluent from a CEPH bleaching sequence (36). The effluent is clarified to reduce suspended solids and pH adjusted with dilute hydrochloric acid. It then passes through an upflow tower filled with granular activated alumina (γAl_2O_3), which removes 96% of the color, 35% of the BOD, and 71% of the COD by a combination of adsorption and precipitation. The aluminum oxide is regenerated by burning off the organics in an indirectly fired rotary kiln. Complete adsorptive capacity of the resin is retained by keeping the kiln temperature less than 800°C. Figure 14-2 represents a schematic of this system.

Wood Shavings

One of the problems with adsorption systems is the cost of the adsorbing material. Resins are expensive to produce and have a limited life. These problems led Poats et al. (37) to look at wood as a possible adsorbing material. Shavings were used as the adsorbent and, when their adsorptive capacity had been reached, would be burned in a hogged fuel boiler. The concept worked but wood shavings proved to have a very low adsorptive capacity, thus requiring large quantities to achieve adequate color removal.

Activated Carbon

Early work demonstrated that activated carbons were capable of removing effluent color by adsorption. Activated carbon is a material characterized by extremely large surface area to unit weight ratios, typically 450–1,800 m^2/g. The large surface area available results in substantial adsorptive capacity. The rate of

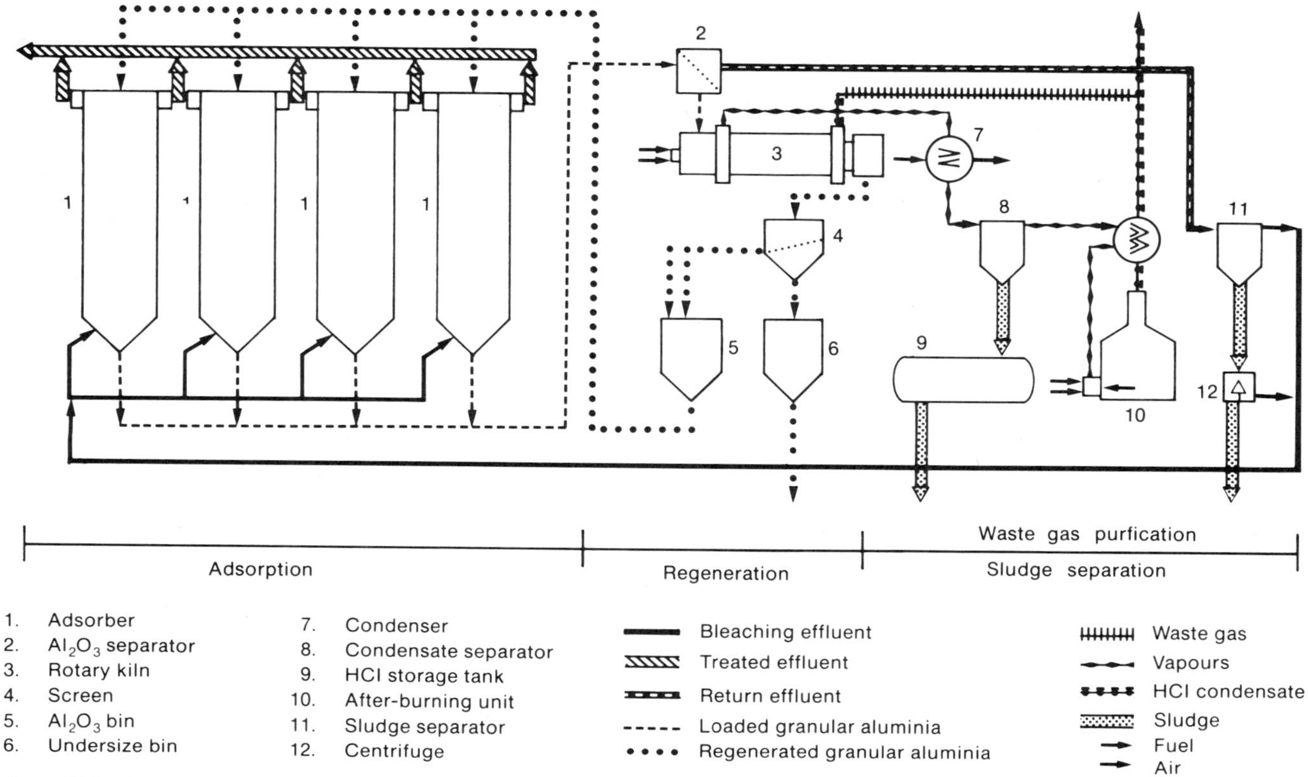

Figure 14-2 Activated alumina color removal system. (From T. W. Joyce and W. H. Petke, *Effluent Decolorization Technologies for the Pulp and Paper Industry*, 1984. Water Resources Research Institute of The University of North Carolina, Report No. 202)

Figure 14-3 Activated carbon adsorption tower system. Copyright © 1981. TAPPI. Reprinted from *Introduction to Environmental Management for the Pulp and Paper Industry*, Study Guide, Volume 6, with permission.

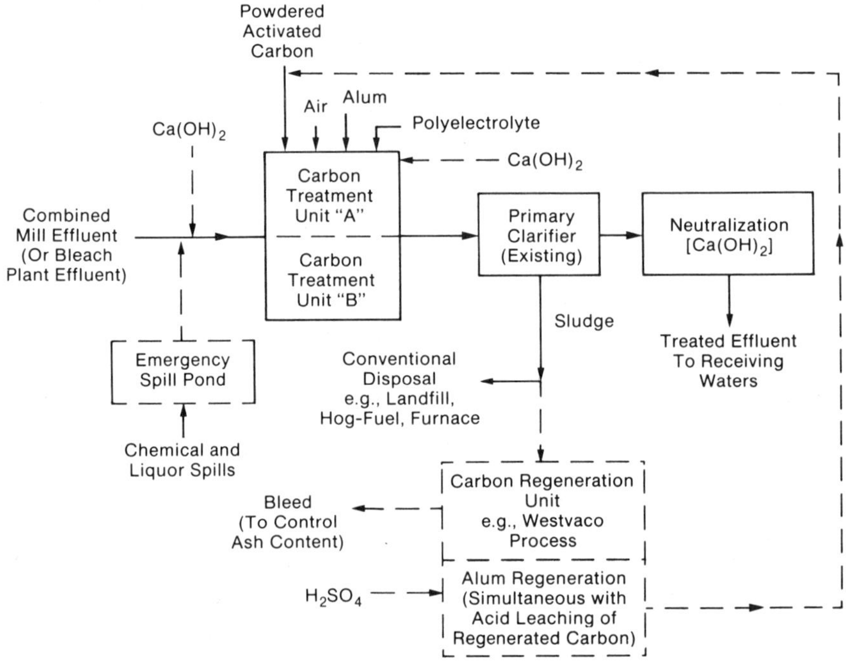

Figure 14-4 Synergistic toxics removal system. (From T. W. Joyce and W. H. Petke, *Effluent Decolorization Technologies for the Pulp and Paper Industry*, 1984. Water Resources Research Institute of the University of North Carolina, Report No. 202.)

adsorption is a function of carbon particle size. Powdered carbon has a faster adsorption rate than granular carbon. The ultimate adsorptive capacity, however, is essentially the same for both. Carbon pore size, particle size, color component molecular weight, and effluent pH (as it affects the solubility and polarity of colored components) are all of significance in determining performance (38). The process had not been commercially applied within the paper industry as of 1984. Figure 14-3 shows the type of equipment used in a granular activated carbon column system.

The high cost of the process and its regeneration requirements have kept it from being applied commercially. Extensive pilot plant work has been done to try to overcome these problems. One idea was to use mill-produced activated carbon on a once-through basis (39). The study evaluated both a column granular carbon process and a multistage, countercurrent, fine activated carbon process (40, 41).

More recently, carbon has been used in combination with other techniques to detoxify effluent. The system involved the synergistic use of powdered activated carbon, air, alum, and polyelectrolyte (42), producing an effect greater than the sum of the effects of each component individually. It effectively detoxified the effluents tested—bleach plant effluent and total mill effluent—and removed 70–85% of the color and 20–40% of the BOD and COD. A schematic of this process is shown in Figure 14-4.

Membrane Processes

There is no question that membrane processes are effective in removing color with any degree of efficiency desired. The problem is to engineer an economical, workable system. At present, there are essentially three types of reverse osmosis (RO) and ultrafiltration systems available on the commercial market:

1 Spirally wound sheet modules
2 Capillary fiber modules
3 Hollow core tube modules

RO is a tight-membrane, high-pressure process, whereas ultrafiltration is a loose-membrane, low-pressure process.

The most common membrane material is cellulose acetate. Membrane life is greatly influenced by the pH and temperature of the wastewaters being treated. In general, a pH less than 7 and a temperature less than 38°C (100°F) are required to minimize unnecessary deterioration of the membrane. A large factor in system cost is membrane life.

Membranes may foul due to buildup of materials on the membrane surface, which causes a marked decrease in flux rate. Fouling is inherent in all existing systems, but may be minimized by pretreatment of the wastewaters to remove particulate and colloidal materials or by pressure pulsing and cleaning of the membrane. For tubular membranes, fouling is minimized at velocities above 3 ft/sec. For pulp and paper wastewaters, the spirally wound and capillary fiber modules have been found to be more prone to invisible fouling problems than are tabular modules.

With tubular membranes, RO has been demonstrated to give 99% color removal on caustic extraction effluent at a flux rate of 7 gal/day/ft^2 at 600 psig (43). The waste was pretreated by filtering through a 40-mesh screen, pH adjusted, and cooled to less than 38°C.

RO hyperfiltration is a high-pressure operation and is probably a more powerful technique than is necessary for color reduction, so recently interest has turned to ultrafiltration. Laboratory studies of the technique have been conducted (44), as has a pilot plant investigation (45). Bleached Kraft mill decker and caustic extract effluents were treated in a 7 gal/min unit that operated at a pressure of 100 psig. The unit consisted of a five-stage, circular-wound cellulose acetate membrane operated at 18–20 GPD/ft^2 and 100–140 psi. Figure 14-5 shows the pilot-scale tubular membrane system used by the Institute of Paper Chemistry to conduct the trials. Color removal from bleach plant caustic extract and from decker effluent was 90–97%, leaving a concentrate of 6–20% solids to be disposed of by incineration (45). The major problem encountered in the operation of the system was particulate collection in the ultrafiltration cartridges that hindered the continuance of a uniform flux rate at a constant pressure loss.

Recently, polysulfone membranes have been developed. Fremont (46) developed and tested a polysulfone membrane coated with polyethyleneimine, which could effect 92% color removal at 44 psi and 94% color removal at 100 psi. A system has been installed in Sweden using polysulfone membranes to treat the alkali extraction stage of a softwood Kraft bleach plant (47), consisting of a cascade of ultrafiltration modules. Cascading results in lower flux from each succeeding module but gives a concentrated final effluent stream. Fibers had to be

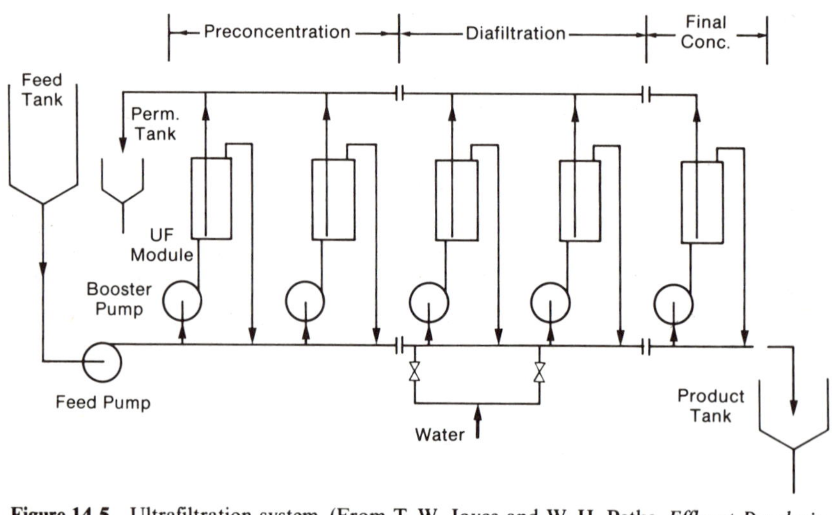

Figure 14-5 Ultrafiltration system. (From T. W. Joyce and W. H. Petke, *Effluent Decolorization Technologies for the Pulp and Paper Industry*, 1984. Water Resources Research Institute of the University of North Carolina, Report No. 202.)

removed to a level of less than 10 ppm before ultrafiltration to prevent clogging the membrane and allow satisfactory operation. This system achieved 86% color reduction, 25% BOD reduction, and 70% COD reduction. Acute toxicity was reduced by 50%.

Electrochemical Processes

The principle behind electrochemical treatment is that chloride in the effluent is converted electrochemically to chlorate, hypochlorite, and chlorine. The chlorine and hypochlorite oxidize the organic compounds in the effluent and by so doing are themselves reduced to chloride again. Thus this is a cyclic process that does not generate sludge and is driven by electrochemical energy. Entire bleached Kraft effluents can be reduced in color by 80% and caustic extraction effluent by more than 90% with this process (48). Many variations of the process have been investigated (49, 50) using different mill streams and electrode types. One variation of the process also used the hydrogen gas bubbles produced to generate a froth that was used to float the solids out of the system and to aid in the removal of toxic substances. Electrochemical systems are effective but high in operating cost.

Other Techniques

A number of other techniques for color and toxicity removal have been investigated, including amine treatment (51–53), precipitation with seawater (54), irradiation (55, 56), and ultrasonic oxidation (57). Paper industry color removal has been under investigation for more than 30 years, and it appears likely that individual specialized solutions, involving elements of the technologies discussed in this chapter, will be developed to solve each mills' individual problem.

APPLICABILITY OF PHYSICOCHEMICAL TECHNOLOGIES

All of the technologies discussed remove BOD and COD as well as color and toxicity, but, as mentioned in the introduction, it is cheaper to remove BOD and COD biologically. Therefore the removal efficiency of the physicochemical processes on these parameters is of secondary interest and has not been discussed. The physicochemical treatment methods may become important as devices to remove toxic pollutants if biological treatment proves inadequate; however, this event seems unlikely at this time.

COST

Cost, rather than technical feasibility, is usually the major issue of color and toxics removal. Tables 14-2 and 14-3 attempt to put costs in perspective.

Table 14-2 Costs of Decolorization Alternatives for the First Caustic Extraction–Stage Effluent

Type	Scale	Costs (1980 $ Basis)			
		Percent Color Removal	Capital Cost ($/ton)	Operating Cost ($/ton)	Total Cost ($/ton)
Ion exchange	Mill (300 tpd)	90	2.8	3.7	6.5
Ultrafiltration	Mill (54.7 tpd)	87	1.7	3.8	5.5
Mini-lime	Mill (625 tpd)	90	0.7	2.7	3.4
Massive-lime	Pilot plant	93	3.1	3.2	6.3
Polymeric adsorbent	Pilot plant	85	0.6	1.3	1.9

Source: From T. W. Joyce and W. H. Petke, *Effluent Decolorization Technologies for the Pulp and Paper Industry*, 1984. Water Resources Research Institute of The University of North Carolina, Report No. 202.

Table 14-3 Costs of Decolorization Alternatives for the Total Mill Effluent

Type	Scale	Costs (1980 $ Basis)			
		Percent Color Removal	Capital Cost ($/ton)	Operating Cost ($/ton)	Total Cost ($/ton)
Ozone	Pilot plant	85	1.0	4.0	5.0
Dispersed air flotation	Laboratory	90	NR[a]	6.9	NR
$Fe_2(SO_4)_3$	Laboratory	91	2.3	10.8	13.1
Alum	Laboratory	93	2.3	10.3	12.6
Activated carbon adsorption	Pilot plant	80	2.2	7.8	10.0

Source: From T. W. Joyce and W. H. Petke, *Effluent Decolorization Technologies for the Pulp and Paper Industry*, 1984. Water Resources Research Institute of The University of North Carolina, Report No. 202.
[a]NR, not reported.

Table 14-2 provides cost estimates (1980 dollars) of $1.90–6.50/ton to treat just the first caustic extraction stage effluent and Table 14-3 of $5–13/ton to treat the total mill effluent. The cost in current dollars would be somewhat higher. This is a considerable cost burden for a product (bleached pulp) whose final market value is $300–400/ton.

REFERENCES

1 T. W. Joyce and W. H. Petke, *Effluent decolorization technologies for the pulp and paper industry*, University of North Carolina Water Resources Institute Project No. 84-07-50113, Raleigh, 1984.

2 R. J. Rush and E. E. Shannon, *Review of color removal technology in the pulp and paper industry*, Environmental Canada Report No. EPS 3-WP-76-5, CPPA, Toronto, Ontario, 1976.

3 T. Obiaga and J. Ganczarezyk, "Biological removal of lignin from Kraft mill effluents—changes in molecular size distribution," *Tappi*, 57:137 (1974).

4 W. A. Maggio, *Experimental chemical treatments for Kraft mill wastes*, Technical Bulletin No. 50, National Council of the Paper Industry for Air and Stream Improvement, New York, 1952.

5 P. Lunar and C. Dence, *The mechanisms of color removal in the treatment of pulping and bleaching effluents with lime. II. Treatment of chlorination stage bleaching effluents*, Technical Bulletin No. 242, National Council of the Paper Industry for Air and Stream Improvement, New York, 1970.

6 H. Edde, *Settleable solids removal practices in the pulp and paper industry*, Technical Bulletin No. 178, National Council of the Paper Industry for Air and Stream Improvement, New York, 1964.

7 *Development document for interim final and proposed effluent limitations guidelines and new source performance standards for the bleached Kraft, groundwood, sulfite, soda, deink, and non-integrated paper mills segment of the pulp, paper, and paperboard mills point source category*, EPA 440/1-76/047a, U.S. Environmental Protection Agency, Washington, D.C., 1976.

8 D. R. Smith and H. F. Berger, *Tappi*, 51 (10):37A (1968).

9 C. C. Davis, *Color removal from Kraft pulping effluent by lime addition*, Interstate Paper Corporation, ENC, Riceboro, Da., December 1977.

10 E. L. Spruill, *Tappi*, 56 (4):98 (1973).

11 M. Gould, (Georgia-Pacific Corp.), U.S. Pat. 3,531,370 (1970).

12 E. L. Spruill, "Long term experience with Continental Can's color removal systems," *Environmental Conference 1974*, Technical Association of the Pulp and Paper Industry, Atlanta, 1974, p. 19.

13 M. Gould, "Color removal from Kraft mill effluent by an improved lime process," *Tappi*, 56:79 (1973).

14 R. L. Schmidt and T. W. Joyce, "An enzymatic pretreatment to enhance the limp precipitability of pulp mill effluents," *Tappi*, 63:63 (1980).

15 M. K. Oldham and R. J. Rush, "Color removal in Kraft mill wastewaters with magnesium," *Journal—Water Pollution Control Federation*, 50 (1):40–45 (January 1978).

16 S. E. Smith and R. F. Christman, *Journal—Water Pollution Control Federation*, 41:222 (1969).

17 H. K. Willard, "Coagulation of pulp and paper aerated lagoon effluents for color and solids removal," *AIChE Symposium Series*, 69:173 (1973).

18 E. J. Middlebrooks et al., *Water & Sewage Works*, 116:IW-7–IW-9 (1969).

19 R. H. Scott, *Pulp & Paper*, 48 (4):82 (1974).

20 H. R. Fuller, U.S. Pat. 3,627,679 (1971).

21 H. D. Bauman and L. R. Lutz, "Ozonation of a Kraft mill effluent," *Tappi*, 57 (5):116 (1974).

22 R. G. Rice and M. E. Browning, *Ozone for industrial water and wastewater treatment: A literature survey*, U.S. EPA Robert S. Kerr Environmental Research Laboratory, Ada, Okla.

23 P. B. Melynk, J. Jadkins, and A. Netzer, *Tappi*, 60 (3):97–100 (1977).

24 L. A. Coburn, A. J. Englande, M. P. Lockwood, and T. Collins, "Kraft bleach plant effluent treatment by ultraviolet/oxidation process," *Proceedings of the TAPPI Environmental Conference*, Savannah, Gao, April 9–11, 1984, pp. 277–286.

25 A. Campbell, *A bench scale evaluation of a process for decolorization of bleach plant effluent using the white rot fungus Phanerochaete chrysosporium*, Ph.D. dissertation, Department of Wood and Paper Science, North Carolina State University, 1983.

26 E. Gerrard, *Some factors affecting fungal decolorization of pulp and paper mill effluents in the MYCOR process,* M.S. thesis, Department of Wood and Paper Science, North Carolina State University, 1983.

27 R. L. Sanks, *Ion exchange color and mineral removal from Kraft bleach wastes*, EPA Report R2-73-255, U.S. Environmental Protection Agency, Washington, D.C., May 1973.

28 S. L. Rock, D. C. Kennedy, and A. Bruner, "Decolorization of Kraft mill effluents with polymeric adsorbents," *Environmental Conference 1974*, Technical Association of the Pulp and Paper Industry, Atlanta, 1974, p. 25.

29 T. A. Chamberlin, G. C. Kolb, S. F. Brown, and D. H. Philp, "Color removal from bleached Kraft effluents," *Environmental Conference, 1975*, Technical Association of the Pulp and Paper Industry, Atlanta, 1975, p. 35.

30 *Decolorization of Kraft pulp bleaching effluents using Amberlite XAD-8 polymeric adsorbent*, Rohm and Haas Co., St Jouis, August 1971.

31 S. L. Rock, A. Bruner, and D. C. Kennedy, "Decolorization of Kraft mill effluents with polymeric adsorbents," *Tappi*, 57 (9):87 (1974).

32 B. G. Broddevall, "Purification of waste water from sulfate pulp bleaching plants," U.S. Pat. 3,990,969 (1976).

33 L. G. Anderson, B. Broddevall, S. Lindberg, and J. Phillips, *Tappi*, 57 (4):102 (1974).

34 S. Lindberg and L. B. Lund, *Tappi*, 63 (3):65 (1980).

35 H. B. Borjeson and S. Lindberg, "Field report on a nonpolluting bleach plant," *Tappi*, 64 (10):89 (1981).

36 H. Ullrich, "Purification of pulp-bleaching waste water with aluminum oxide," *Progress in Water Technology*, 10 (5/6):89 (1978).

37 U. J. P. Poats, G. McKay, and J. J. Healy, "Removal of basic dye from effluent using wood as an adsorbent," *Journal—Water Pollution Control Federation*, 50 (5):926 (1978).

38 P. R. Rankin and A. Benddik, "Lignin adsorption of activated carbon," paper presented at the Fourth Joint AIChE-CSChE Engineering Conference, Vancouver, British Columbia, September 1973.

39 W. Timpe et al., "Activated carbon and other techniques for color removal from Kraft mill effluents, "*Proceedings of the 15th EUCEPA Conference*, Rome, May 1973, pp. 535–557.

40 W. G. Timpe, E. W. Lang, R. R. Haynie, and M. R. Garrett, "The use of activated carbon for water renovation in Kraft pulp and paper mills," *7th Water and Air Conference*, Technical Association of the Pulp and Paper Industry, Atlanta, 1970, p. 75.

41 W. G. Timpe, and E. W. Lang, "Activated carbon treatment of unbleached Kraft effluent for reuse: Pilot plant results," *Environmental Conference 1973*, Technical Association of the Pulp and Paper Industry, Atlanta, 1973, p. 203.

42 A. Wong, T. Tenn, J. Dorica, and S. Prahacs, "Detoxification and decolorization of Kraft pulp mill effluents using activated carbon," *Pulp and Paper Canada*, 78 (4):37 (1977).

43 A. J. Wiley, G. A. Dubey, and I. K. Bansal, *Reverse osmosis concentration of dilute pulp and paper effluents*, EPA 12040 EEL, U.S. Environmental Protection Agency, Washington, D.C., February 1972.

44 H. Berger and W. J. Gillespie, *Tappi*, 53 (5):883 (1970).

45 H. A. Fremont, D. C. Tate, and R. L. Goldsmith, *Color removal from Kraft mill effluents by ultrafiltration*, EPA Technology Transfer Series, EPA-660/2-73-019, U.S. Environmental Protection Agency, Washington, D.C., December 1973.

46 H. A. Fremont, "Color removal from paper and pulp mill aqueous effluents," U.S. Pat. 4,226,673 (1980).

47 H. Lundahl and I. Mansson, "Ultrafiltration for removing color from bleach plant effluent," *Tappi*, 63 (4):97 (1980).

48 O. Klaus, *Journal—Water Pollution Control Federation*, 50:286 (February 1978).

49 *Electrochemical process for treatment of Kraft mill effluent*, Canada Environmental Protection Service CPAR Report No. 523-1, Barringer Research, Ltd., Toronta, Ontario, 1976.

50 D. O. Herea and F. E. Woodard, *Tappi*, 59 (1):134 (1976).

51 *PPI, Pulp & Paper International*, 15 (5):69 (1973).

52 J. J. Svarz, "Removal of color from paper mill waste waters," U.S. Pat. 4,058,458 (1977).

53 *Use of high molecular weight amines for the purification of pulp mill effluents*, CPAR Report No. 1-2, Pulp and Paper Research Institute of Canada, Toranta, Outerio, 1971.

54 B. Rapson, D. P. Sullivan, and J. A. Brothers, *Paper Trade Journal*, 158 (8):23 (1974).

55 K. Sameshina, M. Suminato, and T. Kondo, "The color of waste liquor from pulp industry. VII. Color reduction by ultraviolet irradiation treatment," *Journal of Japan Wood Research Society*, 21 (3):188 (1975).

56 S. Miguro, K. Sameshima, M. Sumimoto, and T. Kondo, "On spent liquor of semichemical pulping. V. Improvement of the color removal efficiency by the photochemical treatment," *Japan Tappi*, 30 (8):44 (1976).

57 L. L. Turai and F. Rosario, "Ultrasonic oxidation of chlorinated phenols found in waste water," *APPITA*, 35 (5):407–11 (March 1982).

15
Conventional Treatment: Suspended Solids Removal

In Chapters 7 through 13, we have discussed the state-of-the-art technologies for conserving water and containing losses in the papermaking, pulping, and bleaching operations. It is a virtual certainty that there will be further developments in these areas as the industry faces more stringent effluent requirements, higher energy costs, and the need to expand production without increasing the water supply. If the problem of eliminating pollution from the process while maintaining product quality and production quantity is approached creatively by the industry's trained personnel—engineers, pulp and paper scientists, and those trained on the job—there is every hope it will be solved.

An example from the municipal water supply field that illustrates what can be accomplished when necessity demands is the problem faced by Chanute, Kansas, in the drought years of 1956 and 1957. The city reservoir impoundment on the Neosho River was scheduled to go dry. To meet its water needs, the city recycled the wastewater from its trickling filter wastewater treatment plant into its water supply reservoir (1, 2). This recycling system was operated for 5 months. The total cycle time was about 20 days from reservoir to treatment plant, so all of the water in the system was reused approximately 8 to 15 times. The water gradually developed a pale yellow color and an unpleasant taste and odor, so when the drought ended the system was abandoned.

The National Academy of Sciences has studied the problem of water recycling for drinking water purposes and has recommended against the practice because of the potential buildup of trace contaminants that might affect health. The example shows, however, what can be done when necessary. Recycling processes in the pulp and paper industry are not as sensitive as those used for drinking water. Thus the main question is not what is technologically possible, but what is desirable. Not all of the technical problems have been solved, but it seems reasonable to believe that they could be whenever it becomes desirable to do so.

Figure 5-1 (Chapter 5) illustrated that a mill should try to operate at the combination of external and internal treatments that produce the least cost, and this is a good principle. However, costs are relative to a particular point in time and hinge on the state of development of technology. Thus we will briefly consider the relative merits of developing internal process technology, external treatment technology, or both together.

ADVANTAGES OF EXTERNAL TREATMENT

In 1977, when the industry had to comply with the requirements of Best Practical Treatment Currently Available (BPTCA), most mills chose to use external biological treatment because the technology was well developed and because external treatment did not have the potential for interference with the manufacturing process. Most mills now have some form of biological treatment or jointly make use of a municipal treatment system. Therefore, it makes sense to continue working to refine the operation of biological treatment systems. However, the one major drawback of external treatment is that it is a direct add-on cost item that cannot yield a return on investment and that requires additional personnel to operate.

ADVANTAGES OF INTERNAL PROCESS MODIFICATIONS FOR POLLUTION CONTROL

The only means available to make waste treatment a profitable venture is to integrate it into the manufacturing process and gain the benefits of conservation of raw materials (fiber, filler, and additives), of saving the cost of treating the process water, and of saving thermal energy. Thermal energy is saved because it is not discharged with the effluent, which should yield a decrease in fuel usage in the power boiler. If the internal systems are carefully designed, it should be possible to operate them with existing personnel. Modifying the pulping and papermaking processes to make them nonpolluting will probably make them more complex and interactive. Happenings in one part of the plant will then more critically affect those in other parts. A more sophisticated control system may be required.

Analyses of the costs of internal versus external pollution control generally show that internal modifications are less capital intensive (3, 4). This fact, plus the consideration that internal controls are the only means to make waste treatment profitable, dictate that technologies for internal pollution alleviation must be further developed.

We are left with the necessity to develop both internal and external technologies. It is likely that existing mills will learn to operate their present external treatment plants more effectively and will update them to meet future requirements. This action will be coupled with internal improvements to reduce the load on the treatment plants and to allow the expansion of production with existing facilities. New plants, on the other hand, will be built to incorporate the best state-of-the-art internal control and will need only a small external treatment plant or none at all.

SUSPENDED SOLIDS

Suspended matter present in pulp and paper wastes consists mainly of bark particles, fiber, fiber debris, and filler and coating materials such as clay, calcium carbonate, and titanium dioxide.

Three general methods for removing these materials are sedimentation, flotation, and screening. Because of the blinding and plugging propensities of pulp and paper wastes, screening systems such as fine screens, microstrainers, and pressure filters are not commonly employed. In addition, adequate screening systems cost approximately the same as equivalent clarifier systems and have more inherent problems (5). In some instances, it may be advantageous to pretreat selected streams with a static DSM screen or microsieve screen in order to recover the long fiber fraction (6), but usually these streams are sent to the saveall. The two most commonly used alternatives for suspended solids removal are gravity sedimentation clarification and froth flotation, with sedimentation largely dominating the industry.

SEDIMENTATION CLARIFIERS

The sedimentation clarifier is usually preceded by a grit chamber and a bar screen. The grit chamber is designed to separate the organics from the inorganics. The aim is to remove all grit particles larger than 65 mesh (>0.15 mm in diameter), which is accomplished by maintaining an average velocity of less than 2.5 ft/min. The self-cleaning traveling trash screen should be designed with approximately 5/8-in. slots if mechanical operational problems are to be avoided (7).

Most U.S. pulp and paper mills have primary clarifiers in operation. Circular, mechanically cleaned clarifiers are the most common type of settling unit employed, followed by reactor clarifiers and then by rectangular, mechanically cleaned basins. Most mills do not add flocculents or coagulants to their primary clarification system, since the solids have good settling characteristics without flocculation. The removal efficiency experienced varies with the type of paper production, as shown in Figure 15-1. Removal of 80–90% of suspended solids is achieved at most mills (8). This action represents removal of 95–100% of the settleable solids, which are the only solids affected by a sedimentation clarifier, since not all suspended matter is settleable. Differences in performance are due to the nature of the solids discharged from the different mills. Those that have highly effective fiber recovery systems have difficulty attaining maximum efficiency because of a preponderance of short fibers and bark particles. Secondary fiber, filled-grade, and coated-grade mills have high-ash suspended solids due to the presence of clay and other filler. Even if a high percentage of the suspended solids are removed, the effluent from such mills may still be turbid due to the high scattering power of even a small amount of suspended filler material. Surface active agents such as starch can peptize some of the filler, making it very difficult

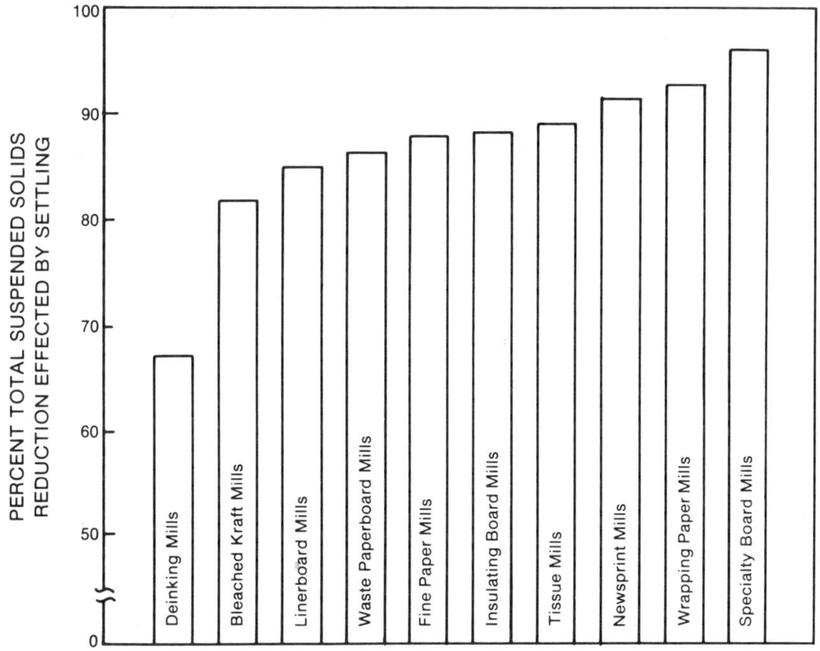

Figure 15-1 Percentage of total suspended solids reduction effected by settling

to remove. This action is one cause of the low removal efficiencies experienced by deinking plants.

The suspended matter removed contains a certain amount of associated BOD_5, depending on the type of mill involved, as shown in Figure 15-2. Specialty board mills, wrapping paper mills, and tissue mills have almost all of their BOD_5 associated with the suspended solids, so high removal efficiency of BOD is experienced in the primary clarifier. In contrast, integrated operations experience low BOD removal through settling because most of their BOD is dissolved.

A mechanically cleaned clarifier commonly consists of a large circular concrete tank with a rotating sludge scraper mechanism mounted in the center. Figure 15-3 presents a schematic of such a circular clarifier. Effluent usually enters the tank through a well located on a center pier, this configuration is termed "center feed and peripheral take-off." The opposite configuration ("peripheral feed and center take-off") is also possible but less common. Settled sludge is raked to a center sump or concentric hopper and is conveyed by solids-handling pumps to areas of further concentration or disposal. The sludge rake must have sufficient torque to operate under heavy load conditions, and reliable torque cutout alarms are needed to protect the drive mechanism. The scraper must be close to the bottom to prevent accumulation of sludge that might then decompose anaerobically. The bottom of the clarifier commonly slopes toward the center at a 1:12

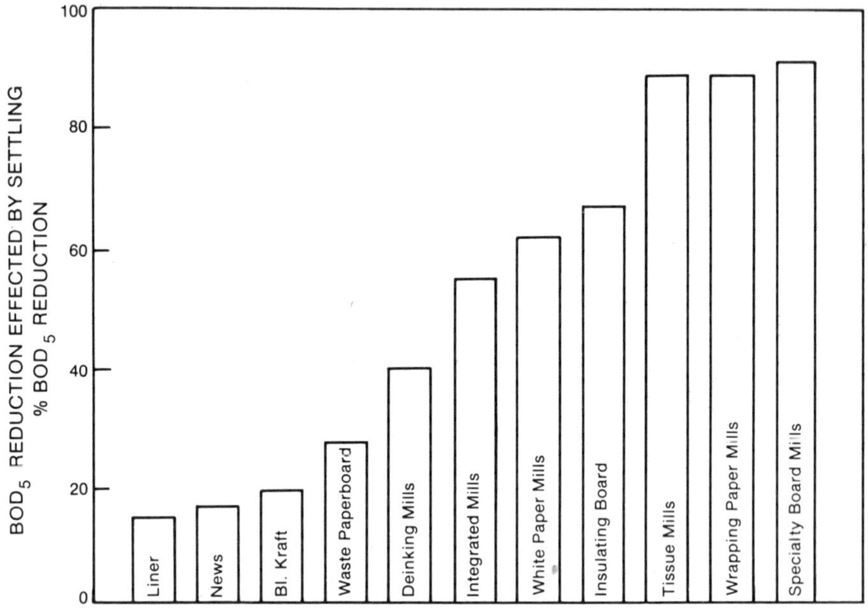

Figure 15-2 BOD_5 reduction effected by settling

ratio. Floating material is collected by a surface skimmer attached to the rotating mechanism and is discharged to a hopper through a line at least 6 in. in diameter. A sound foundation is essential to prevent subsidence of the clarifier, heavy pilings should be used if there is any doubt. Care is also essential where the water table is high, since groundwater can uproot the unit when it is empty.

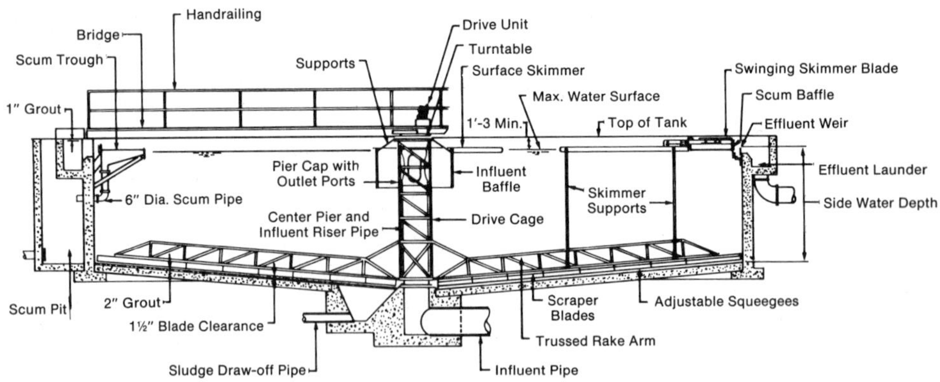

Figure 15-3 Schematic diagram of a circular clarifier. Copyright © 1981. TAPPI. Reprinted from *Introduction to Environmental Management for the Pulp and Paper Industry*, Study Guide, Volume 6, with permission

If multiple clarifiers are required, rectangular units fit closer to one another and use the land more efficiently. Multiple units are useful in that some clarification can be accomplished even if one unit is down for maintenance.

In practice, clarifier overflow rates are 8–50 $m^3/day/m^2$ (200–1,200 gpd/ft^2), with common design practice giving 25–32 $m^2/day/m^2$ (600–800 gpd/ft^2) (9). A typical mill would have a suspended solids content of 3–5 lb/1,000 gal (350–600 ppm) entering the primary clarifier and, at 90% removal efficiency, have 0.3–0.5 lb/1,000 gal (35–60 ppm) leaving the clarifier in the overflow.

SETTLING THEORY

The most critical variable in the design of a clarifier is the overflow rate, measured in $m^3/day/m^2$ ($gal/day/ft^2$), which is synonymous with the critical settling velocity. In a clarifier, three types of settling occur—discrete particle settling, flocculated particle settling, and zone settling. In discrete particle settling, individual particles settle with no interaction between them. In flocculated particle settling, the particles agglomerate and change in density as settling occurs. There is also a limited amount of interaction between the flocs, because those of higher density and larger diameter will settle faster and, in the process, may interact with slower-settling flocs. In zone settling, the flocs form a continuous lattice and the settling rate is the rate of collapse of the lattice structure. In a normal clarifier, discrete particle settling and flocculated settling occur in the upper region and zone settling in the lower region. Either region can control the ultimate settling rate because all of the material being settled must pass through each region. The theory for discrete particle settling is well developed, but for flocculated settling and zone settling experimental procedures need to be used. The theory for discrete particle settling will be discussed in depth because it provides a qualitative understanding of the important factors in sedimentation clarification. For information on the analysis of flocculated settling and zone settling, references 10 through 12 can be consulted. Both flocculated and zone settling are unsteady-state, time- and composition-dependent phenomena for which it is difficult to develop quantitative theoretical models.

Discrete Particle Settling

For a particle in an aqueous suspension, the force causing it to settle is its weight and the force resisting settling is the hydraulic drag on the particle. The weight of the particle is less than that in a vacuum because of the buoyant force. The weight of the particle can be expressed mathematically as follows:

$$\text{Particle weight} = v\rho_{pg} - v\rho_L g = vg(\rho_p - \rho_L) \tag{15-1}$$

where ρ_p is the density of the particle, ρ_L is the density of the liquid, v is the particle volume, and g is the acceleration due to gravity.

The drag force can be expressed mathematically as follows:

$$F_D = C_D A(\rho_L V^2/2) \tag{15-2}$$

where C_D is the drag coefficient, A is the projected cross-sectional area of the particle, and V is the velocity of the particle.

Upon entering a liquid, a particle will accelerate in velocity until its weight is exactly offset by the drag force. The velocity reached is the terminal settling velocity. Equating particle weight to drag force enables solution of these equations for the settling velocity (V_s):

$$vg(\rho_p - \rho_L) = C_D A(\rho_L V_s^2/2)$$

$$V_s = \sqrt{2vg(\rho_p - \rho_L)/C_D A \rho_L} \tag{15-3}$$

If a spherical particle is assumed, the volume (v) is $1/6\pi d^3$ and the projected area (A) is $1/4\pi d^2$. Therefore the formula for settling velocity can be modified to:

$$V_s = \sqrt{4/3(g/C_D)[(\rho_p - \rho_L)/\rho_L]d} \tag{15-4}$$

Thus the terminal velocity can be determined if the drag coefficient can be obtained. The drag coefficient is a function of the Reynolds number (N_{Re}), given by the following expression in which μ represents the viscosity of the liquid:

$$\text{Reynolds number} = dV_s\rho_L/\mu$$

For flow past a sphere, the drag coefficient (C_D) is $24/N_{Re}$ for $N_{Re} < 2$. This is the creeping flow region, and this expression is often termed "Stoke's law." For Reynolds numbers > 2 but < 500, $C_D = 18.5/(N_{Re})^{0.6}$, and for Reynolds numbers > 500, $C_D = 0.44$. Most particles settling in a clarifier would be governed by Stoke's law or would fall in the transition zone.

In any region, it is possible to arrive at an expression for the settling velocity, but for illustration and ease of mathematical manipulation the creeping flow region will be considered. Inserting $C_D = 24/dV_s\rho_L/\mu$ into Equation (15-4) and solving for the settling velocity gives

$$V_s = 1/18[(\rho_p - \rho_L)/\mu]gd^2 \tag{15-5}$$

This formula is useful in showing the important parameters that can be manipulated to improve settling. The parameter with the dominant effect is the particle diameter, since it appears squared in the formula. Flocculation is a means of increasing particle diameter and may also have a minor impact on particle density. The density and viscosity of water are determined by the temperature of the effluent being settled. It is interesting to note that increasing the temperature should decrease the liquid density and the viscosity; both of these changes should increase the settling velocity.

Theoretical Sedimentation Tank

Let us now idealize both a rectangular and a circular sedimentation tank to gain a basic understanding of their important design parameters. This discussion is based on original concepts presented in papers by Hazen (13) and Camp (14). We will start with a theoretical rectangular tank, as shown in Figure 15-4. There are

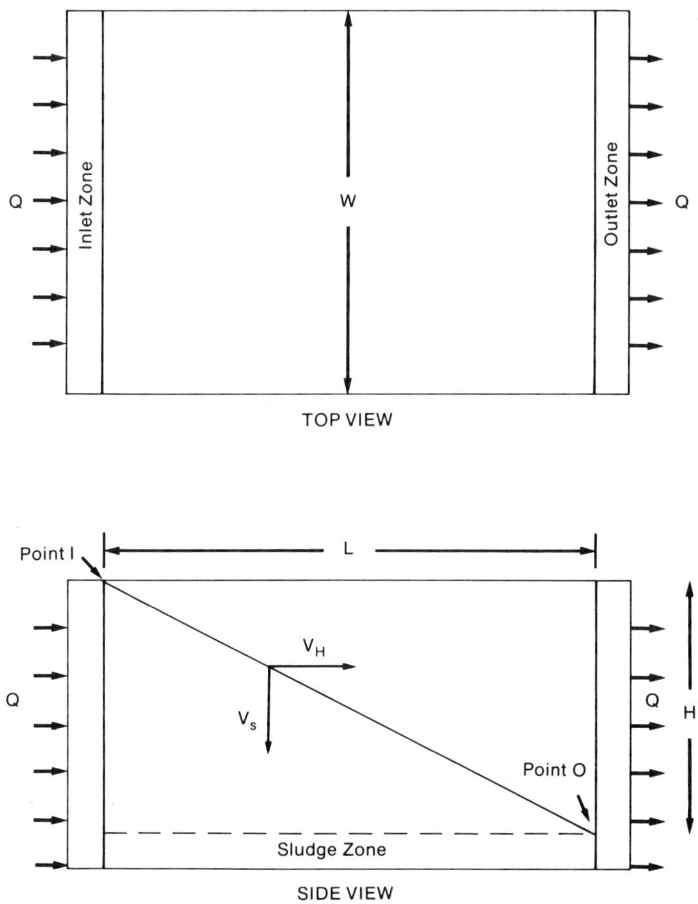

Figure 15-4 Theoretical rectangular settling tank.

three zones in the tank that are not considered part of the settling process: an entrance zone, an outlet zone, and a sludge zone. In the entrance and exit zones, the flow is uniformly distributed over the entire cross section of the tank. In the sludge zone, particles accumulate for disposal but do not enter the outlet zone. The dimensions of the tank are length *L*, height *H*, and width *W*. Considering a discrete particle entering the settling zone, its worst position for collection would be point *I*. A particle entering at point *I* has the longest distance to settle (*H*) over its residence time in the system. If the particle is to be collected, it must follow a straight-line trajectory, as illustrated, or one of even steeper slope. The settling velocity will be V_s, the terminal velocity previously discussed for discrete particle settling. If flocculation occurs, the velocity would increase and the path followed by the particle would be a curve. The minimum detention time to collect this worst-position discrete particle is:

$$t = H/V_s \tag{15-6}$$

This is also equal to the horizontal velocity divided by the length of the tank:

$$t = L/V_H \tag{15-7}$$

The horizontal velocity is caused by flow through the tank and is equal to the flow divided by the cross-sectional area of the tank:

$$V_H = Q/HW \tag{15-8}$$

Combining Equations (15-8) and (15-7) gives:

$$t = LHW/Q \tag{15-9}$$

Observe that *LHW* is equal to the settling volume of the basin. Equating Equations (15-6) and (15-9) gives:

$$LHW/Q = H/V_s \tag{15-10}$$

or

$$V_s = Q/LW = Q/\text{surface area of tank} \tag{15-11}$$

LW is the surface area available for settling, and this formula states the important relationship between the settling velocity and the surface loading rate, which is expressed in $m^3/day/m^2$ ($gal/day/ft^2$). This is the most important design parameter for the clarifier. The depth of a clarifier is determined after the surface area is chosen by selecting a critical settling velocity. The depth is fixed by the desired detention time.

The fact that the surface loading rate is synonymous with the critical settling

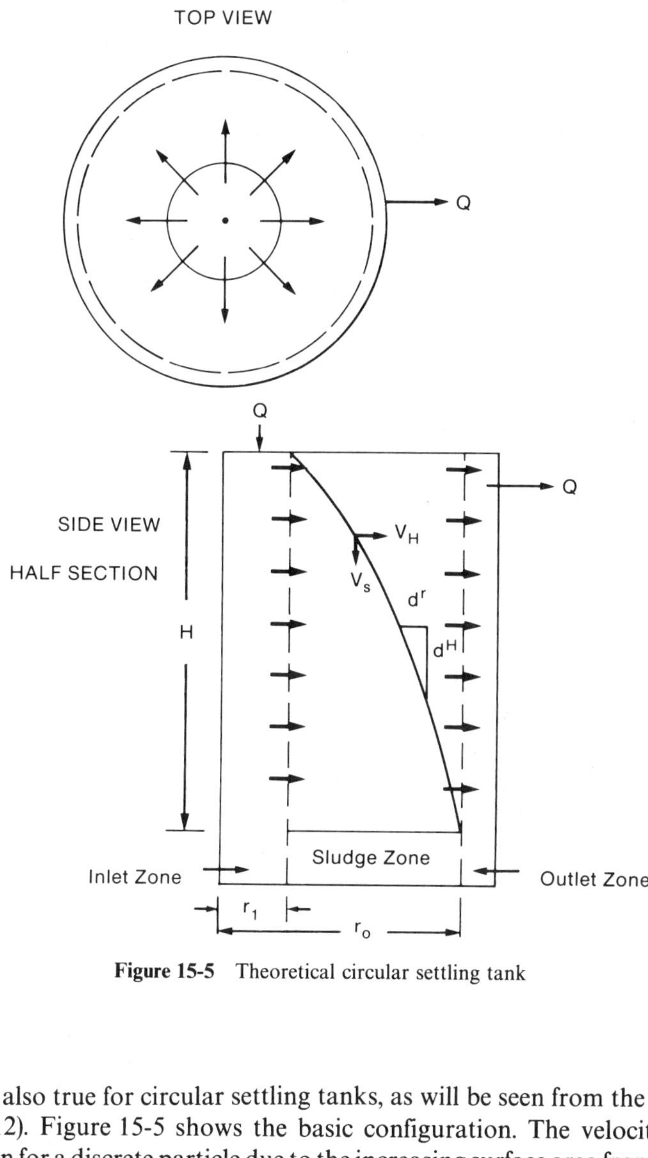

Figure 15-5 Theoretical circular settling tank

velocity is also true for circular settling tanks, as will be seen from the following analysis (12). Figure 15-5 shows the basic configuration. The velocity path is curved even for a discrete particle due to the increasing surface area from center to edge. From a mass balance, the horizontal velocity is:

$$V_H = Q/2\pi rH \tag{15-12}$$

By inspection, the slope of the particle path curve is related to the ratio of settling velocity to horizontal velocity:

$$dH/dr = V_s/V_H \tag{15-13}$$

Combining Equations (15-13) and (15-12) gives:

$$dH/dr = 2\phi rHV_s/Q \tag{15-14}$$

This equation can be rearranged and integrated over the height and radius of the clarifier:

$$\int_0^H \mathrm{d}H = 2\pi HV_s/Q \int_{r_i}^{r_0} r\,\mathrm{d}r \tag{15-15}$$

$$H = 2\pi HV_s/Q(r_0^2 - r_i^2) \tag{15-16}$$

But $2\pi(r_0^2 - r_i^2)$ is the surface area. Therefore:

$$H = HV_s \text{ surface area}/Q$$

or

$$V_s = Q/\text{surface area} \tag{15-17}$$

Thus, for a circular clarifier as for a rectangular clarifier, surface loading rate is synonymous with critical settling velocity.

Any particle entering at the top of the ideal clarifier with a settling velocity less than the critical velocity determined by surface loading will escape capture. However, those particles that have a velocity less than the critical settling velocity but enter below this maximum height have a certain probability of being captured. The fraction removed can be calculated as follows (13):

$$\text{Fraction removed} = (1 - x_s) + \int_0^{x_s} V/V_s dx_s$$

where $(1 - x_s)$ is the fraction of particles captured with velocity V greater than V_s and $\int_0^{x_s} V/V_s dx_s$ is the fraction of particles captured with velocity less than V_s.

In designing a clarifier from a fundamental perspective, the two critical parameters that must be determined are the critical settling velocity and the detention time. These parameters will determine the size of the clarifier. However, from an engineering perspective, many other decisions must also be made before construction. Clarifiers are sensitive to surges, entrained air, and convection currents caused by temperature differentials. Common causes of air entrainment are faulty transfer pumps and the education effects of badly designed piping runs.

FLOTATION CLARIFIERS

In a few mills, dissolved air flotation is used as a primary clarification technique. In certain instances, up to 98% of the suspended solids can be removed (15). Figure 15-6 presents a schematic of a flotation clarifier. Air under pressure is

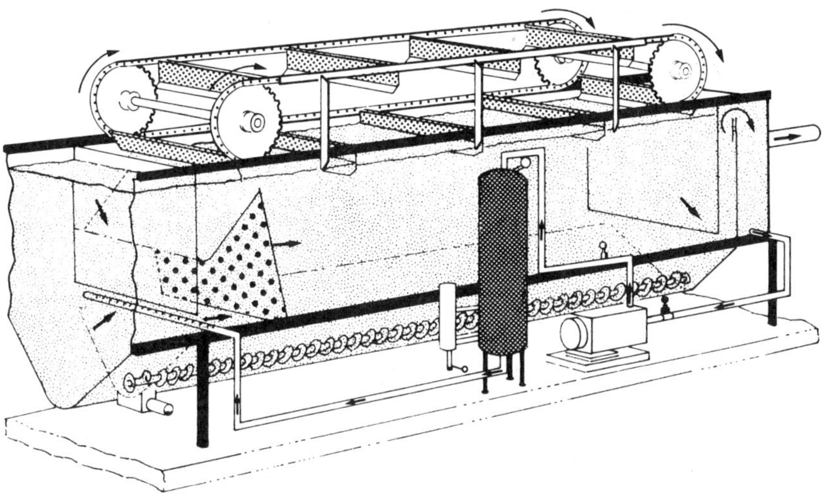

Figure 15-6 Schematic diagram of a floation clarifier

introduced to the effluent entering the bottom of an upflow unit. Supersaturated air is released in the form of very fine air bubbles that become attached to suspended solids, carrying them to the surface. A mechanical scraper skims off the floating floc, separating it from the clarified effluent that spills over the top of the unit. This basic device has been used for many years as a saveall, as discussed in Chapter 7.

A new modification is the German-developed Favair device, which requires less space and has internal lamellae that permit higher flow rates.

To get acceptable results from a flotation unit, a flocculent (alum with or without a polyelectrolyte) must be added. Air flotation is not sensitive to temperature but is adversely affected by large fluctuations in flow and in solids loading. Flow sensitivity can be eliminated by oversizing the unit and operating at fixed flow. Part of the treated water is automatically returned to the inlet to maintain optimum throughput.

Flotation is very effective for removing materials with a specific gravity close to unity but not those of greater density. Flotation units therefore have been developed that include a bottom scraper to remove material that settles rather than floats.

Because flotation equipment is relatively high in cost, requires flocculation chemicals, has high power requirements, and is complex mechanically, it is unsuitable for application other than as a saveall, except where space is at a premium.

NEW DEVELOPMENT

Tube clarifiers (also called "floc barriers") have been developed to improve the efficiency of gravity clarifiers or to increase the allowable surface loading rate

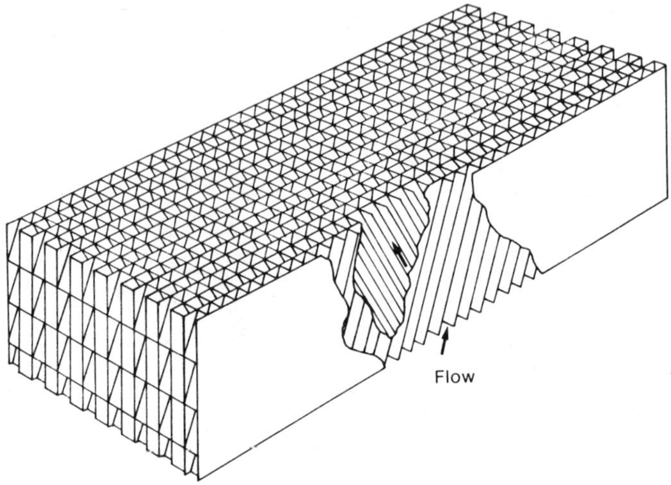

Figure 15-7 Module of a tube settler (floc barrier)

while maintaining efficiency. A tube clarifier is illustrated in Figure 15-7. It is a matrix of parallel plastic tubes, 2 in. square or chevron-shaped in cross section, inclined 45–60° to the vertical. Since the particles have a longer hydraulic path to negotiate and since convection currents are largely eliminated, more efficient settling takes place. Floc barriers can be installed in existing gravity clarifiers or incorporated into the design of new units. Figure 15-8 illustrates how they are installed. The overflow rate can be increased by a factor of 1.5 to 4.0, depending on the type of suspended solids, while maintaining the same removal efficiency.

The advantages of a tube settler can be seen from ideal settling theory. If we use V_s as the settling velocity, V_H as the horizontal velocity, D as the depth of the floc barrier chamber, and L as the length of the chamber, then the particle has to settle a distance H^1 to be collected in a sludge layer (see Fig. 15-9). This distance is much less than the depth of a conventional clarifier. From ideal settling theory for

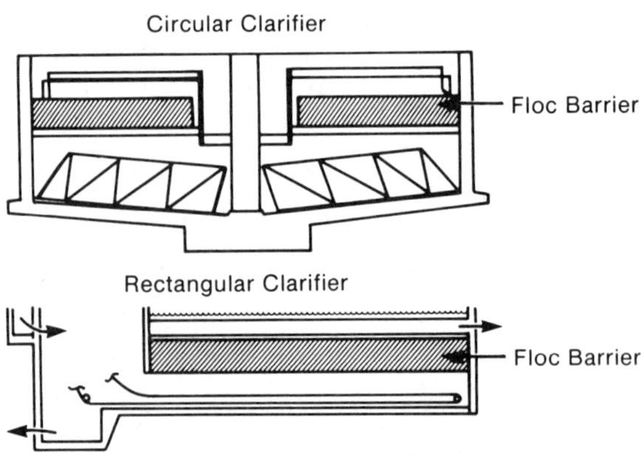

Figure 15-8 Tube settler in an existing clarifier

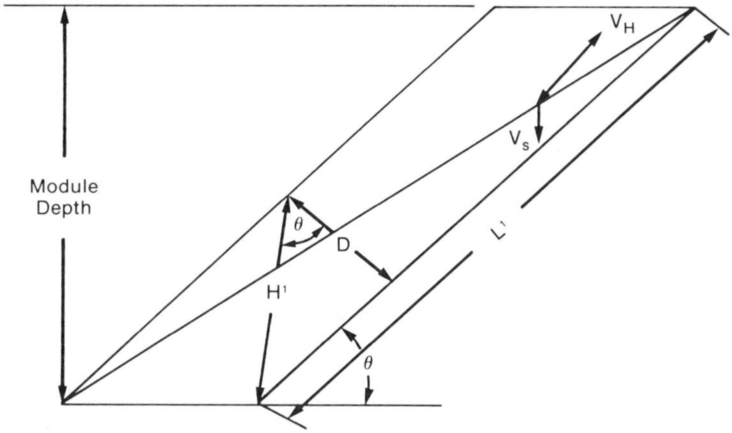

Figure 15-9 Single tube in a floc barrier

a conventional clarifier:

$$V_s/H = V_H/L \tag{15-18}$$

For typical dimensions of H (3.35 m or 11 ft) and L (18.3 m or 60 ft), and assuming a critical settling velocity of 1.5 m/hr (5 ft/hr), the horizontal velocity of the system would be:

$$\begin{aligned} V_H = V_s L/H &= 5\,\text{ft/hr}(60/11) = 27\,\text{ft/hr} \\ &= 1.5\,\text{m/hr}(18.3/3.35) = 8.23\,\text{m/hr} \end{aligned} \tag{15-19}$$

Now let's calculate what flow velocity would be tolerated in a tube settler with the same material being settled. For the tube settler chamber:

$$V_H/L' = V_s/H^1 \tag{15-20}$$

$$H^1 = D/\cos\theta \tag{15-21}$$

Assume a depth of 2 in. (5 cm) and a 45° incline for the chamber:

$$\begin{aligned} H^1 &= (2/12)\,\text{ft}/\cos 45^\circ = 0.236\,\text{ft} \\ &= 0.0508\,\text{m}/\cos 45^\circ = 0.0719\,\text{m} \end{aligned} \tag{15-22}$$

The value of L' is the module depth divided by the sine of the angle, since that is the critical path length. If a 3-ft-thick module is assumed:

$$\begin{aligned} L' &= 3\,\text{ft}/\sin 45^\circ = 4.24\,\text{ft} \\ &= 0.914\,\text{m}/\sin 45^\circ = 1.29\,\text{m} \end{aligned} \tag{15-23}$$

Therefore, the horizontal velocity that can be tolerated is:

$$V'_H = (L'/H')V_s = (4.24\ \text{ft}/0.236\ \text{ft})5\ \text{ft/hr}$$
$$= 89.8\ \text{ft/hr} \tag{15-24}$$

or

$$= (1.29\ \text{m}/0.0719\ \text{m})1.5\ \text{m/hr}$$
$$= 27.4\ \text{m/hr}$$

$$V'_H/V_H = 89.8/27 = 3.3 \tag{15-25}$$
$$= 27.4/8.23 = 3.3$$

Thus 3.3 times as much flow may theoretically be put through a unit with a floc barrier present without affecting clarifier efficiency. This agrees well with the figure of 1.5 to 4.0 cited above and gives the theoretical basis for the interest in floc barriers or tube clarifiers. These devices will find application in existing clarifiers that are overloaded or where space is at a premium. However, they are sensitive to plugging problems.

An improvement on the tube clarifier is the Swedish-developed lamella separator. In this unit, the tubes are replaced by inclined plates. It is a downflow unit, whereas the tube system is an upflow system. Liquids and solids pass in the same direction down between the plates. The clarified effluent returns to the surface through vertical takeoff pipes and overflows. The sludge slides down the inclined planes to the bottom of the unit.

Both of these new developments appear to have technical advantages, but careful pilot plant testing should be done on site to verify that the advantages can be realized in the particular application involved.

FLOW EQUALIZATION

After primary clarification, the effluent flows to secondary biological treatment. If the flow from the mill is highly variable in strength, it might be advisable to install an equalization basin between the primary clarifier and the biological treatment system, particularly if a high-rate biotreatment system is used. The detention time in the completely mixed equalization basin should be at least double the time period associated with the frequency of the strength variation, if possible.

REFERENCES

1 P. F. Metzler et al., *Journal—American Water Works Association*, 51:1021 (August 1958).

2 P. F. Metzler, *Public Works*, 90:117 (May 1959).

3 A. M. Springer and R. C. Peterson, "A systems analysis approach to process water reuse in

paper and board manufacturing operations," paper presented at the AIChE Annual Meeting, San Francisco, September 25–28, 1979. *AIChE Symposium Series* 200.

4 D. W. Reeve, G. Rowlandson, J. O. Kramer, and W. H. Rapson, *Tappi*, 62(8):51–54 (1979).

5 R. S. Fuller, *Tappi*, 56(6):104 (1973).

6 D. S. Diehl, "Fiber savings by pressure-fed static screens," *Chemical Engineering Progress*, 79:23–26 (September 1983).

7 H. S. Azad, ed., *Industrial Wastewater Management Handbook*, McGraw-Hill, New York, 1976.

8 C. A. Knapp, F. P. Coughlan, and J. J. Baffa, *Journal of the Sanitary Engineering Division, American Society of Civil Engineers*, 90:41 (1964).

9 H. R. Jones, ed., *Pollution Control and Chemical Recovery in the Pulp and Paper Industry*, Noyes Data, Park Ridge, N. J., 1973.

10 Metcalf & Eddy Inc., *Wastewater Engineering: Treatment/-Disposal/Reuse*, 2nd ed., McGraw-Hill, New York, 1979, pp. 201–221.

11 R. S. Ramalho, *Introduction to Wastewater Treatment Processes*, 2nd ed., Academic Press, New York, 1983, pp. 96–125.

12 T. D. Reynolds, *Unit Operations and Processes in Environmental Engineering*, Brooks/Cole Engineering Division, Monterey, Calif., 1982, pp. 82–95.

13 A. Hazen, *Transactions of the American Society of Civil Engineers*, 53:45 (1904).

14 T. R. Camp, *Transactions of the American Society of Civil Engineers*, 111:909 (1946).

15 M. Gould and J. Walzer, *Chem 26/Paper Processing*, 8(11):50–52 (November 1972).

16
Secondary Biological Treatment

All weak wastes generated in the paper industry, that is, those less than 4,000 mg/l, are susceptible to biological waste treatment for BOD_5 reduction. In addition, other benefits are obtained, such as destruction of toxicity to aquatic life (1–7), reduction in foaming tendency (8), reduction in dispersed turbidity due to inorganic coating additives, and elimination of slime production in the receiving waters (9). The biological treatment system has little effect on effluent color.

The most common biological treatment systems used in the pulp and paper industry include:

1 Large storage oxidation basins
2 Aerated stabilization basins
3 Various modifications of the activated sludge process
4 Land disposal

FUNDAMENTALS OF BIOLOGICAL TREATMENT

All of the methods listed are based on a mixed culture of living organisms. The organisms that are involved in biological treatment are bacteria, fungi, algae, protozoa, rotifers, yeasts, and, inadvertently, viruses. The systems are living, and must be operated and controlled with this fact in mind.

Organisms are generally classified as either plants, animals, or protists. The majority of the organisms found in waste water treatment plants are protists. Rotifers are an exception, for they are classified as animals. Some of the organisms, such as bacteria, viruses, protozoa, and yeasts, are unicellular, while others, such as fungi, algae, and rotifers, are multicellular.

Figure 16-1 shows a generalized bacterial cell. The cell wall is the outermost structure, which serves to protect the cell against osmotic pressure changes and to determine the appearance of the cell. The cell wall makes up 10–40% of the total weight of the cell. The material contained within the cell wall is cytoplasm, a colloidal suspension of proteins, lipids, and carbohydrates. The cytoplasm is granular in appearance because it contains small structures called "ribosomes," which are the site of protein synthesis. Sometimes it also contains granules of glycogen, lipid, and sulfur. The cytoplasm is contained within the cell wall by the

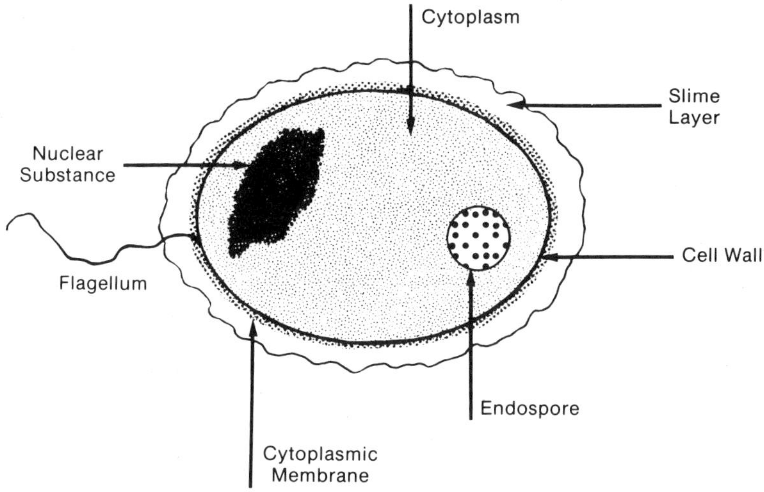

Figure 16-1 Generalized bacterial cell

cytoplasmic membrane, a lipoprotein complex that controls the passage of material into and out of the cell. Passage of material can occur by either active or passive transport. Some bacteria contain endospores, which give them the ability to transform themselves into small spore-like forms that are highly resistant to environmental factors such as heat, light, and lack of moisture. They can then regenerate into active cells when conditions are more favorable. Within the cell there is also a chromatinic area that contains the DNA, the cell's genetic information. For motility some cells have flagella, which can be arranged at only one end, as in the figure, or all around the cell. Lastly, the cell may be surrounded by a slime layer of polysaccharides too viscous to diffuse away from the cell.

Bacteria are the primary organism that serve to stabilize wastewater in biological treatment systems; they will be discussed in more depth. Fungi occur in liquid waste treatment systems most often when nitrogen is deficient or the pH is low. Fungi are usually undesirable, since they cause problems during separation of biomass from the effluent. In some instances, however, they are needed to degrade certain large molecules. Algae are important in stabilization ponds where conditions are quiescent enough to permit their growth, but such ponds are now little used in the pulp and paper industry except in final polishing applications.

Protozoa and rotifers feed on the bacteria and are an indication of a healthy population in a biological treatment system. The maturity of a system can be determined by the types of rotifers and protozoa present. As a system progresses toward the desired state of maturity, the populations of these organisms progress through the following stages: amoebae, phyloflagellates, zooflagellates, free-swimming ciliates, suctoria, stalked ciliates, and rotifers (10). These organisms can be seen with a low-power microscope.

BACTERIA

The main organism in a treatment system is a dynamic population of bacteria. Bacteria are single-celled organisms that can assume one of three shapes. Cocci are spherical, bacilli are rod-shaped, and spirilla are spiral-shaped. Bacteria are microscopic in size and range from 0.2 to 15 μm (0.5–1.0 μm for cocci, 0.5–1.0 μm in width by 1.5–3.0 μm in length for bacilli, and 0.5–5.0 μm in width by 6–15 μm in length for spirilla). They multiply through a process called "binary fission," in which the parent cell divides into two new cells. The time between divisions is called the "generation time" and varies from species to species, ranging from about 10 minutes to as long as 3 days. For a generation time of 30 minutes, one bacterium would yield 16,777,216 bacteria in 12 hours if not limited by some environmental factor.

Bacteria grown under laboratory culture conditions exhibit the classic growth curve shown in Figure 16-2. The distinct portions of this curve are termed the "lag phase", "log growth phase," "declining growth phase," "stationary phase," "increaing death phase," and lastly, the "log death phase." During the initial lag phase, the bacteria are adjusting to their new environmental and recovering from the injuries caused by inoculation into the growth medium. The larger the inoculum used, the shorter will be the lag phase. Once adjusted to the medium, the cells begin to grow and divide at the maximum rate possible for the particular species. They are not limited by either substrate or nutrients. In this condition, the number of cells increases exponentially and the log of the

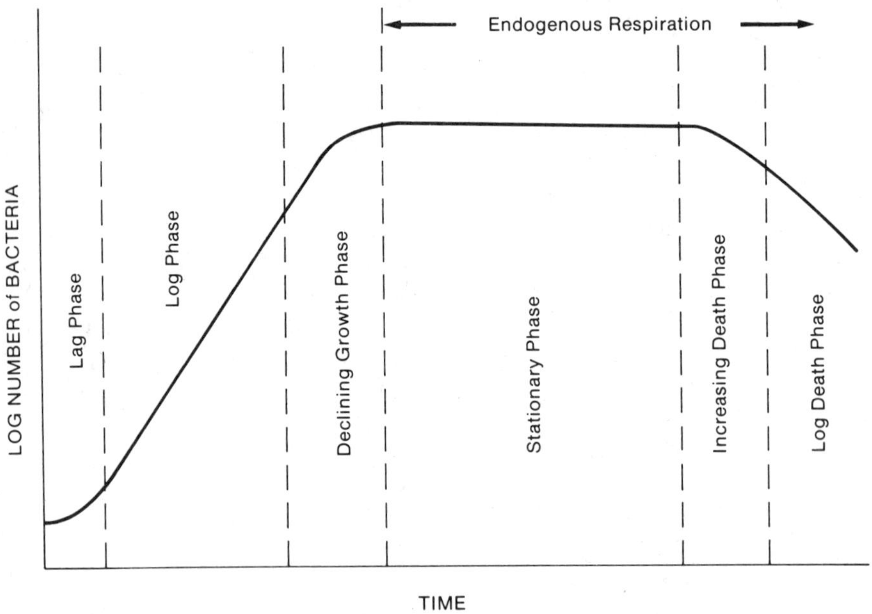

Figure 16-2 Bacterial growth curve

number of cells increases linearly with time. Near the end of this phase, cell growth begins to decline because some factor becomes limiting, bringing on the declining growth phase. The factor could be the substrate, nutrients, or toxic waste products produced by the organisms. In most instances in wastewater treatment, the declining growth phase is due to a depletion of substrate and not to the buildup of toxic end products. With the decrease in growth rate, a point is reached where the death rate of cells is equal to the fission rate—the stationary phase. With time the environment becomes more and more unfavourable, the death rate becomes larger than the growth rate, and the increasing death phase begins. Cells are reproducing very slowly, and death of cells becomes the dominant factor.

These last two phases represent endogenous respiration. In endogenous respiration, the microbes use as food their own stored resources and portions of dead cells in the environment. The net effect of endogenous respiration is to decrease the total amount of cell mass. The rate at which cell mass endogenously decays is a relatively constant value of 5–20% per day. Endogenous respiration occurs throughout cell growth, but it becomes dominant in the latter phases because growth has stopped.

In a wastewater treatment system, the organic matter on which the bacteria feed is the factor by which growth should eventually be limited. The way in which a biotreatment plant is operated will determine what phase of growth is dominant in the plant, but it is common for some fraction of the bacteria to be in a state of endogenous respiration.

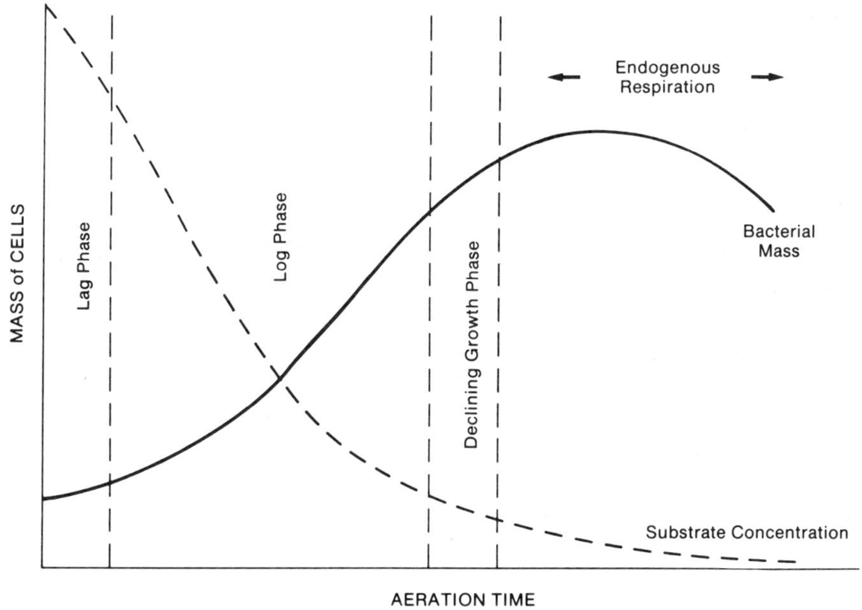

Figure 16-3 Mass mixed-culture bacterial growth curve

If a mixed culture of bacteria is put into a fixed volume of waste and the sample is aerated for an extended period of time, the growth curve of Figure 16-3 results. The difference in shape between Figures 16-2 and 16-3 is due to the fact that Figure 16-2 is a log plot based on the number of cells and Figure 16-3 is a linear plot based on the mass of cells. The log plot shows a more distinctive pattern of phases than does the linear plot based on cell mass.

The function of the bacteria in the biological waste treatment system is to convert the soluble organic compounds into cellular material. To accomplish this conversion, the following must occur:

1 Transfer of organic matter through the cell wall and cytoplasmic membrane into the cytoplasm
2 Production of energy by cellular respiration. The oxidation of organic matter results in the formation of high-energy phosphate compounds, which then act as the energy source for synthesis of cell components
3 Portions of the substrate are transformed into proteins and other cellular matter. Waste products from the cytoplasm are transferred out of the cell

These processes are accomplished through a multitude of biochemical reactions that are interrelated through complex biochemical pathways. Involved in these reactions are organic catalysts called "enzymes." The enzymes are extremely specific and may catalyze only one simple reaction in the pathway. The exact nature of these reactions is beyond the scope of this discussion. It is interesting to observe, however, that the kinetics of bacterial reactions are very similar to enzyme kinetics.

BACTERIAL KINETICS

The most common equation used to describe bacterial kinetics is the Monod equation, which is similar to Michaelis Menton enzymatic kinetic (11). According to the Monod equation:

$$dS/dt = k_s SX/(K_s + S) \qquad (16\text{-}1)$$

where dS/dt = rate of substrate use (mass/volume time), k_s = maximum rate of substrate use [mass/(mass microbes)(time)] = rate constant, K_s = substrate concentration when the rate of use is half the maximum rate (mass/volume), X = mass of microbes (mass), and S = substrate concentration (mass/volume).

This is a variable-order kinetic expression that depends on the magnitude of the substrate concentration in comparison to K_s. If S is large in comparison to K_s, the S in the numerator cancels the S in the denominator, the reaction is zero order, and Equation (16-1) reduces to:

$$dS/dt = Xk_s \qquad (16\text{-}2)$$

Table 16-1 Pseudo First-Order Rate Constants for Selected Wastewaters

Type of Waste	Reaction Rate K_1 TOC/(g MLVSS) (hr) at 25°C
Pulp and paper	0.375
Pulp and paper	0.528
Chemical manufacturing	0.479
Chemical manufacturing	0.601
Oil refining	0.504
Oil refining	0.660
Petrochemical	0.686
Petrochemical	0.713
Municipal (domestic)	1.717

Source: From *Unit Operations and Processes in Environmental Engineering* by T. D. Reynolds. Copyright © 1982 by Wadsworth, Inc. Reprinted by permission of Brooks/Cole Engineering Division, Monterey, California 93940.

If, on the other hand, K_s is large in comparison to S, the S in the denominator can be neglected and Equation (16-1) reduces to:

$$dS/dt = (k_s/K_0)XS = K_1 S \tag{16-3}$$

which is a first-order rate expression in terms of the substrate. Table 16-1 provides some representative values for this rate constant K_1.

This Monod variable-order kinetic model can be combined with hydrodynamic equations for reactor systems to produce design equations for wastewater treatment plants. First, we will examine the important bacterial parameters involved in the operation of a wastewater treatment plant.

FACTORS AFFECTING BACTERIAL GROWTH

Nutrients and Oxygen

Chemical analysis shows that 80–90% of the total weight of a bacterial cell is water. Of the remaining dry mass (10–20% of the total), about 95% consists of the six elements carbon, hydrogen, oxygen, nitrogen, phosphorus, and sulfur. Also present are significant quantities of sodium, potassium, magnesium, calcium, and iron. In addition, minute quantities of manganese, cobalt, copper, zinc, and molybdenum can be identified.

This analysis implies that the nutrient source for bacterial growth must include a large amount of carbon. In wastewater treatment, the organic matter to be removed serves as the carbon source, and bacteria that use such a source are termed "heterotrophs." Bacteria that use carbon dioxide as their carbon source are termed "autotrophs."

Both the hydrogen and oxygen required may come from the water itself, but the oxygen requirement is largely satisfied by the oxygen dissolved in the wastewater through aeration. Bacteria can be classified as aerobic, those that can live only when oxygen is present; anaerobic, which require the absence of oxygen; and facultative, which can live either in the presence or the absence of oxygen.

For proper operation of a waste treatment plant, significant quantities of nitrogen and phosphorus must be added to the wastewater stream from a pulp and paper mill. Sufficient quantities of sulfur, sodium, potassium, magnesium, calcium, iron, manganese, cobalt, copper, zinc, and molybdenum are usually present to satisfy the requirements. Some species of bacteria have a vitamin requirement, but most can manufacture what they need.

Temperature

Bacteria are generally classified as psychophiles, mesophiles, or thermophiles. Psychophiles can exist at a temperature range of 7–35°C, with the optimum being 20–30°C. Mesophiles can exist at 25–40°C, with the optimum being around 30°C. Thermophiles can exist at 25–75°C, with the optimum being 40–60°C. Most bacteria in wastewater treatment plants are mesophilic, and the plant will experience operating difficulties if the temperature falls below 15°C or increases to above 37°C. Over time the bacterial population in a particular plant adjusts to the temperature variations experienced in that plant.

pH

The normal bacterial population in wastewater treatment can survive a range of pH from about 5 to 9, but optimum operation will occur if the pH is around neutral, about 6.5–7.5. The pH of the environment immediately surrounding the bacteria is of importance, not the pH of the waste being fed to treatment. To some extent, the bacterial systems are self-buffering.

Toxicants

Toxic materials can, of course, destroy a bacterial population. Toxicity may be caused by compounds such as phenols, chloro-organics, cyanide, sulfide, and heavy metals (lead, copper, nickel, and zinc). Bacterial growth will be inhibited by a toxic material above a certain concentration, which can be determined experimentally. In most cases, a microbial population can acclimate to the presence of a toxicant and learn to tolerate a higher concentration without inhibition.

MATHEMATICAL MODELS FOR WASTEWATER TREATMENT SYSTEMS

In general, there are two broad classes of biological wastewater treatment systems—fixed-film systems such as trickling filters and the rotating biological surface, and dispersed-growth systems such as aerated stabilization basins and activated sludge systems. Only simple models for dispersed growth systems will be developed, because such systems are the major type used in the pulp and paper industry.

The kinetic expression most often used is that of Monod, as given previously in Equation (16-1). The mass of organisms present in a waste treatment system is the result of their production by feeding on the organic matter in the waste less the amount of endogenous respiration that occurs. Expressing this relationship mathematically gives:

$$dX/dt = Y(dS/dt) - k_d X \tag{16-4}$$

where dX/dt = rate of generation of organisms (mass/volume time), dS/dt = rate of substrate use (mass/volume time), Y= yield (mass of organisms/mass of substrate), X = mass of organisms (mass/volume), and k_d = endogenous respiration constant (1/time).

If both sides of Equation (16-4) are divided by the mass of organisms, the resulting expression is:

$$\frac{dX/dt}{X} = \frac{Y dS/dt}{X} - k_d \tag{16-5}$$

But $X/(dX/dt) = \theta_c$, which is the mean cell residence time of microorganisms in a system, or the age of the sludge [(mass of cells in the aeration tank/volume)/mass of cells wasted per day]. And $(dS/dt)/X = U$, which is the ratio of food to microorganisms or the mass of food used per unit time per organism. Therefore Equation (16-5) can be expressed as:

$$l/\theta_c = YU - k_d \tag{16-6}$$

This equation applies to all dispersed-growth biological treatment systems. It shows that the two controlling parameters—the food/microorganism ratio and the mean cell residence time—are interrelated in a fixed manner, and that a change in one will affect the other. A waste treatment technique can be based on either of these parameters but not on both of them independently. The food/

microorganism ratio and the mean cell residence time are determined by the hydraulic interconnections of the system.

SIMPLE MATHEMATICAL MODEL FOR AERATED STABILIZATION BASINS

The simplest hydraulics for a system are an in-and-out configuration that assumes complete mixing. Such a system is illustrated in Figure 16-4.

Let us define V as the tank Volume, X as the mass of organisms, S_0 as the substrate concentration, and Q as the flow rate. The liquid retention time is Θ, which equals V/Q. The mean cell residence time for the system is Θ_c:

$$\Theta_c = \text{mass of cells in reactor/rate of cell wastage} = VX/QX \qquad (16\text{-}7)$$

So, for this system, $\Theta = \Theta_c$ and the hydraulic retention time is equal to the mean cell residence time. We will see that this is a unique case for this type of system and not true for other types of biological treatment systems. Equation (16-4) applies to all dispersed-growth systems and is applicable in this instance. Taking a mass balance on the organisms:

Rate of change of mass of organisms = net growth − rate of outflow of organisms or, mathematically:

$$V dX/dt = (Y dS/dt - k_d X)V - QX \qquad (16\text{-}8)$$

The goal is to operate waste treatment processes at a steady state, so $dX/dt = 0$ and:

$$(Y - dS/dt - k_d X)V = QX \qquad (16\text{-}9)$$

$$Q/V = [Y(dS/dt)/X] - k_d \qquad (16\text{-}10)$$

$$1/\Theta = YU - k_d \qquad (16\text{-}6)$$

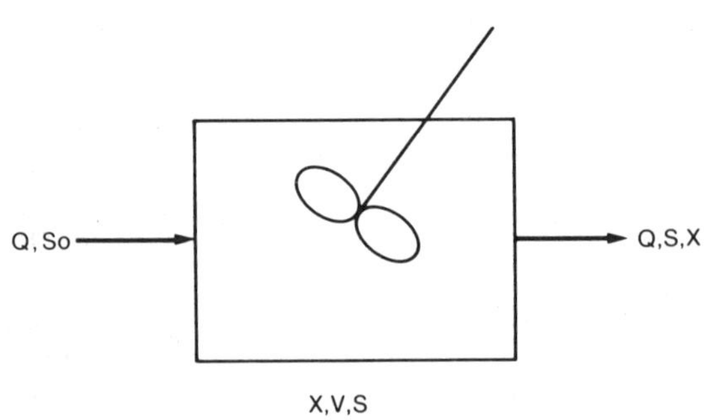

Figure 16-4 Complete mix ASB model

which is the same as Equation (16-6) previously derived, thus showing that it definitely applies to this type of system.

Let's relate these concepts to system efficiency. Define removal efficiency E as follows:

$$E = (S_0 - S)/S_0 \tag{16-11}$$

where S_0 is the influent concentration and S is the effluent concentration. From the Monod kinetic expression:

$$dS/dt = kXS/(K_s + S) \tag{16-1}$$

$$(dS/dt)/X = U = kS/(K_s + S) \tag{16-12}$$

Solving for S gives:

$$S = UK_s/(k - U) \tag{16-13}$$

Inserting (16-13) into (16-11) gives an expression for system efficiency in terms of the food/microorganism ratio:

$$E = 100[S_0 - Uk_s/(k - U)]/S_0 \tag{16-14}$$

Equation (16-6) can be used to substitute for U in terms of Θ_c, thereby transforming equation (16-14) into:

$$E = 100\ \frac{S_0 - \dfrac{K_s(1 + k\Theta_c)}{\Theta_c(Yk - k_d) - 1}}{S_0} \tag{16-15}$$

Equation (16-15) clearly shows that the efficiency of the system depends on its mean cell residence time, which is synonymous with its hydraulic detention time. If the kinetic coefficients and the required removal efficiency are known, the system can be sized.

The quantity of sludge produced per unit time can be calculated. For this system, because no clarification is usually employed, the effluent suspended solids concentration can also be calculated.

For a finite time (Δt):

$$\Delta S/\Delta t = Q(S_0 - S)/V = (S_0 - S)/\Theta_c \tag{16-16}$$

From Equation (16-4):

$$1/\Theta_c = [(YdS/dt)/X] - k_d$$

Multiplying through by Θ_c gives:

$$1 = [(YdS/dt)/X]\Theta_c - k_d\Theta_c \qquad (16\text{-}17)$$

Multiplying through by X gives:

$$X = Y(dS/dt)\Theta_c - k_d\Theta_c X \qquad (16\text{-}18)$$

Solving for X:

$$X = [Y(dS/dt)\Theta_c]/(1 + k_d\Theta_c) \qquad (16\text{-}19)$$

Substituting $\Delta S/\Delta t$ *for* dS/dt from Equation (16-16):

$$X = Y(S_0 - S)/(1 + k_d\Theta_c) \qquad 16\text{-}20)$$

Equation (16-20) gives an expression for the mass of organisms in the effluent and, if multiplied by the flow rate, gives the total mass of sludge produced. The amount of sludge produced is a function of yield, amount of substrate removed, endogenous respiration, and residence time in the system, as Equation (16-20) shows.

Equations similar to these developed here are used as part of the design procedure for aerated stabilization basins (13, 14). More sophisticated models divide the basin into segments rather than assume complete mixing throughout. Table 16-2 gives the kinetic coefficients obtained for a group of integrated bleached Kraft mills. These coefficients were obtained by calibrating the model used with existing performance data from six integrated mills. The absolute values of the kinetic coefficients depend to some extent on how the basin is operated, which affects which populations of bacteria grow and in what growth phase.

The rate constant can be adjusted for temperature in the approximate range of 15–37°C by the following relation, in which $\Theta = 1.072$ (15):

$$k = k_{20}\Theta^{T-20}$$

Table 16-2 Kinetic Coefficients for an Integrated Bleached Kraft Mill

k	2.2/day
k_d	0.12/day
Y	0.4 mg/mg
K_s	150.0 mg/l

SIMPLE MATHEMATICAL MODEL FOR AN ACTIVATED SLUDGE SYSTEM

The hydraulics of an activated sludge system differ from those of an aerated stabilization basin in that solids are separated in a secondary clarifier and recycled to the aeration tank. Figure 16-5 is a flow diagram of the simplest type of activated sludge system, from which we will develop a basic design equation and an equation to predict the concentration of organisms in the system.

The definitions of terms are the same as those previously used. The subscript r refers to the recycle stream, w to the wasting stream, and e to the effluent stream. The system is more complicated because a settling tank is involved. We will assume that the only function of the settling tank is to perform the solids separation and will ignore any treatment for substrate removal that might occur there. In calculating the mean cell residence time in the system, only the volume of the aeration tank will be considered. The settling tank is just a reservoir from which solids are returned to maintain a fixed solids level in the aeration tank. These assumptions may break down if large amounts of solids are held in the clarifier, such as occurs in a UNOX system.

The hydraulic detention time in the system is Θ_s:

$$\Theta_s = (V_s + V)/Q = V_t/Q \tag{16-21}$$

where V_t = total system volume and $\Theta = V/Q$ = residence time in the reactor.

The mean cell residence time is given by a mass balance:

$$\Theta_c = VX/[Q_w X + (Q - Q_w)X_e] \tag{16-22}$$

where Q_w = quantity of mixed liquor wasted from the reactor and X_e = micro-

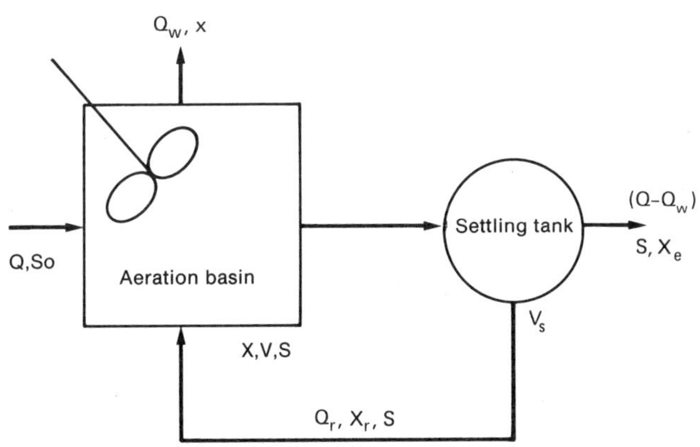

Figure 16-5 Simplified activated sludge system

organism concentration in the effluent. For simplicity of calculation, we will assume that mixed liquor rather than sludge is wasted, which is a common assumption. The sludge wasting calculation reaches the same conclusions, so for clarity the simpler calculations are chosen for illustration. If the settler is operating perfectly, $X_e = 0$ and $\Theta_c = V/Q_w$.

In this system, as diagrammed in Figure 16-5, theoretically Θ_c and Θ can be controlled separately. In practice there are limits on their variation, but the statement is reasonably correct.

A mass balance on the system begins the process of relating proess efficiency to kinetic coefficients, hydraulic detention time, and mean cell residence time:

Rate of accumulation of bacterial mass	=	rate of flow of bacteria in	−	rate of flow of bacteria out	+	net growth within the system

Expressing this relationship mathematically gives:

$$V\,dX/dt = QX_0 - [Q_wX + (Q - Q_w)X_e] + Vr_g \tag{16-23}$$

Assuming that no organisms flow in ($X_0 = 0$) and that steady state conditions prevail, so that the rate of wastage equals the rate of growth (r_g), $dX/dt = 0$ and:

$$Q_wX + (Q - Q_w)X_e = Vr_g \tag{16-24}$$

$$r_g = Y(dS/dt) - k_dX \tag{16-25}$$

$$[Q_wX + (Q - Q_w)X_e]/V = Y(dS/dt) - k_dX \tag{16-26}$$

Dividing through by X gives:

$$[Q_wX + (Q - Q_w)X_e]/VX = [Y(dS/dt)]/X - k_d \tag{16-27}$$

Observe that the left-hand side is equal to $1/\Theta_c$, according to Equation (16-22),. and that $(dS/dt)/X$ equals U. Thus this expression reduces to:

$$1/\Theta_c = YU - k_d \tag{16-6}$$

Therefore this concept also applies to the activated sludge system.

$$dS/dt \cong \Delta S/\Delta t = (S_0 - S)/\Theta \tag{16-28}$$

Substituting the definition of U into Equation (16-6) gives:

$$1/\Theta_c = [(Y(dS/dt)]/X - k_d \tag{16-29}$$

Substituting Equation (16-28) for dS/dt in Equation (16-29) yields:

$$1/\Theta_c = (Y/X)[(S - S_0)/\Theta] - k_d \tag{16-30}$$

Solving for $(S - S_0)$ and dividing by S_0 gives an expression for the fractional treatment efficiency:

$$(S - S_0)/S_0 = (X\Theta/S_0 Y)[(1/\Theta_c) + k_d] = E/100 \tag{16-31}$$

In this system, the removal efficiency depends on the hydraulic detention time as well as the mean cell residence time. It is possible to get high removal efficiencies at low hydraulic detention times by increasing the mass of organisms to a high level. When the system is at steady state, the mass of organisms in it is constant and the quantity generated is the quantity that must be wasted. These systems are operated at steady state, so the quantity wasted must equal the quantity generated.
Quantity grown = rate of removal

$$[Y(dS/dt) - k_d X_m]V = QX_m \tag{16-32}$$

where X_m is the concentration of new solids grown in this pass through the reactor. Solve for X_m and recognize that $V/Q = \Theta$:

$$Y(dS/dt) - k_d X_m = X_m/\Theta \tag{16-33}$$

$$X_m = [Y(dS/dt)\Theta]/(1 + k_d\Theta) \tag{16-34}$$

Here dS/dt is approximately equal to $(S_0 - S)/\Theta$. Therefore the concentration of new solids generated can also be expressed by:

$$X_m = [Y(S_0 - S)]/(1 + k_d\Theta) \tag{16-35}$$

and the mass that must be wasted by:

$$X_m Q = [Y(S_0 - S)Q]/(1 + k_d\Theta) \tag{16-36}$$

If the solids concentration in the reactor is to remain constant, the quantity $X_m V$ that has been produced must be wasted. Thus Equation (16-36) represents the magnitude of the sludge disposal problem.

Both aerated stabilization basin (ASB) and activated sludge systems can be operated in a plug flow mode rather than in the complete mix mode. The treatment efficiency for a given detention time is higher for a plug flow system than for a complete mix system. The problem with plug flow systems is their susceptibility to shock loads. Thus it is desirable to build in as much plug flow character as is compatible with the variable nature of the waste. The mathematics for plug flow systems are more complicated but have been solved (11) using Monod kinetics with an average microorganism mass. The resulting equation, which is analogous to Equation (16-6), is:

$$1/\Theta_c = [Yk(S_0 - S)]/[(S_0 - S) + (1 + \alpha)K_s \ln(S_i/S)] - k_d \qquad (16\text{-}37)$$

where S_0 = influent concentration, S = effluent concentration, S_i = concentration to the plug flow reactor, which is a blend of influent plus recycle, or $S_i = (S_0 + \alpha S)/(1 + \alpha)$, and α = recycle ratio. For more details on the modeling of plug flow systems, references 11, 12, and 16 can be consulted.

To achieve some plug flow character in systems, it is common to resort to staging, since an infinite number of stages would be equivalent to a complete plug flow reactor. More complex models than those presented here can account for plug flow character by breaking the aeration basin into segments (14).

One of the assumptions underlying all modeling exercises is that neither mixing nor oxygen transfer is limiting. Normally, the power required to achieve aeration provides more than enough agitation, so that mixing is not a major isssue. The kinetic coefficients required for different models can be determined from treatability studies on the waste by a classic set of experiments described in many wastewater engineering texts (12, 16, 17).

OXYGEN TRANSFER

The simplest way to characterize mass transfer in this application is the two-film theory of Lewis and Whitman (18). It may not be exactly correct conceptually, but it predicts trends correctly in 95% of cases. Figure 16-6 illustrates the concepts involved. It is postulated that to get transfer from a gas to a liquid, the gas molecules must pass through a laminar film in the gas phase and a laminar film in the liquid phase in order to transfer into the turbulent bulk of the liquid. These two laminar films are the resistances to mass transfer. If the gas is highly soluble in the liquid, then the gas laminar film would be the major resistance. If the gas is only slightly soluble in the liquid, then the liquid laminar film would be the major resistance. In cases of intermediate solubility, both films are important.

The rate of mass transfer is proportional to the concentration difference between the saturation concentration and the concentration actually existing in the liquid. The saturation concentration is determined by the solubility of the gas in the liquid:

$$dm/dt = KA(C_s - C) \qquad (16\text{-}38)$$

where dm/dt = rate of mass transfer, A = surface area of diffusion, C_s = saturation solubility, and C = present state of solubility. Here $dm/dt = VdC/dt$, so Equation

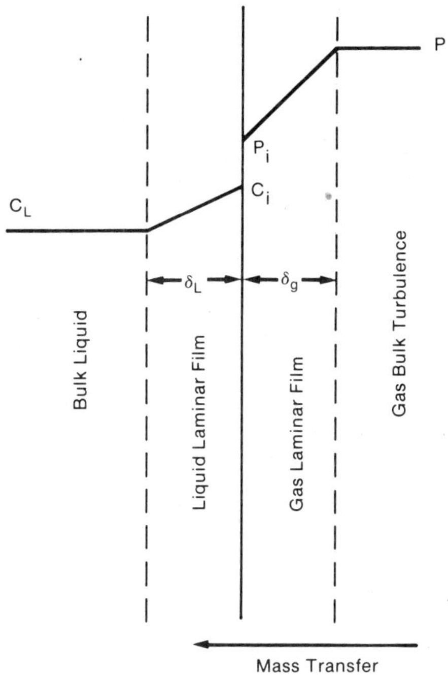

Figure 16-6 Two-film theory of mass transfer

(16-38) can be transformed to:

$$dC/dt = K(A/V)(C_s - C) \tag{16-39}$$

We can define KA/V as $K_L a$, giving the normal form of the equation as found in the literature:

$$dC/dt = K_L a(C_s - C) \tag{16-40}$$

where dC/dt = change in concentration (in mg/l/sec), $K_L a$ = overall mass transfer coefficient (in $\sec^{-1}$), C_s = saturation concentration of the gas in solution (in mg/l), and C = present concentration of the gas in solution (mg/l).

If the mass transfer coefficient is assumed to be constant, this equation is easily integrated from $C = C_0$ at $t = 0$ to $C = C$ at $t = t$:

$$\int_{C_0}^{C} dC/(C_s - C) = K_L a \int_0^t dt \tag{16-41}$$

$$(C_s - C_1)/(C_s - C_0) = e^{-K_L a T} \tag{16-42}$$

It is therefore easy to monitor the oxygen concentration as a function of time and to construct the ratio $(C_s - C_t)/(C_s - C_0)$. This ratio can then be plotted against time on a semilogarithmic plot, and the slope of the line should be the mass transfer coefficient $K_L a$:

$$1n(C_s - C_t)/(C_s - C_0) = -K_L a \tau \qquad (16\text{-}43)$$

This is a linear equation with slope $K_L a$.

The expression "a" represents the surface area of bubbles per unit volume of gas, which is very appropriate for aeration. For the same volume of gas, the smaller the bubble size the higher the mass transfer provided, the bubbles are big enough to rise through the liquid at a reasonable rate. Values of $K_L a$ are temperature sensitive, increasing with increasing temperature, but overall mass transfer may decrease with increasing temperature because C_s decreases. $K_L a$ can be corrected for temperature (16) by:

$$(K_L a)_T = (K_L a)_{20}(1.024)^{T-20} \qquad (16\text{-}44)$$

Surface active materials can also affect the value of $K_L a$, so it is important that it be determined by tests on the wastewater of interest.

AERATORS

Two different classes of aerators are available for providing aeration to biological waste treatment systems—diffusers and surface aerators. Diffuser systems are made to produce either coarse or fine bubbles. Coarse bubble systems transfer 4–8% of their air volume to the liquid, while fine bubble systems transfer 8–12% of their air volume to the liquid. The air flow per diffuser head would be 4–16 scfm (at 20°C and 1 atm) (12).

Table 16-3 Aeration Equipment Efficiency

Equipment	Power	Average Efficiency
Direct-drive surface aerator	100 HP	92
	10 HP	85
Low-speed, gear-reduced surface aerator	Two-stage	97
	Three-stage	95
Diffused aeration	Centrifugal blower	75
Diffused aeration	Positive-displacement blower	70

Table 16-3 shows a comparison of aeration equipment efficiency for diffusers and surface aerators (19). Even with fine bubble diffuser systems, the efficiency of diffused systems is much less than that of surface aerators, so most pulp and paper mill installations have chosen surface aerators.

Surface aerators are often characterized in terms of their N value (19):

$$N = N_0[(C_w - C_L)/9.17]1.02^{T-20}\alpha$$

where N = rate of oxygen transfer at operating conditions (in lb/HP hr), N_0 = oxygen rating of aerator (in lb/HP hr), C_w = saturation oxygen conditions for the waste, C_L = present oxygen conditions in the waste, and $\alpha = K_La$ (wastewater)/K_La (tap water). Common N values for pulp and paper wastes are 1.5–2.2 lb/HP hr.

Figures 16-7 to 16-9 illustrate different types of aeration equipment used in selected segments of the pulp and paper industry. The most common types of equipment used are low- or high-speed surface aerators. Low-speed surface aerators have been reported by manufacturers to give higher transfer efficiencies, but this result has not occurred under field conditions. Large, slow impellers are more efficient mixers than are high-speed impellers (20). However, it is common for failure to involve the gear boxes, which then have to be repaired in the field due to their size. For this reason, slightly more high-speed aerators seem to be used. Aerator life expectancy is on the order of 5–10 years with normal maintenance.

The brush aerator shown in Figure 16-9 is often used in conjunction with the oxidation ditch configuration, which is discussed later in this chapter. These aerators have high oxygen transfer rates.

Turbine aerators combine diffused aeration with the characteristics of surface aeration. In this configuration, the air is introduced directly below an impeller

Five 20 HP platform mounted aerators in the activated sludge process

Figure 16-7 Low-speed surface aerators

Figure 16-8 High-speed surface aerator. (Courtesy Aqua-Aerobics Systems, Inc.)

Figure 16-9 Brush aerator. (Courtesy Lakeside Equipment Corporation.)

that disperses it throughout the system. It is claimed that turbine aerators require less horsepower and give better mixing.

The same claim is made for the jet aeration system, in which the air and water are pumped together through a nozzle and injected directly into the aeration tank to produce mixing.

The last type of aerator featured here is the static air gun incorporating a helical mixer. These aerators are useful in deep basins (>8 ft) where surface aeration is not effective. They avoid the plugging problems encountered with fine or coarse bubble-diffused air systems. For deep basins, draft tubes can also be installed on surface aerators.

INDUSTRY EXPERIENCE WITH VARIOUS BIOTREATMENT SYSTEMS

Storage Oxidation

The earliest system used in the industry consisted of storage oxidation basins, but they are now largely of historical interest. This system was created in response to the need to limit effluent flow to streams during periods of low river flow. In the southern states, where land was relatively inexpensive, large storage basins were constructed. They provided an environment for the growth of organisms and thus provided biological treatment. Reduction in BOD of 50–90% is obtained at a loading of 50–60 lb/acre/day and a retention time of 20 to over 60 days. The two principal advantages of this type of treatment are that it is capable of suffering accidental spills of strong waste without upset and that it has no mechanical devices to give maintenance problems. Design procedures for storage oxidation systems have been developed (21).

Where storage oxidation treatment is still used, it is now usually preceded by a higher rate system, such as an aerated stabilization basin, to meet the current more stringent water quality standards.

Aerated Stabilization Basin (ASB)

A natural extension of the oxidation basin concept is the ASB, which is capable of higher system loads and a higher rate of reaction. Figure 16-10 shows an ASB with surface aerators. In order to increase the rate of biological activity, it is necessary to increase the rate of aeration by mechanical means and to add nutrients to the system. For high rate-activated sludge systems, it is recommended that nutrients be added at a BOD_5: nitrogen: phosphorus ratio of 100:5:1. Such nutrient addition would be the upper limit for an ASB system. The longer the detention time in the system, the less critical the nutrient level. Many mills do not add phosphorus at all and, as a rule of thumb, add nitrogen at a BOD_5:N ratio of 50:1 for a 4-day detention time and 100:1 for a 10- to 15-day detention time. Ammonia and phosphoric acid are the common substances added as

Figure 16-10 Aerated stabilization basin. (Courtesy Aqua-Aerobics Systems, Inc.)

nutrients. For ASBs in the pulp and paper industry, 5- to 10-day detention times are common, but many systems use periods as long as 15 days (22–24). With such detention times the removal efficiencies are normally in the range of 80–90%, producing an effluent with a BOD_5 of about 30 mg/l. It is important not to allow the flow through the ASB to short-circuit, and diversion curtains have been developed to help prevent this problem. A diversion curtain is a polyvinyl chloride sheet placed across the basin to control the flow pattern.

In order to maintain treatment efficiency, it is necessary to keep a residual dissolved oxygen level of 0.5 ppm in the basin. This level usually requires the input of 1.1–1.3 lb oxygen per pound of BOD removed. The majority of installations use mechanical aeration with either high-speed (direct) or low-speed (geared) aerators to provide this input. If the lagoon has an irregular bottom or is deep, a diffusion aerator system would be needed; but such systems require slightly higher operating horsepower.

The consensus is that there have been fewer maintenance problems with high-speed, direct-drive mechanical aerators than with gear reduction units. A well-designed aerator will dissolve 40–50 lb oxygen per horsepower-day (25–27). The industry trend is toward fewer high-horsepower units rather than many small-horsepower units, with a popular size being 75–100 HP. In addition to satisfying the dissolved oxygen requirement, enough horsepower must be provided to keep the ASB completely mixed so that solids will not settle. Figure 16-11 gives the horsepower requirements for various states of mixing for different basin

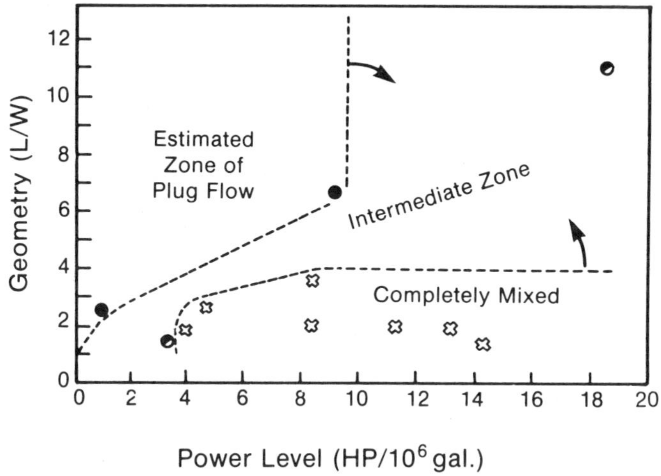

Figure 16-11 Power levels for different characteristics in aerated stabilization basins

length/width ratios (28). The basin may be designed as a completely mixed system, a plug flow system, or some compromise.

Approximately half of the initial capital (installation) cost of an ASB is attributable to the aerators. Aerator power and maintenance comprise a large part of the operating cost of an ASB.

A major issue with the ASB system is biological solids generation. Since most systems are not equipped with a secondary clarifier, no sludge is generated. This is a distinct advantage for the ASB. The minor amount of nonoxidized biological material generated goes out with the effluent. The loading rate of an ASB must be below 2,000 lb/acre/day in order to keep the biological suspended solids generated in the range of 0.1–0.2 lb per pound of BOD removed. Figure 16-12 summarizes data from actual basins that reinforce this conclusion (28). At temperatures between 20 and 43°C, there are no difficulties because the microorganisms are digested by an endogenous respiration process, which usually results in an effluent suspended solids level of about 50 mg/l. At lower temperatures, an increasing amount of biological detritus is generated, which some contend should be removed, since the system may generate a solids amount in excess of NPDES permit requirements.

At one mill at a winter temperature of 18°C (65°F), a BOD removal of 83% and a solids generation rate of 0.33 lb per pound of BOD removed were experienced. Under summer conditions, the same mill had an average temperature of 29°C (85°F), a BOD removal of 87%, and a solids generation rate of 0.22 lb per pound of BOD removed (29). This finding illustrates the general case of a winter solids problem. The plant operated with minimal nutrients in the summer, but maximum amounts were added in the winter.

The biological solids generated in an ASB do not settle satisfactorily in

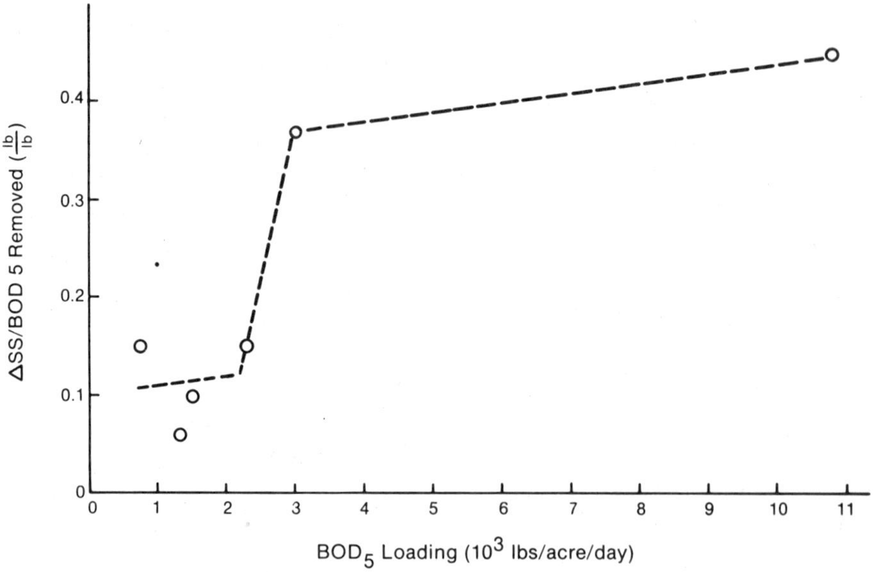

Figure 16-12 Suspended solids generation as related to organic loading

conventional clarification devices, nor are they adequately removed by conventional mechanical filtration devices. In theory it is possible to remove these solids by a combination of chemical coagulation and mechanical filtration, but in practice this procedure would be very costly.

Several mills have looked at the issue of removing these solids. One study (30) on ASB effluents from two integrated bleached Kraft mills demonstrated that the solids could be chemically coagulated and removed by adding 400–700 mg alum per liter plus 3–5 mg nonionic polymer per liter. This system worked but required 280–1,060 lb sodium hydroxide for pH control. The sludge that resulted from removal of the solids was bulky and more difficult to dewater than activated sludge, which itself is notoriously bad. This study concluded that a worse problem was generated than was solved. This same study also examined dual-media filtration, that is, filtration with sand plus anthracite. The dual-media filter removed 37% of the TSS from the ASB effluent. It would not be desirable to couple an ASB—a process that requires little operator attention—to a dual-media filter process requiring a significant amount of operator attention.

The cost–benefit of the required capital and operating costs plus energy required versus the small environmental gain achieved by removing ASB biological solids is highly questionable. Not one case has been reported of biological solids from an ASB causing a problem in the receiving waters. These solids, in fact, are rarely detected beyond the mixing zone at the outfall and appear to be readily assimilated into the natural ecosystem. Many mills use a quiescent pond after the ASB to aid in solids removal.

The BOD removal efficiency decreases at lower system temperatures, becoming as low as 50% at temperatures below 18°C (65°F). This temperature sensitivity is a major drawback to ASB treatment. It has been stated (31) that soluble BOD_5 removal by microorganisms approximates first-order kinetics. Mathematically, a 10°C drop in temperature would require a 35% increases in detention time to keep the effluent BOD_5 constant. Similarly, a 10°C rise in temperature would allow a 25% decrease in detention time. The adverse effect of lower temperatures can be minimized by increased detention time. It can also be lessened by staging the system to keep the first stage at a higher temperature. Deeper basins also experience less heat loss because of less surface area, which helps them to maintain higher temperatures (32).

The design and performance of ASB systems have been extensively discussed in the literature. References 17, 28, and 33 to 36 can be consulted.

Where adequate space is available, ASB waste treatment is generally preferred because it produces no secondary biological solids to be disposed of and because it can absorb mill upset conditions without major disruption of waste treatment. It also has an approximately 2:1 capital and operating cost advantage over a high rate-activated sludge system.

The world's largest ASB system is found within the paper industry. It covers 244 acres and has 43 75-HP aerators, treats flow of 78 million gal/day with a BOD_5 loading of 170,000 lb/day, and achieves an average removal efficiency of 86%.

Activated Sludge

In situations in which space is limited or the aerated stabilization basin will not meet the suspended solids requirement, a logical alternative is the activated sludge process or one of its modifications. This process was borrowed from sanitary waste treatment and is described in textbooks on that subject (16, 17), so comments here will be brief. In contrast to the ASB, in which the basin solids concentration is 50–200 mg/l, basin solids in an activated sludge system are 2,000–5,000 mg/l; the solids are separated in a secondary clarifier and recycled to maintain the high mixed liquor suspended solids. The detention time is 3–8 hours, and the excess biological solids produced are purged from the bottom of the secondary clarifier. These solids create a sludge disposal problem that is a major disadvantage of the process. This sludge is extremely difficult to dewater and usually must be mixed with primary sludge, bark, or fly ash in order to be successfully dewatered. The amount of sludge generated ranges from 0.5 to 0.75 lb per pound of BOD_5 removed. Secondary clarifiers are normally designed at an overflow rate of 25 $m^3/day/m^2$ (600 $gal/day/ft^2$).

The activated sludge system is capable of loadings in excess of 100 lb BOD_5 per 1,000 ft^3 of aeration volume (4) and can achieve BOD_5 removal efficiencies of 80–90%. Such systems are controlled on the basis of sludge age, which is the same as mean cell residence time. System volume is so small that, unlike the aerated stabilization basin, it is insensitive to winter operating temperatures. However, effluents above 37°C (100°F) would have to be cooled before treatment. The short detention time and small system volume make the system susceptible to process upsets, after which up to 2 or 3 weeks may be required to return to optimum operating efficiency. Such system shock or a wide variety of other factors can produce bulking, or the production of nonsettleable secondary sludge. Bulking is extremely troublesome and is the main difficulty of the process.

Bulking is often associated with the presence of filamentous organisms. If the sludge is examined under a microscope and found to have an extended filament length of about 10^7 μm/ml, then problems with bulking will probably occur (37). The usual parameter by which the state of the sludge is monitored is the sludge volume index, which is the volume in milliliters that a gram of sludge occupies after settling for 30 minutes in a 1,000-ml graduated cylinder. For mechanical aeration, a value greater than 300 ml/g may indicate a problem. This is only a general indication, however, since each plant must develop its own set of guidelines based on the sludge volume indexes experienced with the particular waste of interest.

Conditions that favor filamentous growths in conventional activated sludge are food/microorganism (F/M) ratios greater than about 0.8 lb BOD/lb mixed liquor volatile suspended solids (MLVSS)/day, lack of nitrogen and phosphorus, low levels of dissolved oxygen, and loading upsets. pH shocks or the introduction of toxic material can also cause bulking. Limited dissolved oxygen has been noted more frequently than any of the other causes of bulking (16).

Bulking is usually associated with the growth of filamentous organisms, but if this cause is excluded through microscopic examination, the bulking problem is probably due to the bacterial cells binding water. The causes of water binding have not been delineated but the probably similar to those causing filamentous growth. Something has changed in the nature of the waste received by the plant or in the way in which the plant is operated to cause the bulking problem to occur. It would be useful to develop a checklist of plant operating parameters to verify that all are in the proper range. Parameters not within their proper range might provide a clue to the cause of the bulking.

Figure 16-13 illustrates the commonly perceived way in which sludge settleability, measured by the sludge volume index, varies with organic loading. Either underloading or overloading can cause problems with settleability. Overloading causes filamentous growth, while underloading causes dispersed organisms. The normal loading rate for activated sludge is 0.2–0.5 kg BOD_5/kg MLVSS/day. Increasing the rate to 0.8 will likely cause a problem. The normal loading rate should correspond to a sludge age of 5–10 days.

While the cause of the bulking problem is being sought, the plant can be treated with a strong oxidizing agent such as chlorine or hydrogen peroxide. The surface area of the filamentous organisms is so much greater that they will be preferentially attacked. A common application point for this treatment is the return sludge.

An activated sludge system requires supplemental nutrients at a BOD_5:nitrogen:phosphorus ratio of 100:5:1. Approximately 1 lb of oxygen is required per pound of BOD_5 removed, usually supplied by surface aerators. The

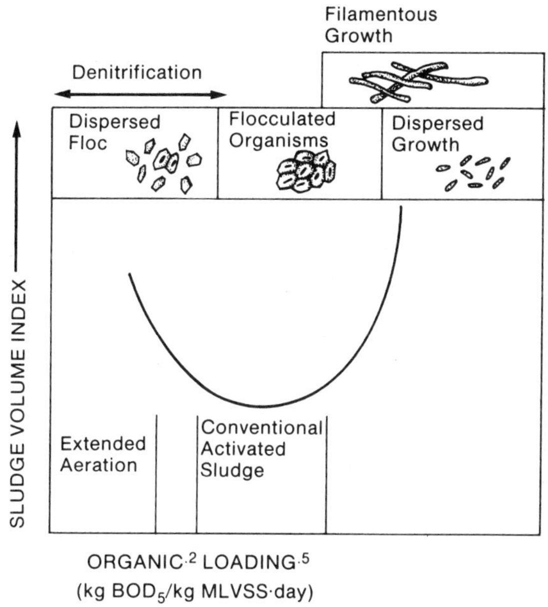

Figure 16-13 Sludge settleability as a function of the F/M ratio

first large plants were installed in the paper industry in 1953. The design and operation of these first installations (38,39) and later ones (40–43) have been described in the literature.

Two traditional modifications of the activated sludge process have been tried

Table 16-4 Intermediate Rate-Activated Sludge System Operating Parameters

System	Parameters
Design flow	4.0 MGD
Actual flow	2.5 MGD
F/M	0.2
MLVSS	700–800 mg/l
Recycling of secondary solids	75%
Sludge volume index	150
Dissolved oxygen in aeration	1.5–2.0 ppm
pH	7.0
Residual phosphorus	1 mg/l
Residual nitrogen	1 mg/l
Sludge generation	0.5 kg/kg BOD removed
Primary-clarifier surface loading	210 GPD/ft^2
Secondary-clarifier loading	150 GPD/ft^2

in the industry—contact stabilization (44) and extended aeration (40). Neither has proven superior to the traditional process used for pulp and paper wastes, so these modifications will not be discussed further.

In addition, some mills have experimented with intermediate-rate activated sludge systems with retention times of about 1 day (24 hours) in the aeration chamber (45, 46). The concept is to gain stability through the longer aeration time and also to generate less biological solids through endogenous respiration. Table 16-4 lists the parameters for such a system operating on waste from a groundwood-coated fine paper mill (47). The system seems to have merit, but may be more susceptible to the production of nonsettling secondary biological solids (46).

The main deterrent to more extensive application of the activated sludge process is its high capital and operating costs, which are usually twice as great as those of aerated lagoons.

Pure Oxygen–Activated Sludge

With the development of the pressure swing absorption system (molecular sieve) for producing pure oxygen in small quantities at an economical cost (48) has come renewed interest in pure oxygen–activated sludge. As of 1984, 16 such plants were in operation in the paper industry. The paper industry is the largest industrial user of the process (49). Figure 16-14 shows a schematic of one such process. In practice, the aeration tanks are covered and the oxygen introduced into the aeration tank is recirculated. More than 90% use of oxygen is claimed. Such plants operate at very high mixed liquor suspended solids levels of 5,000–7,000 mg/l and at high dissolved oxygen concentrations > 5 ppm. They achieve 87–97% BOD removal, with an effluent BOD frequently less than 30 mg/l. Detention times are on the order of 1.5–3.0 hours, less than for activated sludge. The F/M ratio is in the range 0.4–1.0, and the volumetric loading rate is 2–5 kg BOD_5/m^3/day. These are high loading rates. Of the sludge produced, 30–50% is recycled back to the aerator. The BOD and suspended solids content of the final effluent is similar to that from activated sludge.

The process claims a number of advantages, including less susceptibility to shock loading, reduced aeration tank volume, improved sludge settling, and improved sludge dewaterability (50–54). Although some of these advantages are real, a comparison study of paper industry installations did not demonstrate a definite superiority to conventional activated sludge (49).

One problem that has developed at some UNOX plants is toxicity of the effluent due to high concentrations of CO_2. One plant had 250–400 ppm CO_2 dissolved in the effluent (55). The available literature indicates that a CO_2 concentration of 100–200 ppm is immediately fatal to fish, and a concentration of 50–100 ppm is fatal in a matter of hours (56). This effluent had an LC_{50} of 59% for bluegill fry. Because the UNOX process is enclosed and under pressure, carbon dioxide cannot escape. The problem was solved at this mill by pH adjustment and

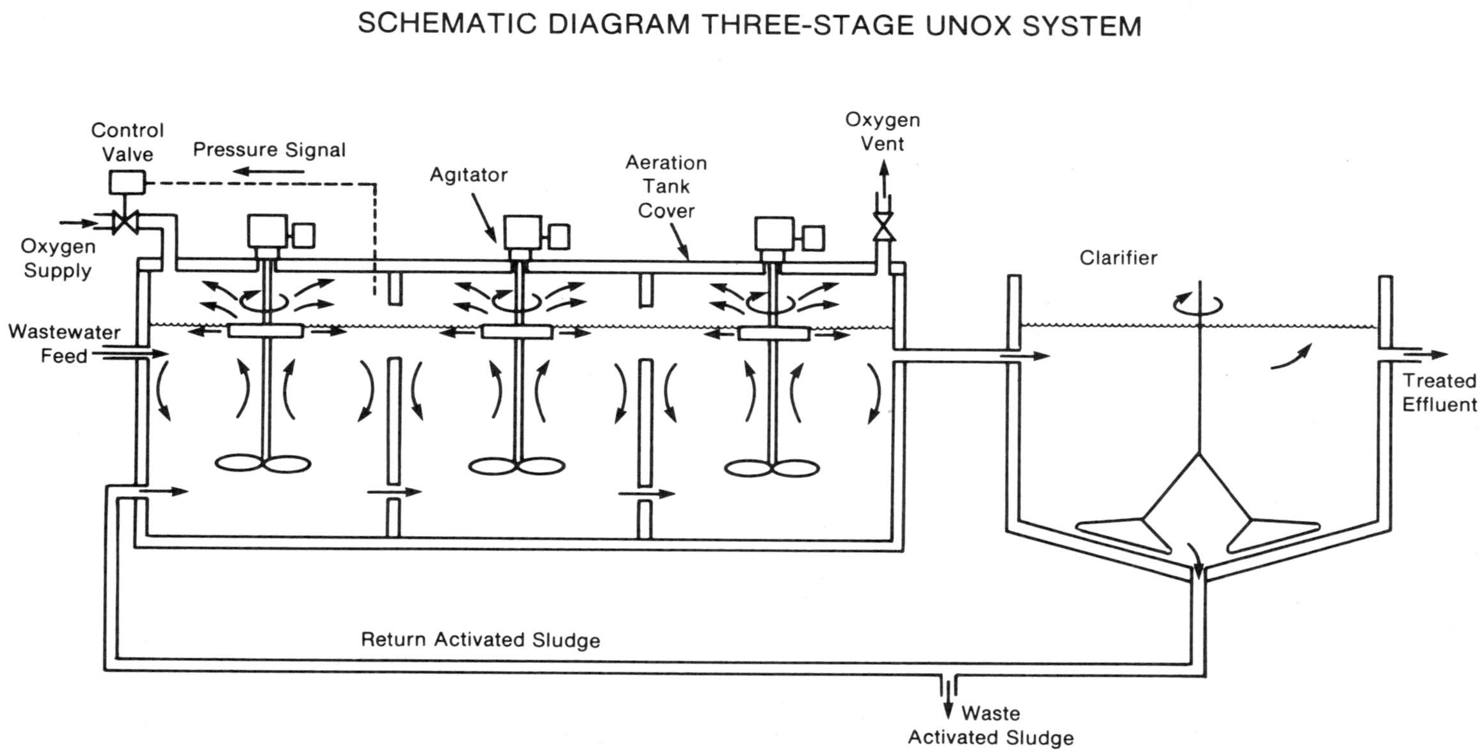

Figure 16-14 Three-stage oxygen-activated sludge system

air stripping, which decreased the CO_2 concentration to 23 ppm or less and thereby rendered the effluent nontoxic to fish.

Besides UNOX, there are other oxygen-activated sludge processes commercially available, such as the OASIS system from Air Products and the MAROX system from FMC. Neither of these systems has so far been applied within the paper industry.

Two-Stage Biotreatment Systems

A wide variety of possibilities exist for staging biological treatment systems, but in most instances the costs outweigh the benefits. Some exceptions within the industry are two-stage aerated stabilization basins, aerated stabilization basins followed by oxidation ponds, trickling filters followed by aerated stabilization ponds, and the Zurn Attisholz (ZA) process.

The Zurn Attisholz process has been used in several secondary-fiber fine paper and tissue mills (57). Basically, it consists of a specialized two-stage activated sludge system with different operating conditions in each stage. The ZA system maintains a very high mixed liquor suspended solids concentration of 6,000–8,000 mg/l in the first stage, and the sludge is recycled at a rate of 200%. The dissolved oxygen level is 0.1–0.5 ppm. In the second stage the mixed liquor suspended solids are 2,000 mg/l, sludge recycle is 150% of incoming flow, and dissolved oxygen is 2 ppm. The system is capable of BOD_5 removal efficiencies of about 95% and is reportedly very stable in operation. Its cost is reported to be slightly less than that of conventional activated sludge, which is surprising in view of the duplication of aeration tanks and secondary clarifiers required (57).

Oxidation Ditch

The oxidation ditch treatment process is essentially an extended aeration activated sludge process providing aeration for a time period in excess of 24 hours (58). The primary clarifier is often eliminated. For economy and simplicity, an earthern ditch in the shape of a racetrack is used for the aeration tank, and air is provided by a horizontally mounted aeration rotor or brush aerator. Jet aeration has been used in one installation (59). The oxygen requirement is 1.5 lb per pound of BOD removed. Figure 16-15 is a diagram of an oxidation ditch system. The ditch overflow is clarified and the settled sludge returned to the aeration basin.

There are advantages and disadvantages to eliminating the primary clarifier. Such elimination increases the load, requiring a larger aeration basin. The mixed liquor suspended solids are maintained at a high level, in the range of 3,000–8,000 mg/l. This gives a very low F/M ratio of 0.1–0.2. The high solids level gives the system good shock load resistance. The solids retention times are on the order of 25 days, giving a lot of time for endogenous respiration, so the solids production rate is only 0.15–0.3 kg VSS/kg BOD removed (VSS, volatile suspended solids). The capital cost is about 20–33% lower than that of

Figure 16-15 Oxidation ditch. (Courtesy Lakeside Equipment Corporation.)

conventional activated sludge. The system is capable of BOD removal efficiencies of 80–94% and is simple to operate.

Trickling Filter

Trickling filters have been tried at a couple of mills in the pulp and paper industry. Experience with them has shown that they are unable to accomplish high degrees of BOD_5 removal (60). Their removal efficiency is normally 50% or less, which is far lower than that needed to meet regulatory requirements. The development of plastic media for these filters has helped overcome this deficiency, but not enough. Trickling filters can be used by the industry as a pretreatment device and possibly as a cooling stage for treating hot effluents before the application of other methods of oxidation.

Land Disposal

If we consider biological treatment in the broadest sense of the term, then land disposal must also be included. BOD is removed by a biological mechanism, and one of the advantages of land disposal is that BOD reduction may approach 100%. The following types of land disposal have been used:

1 Seepage ponds
2 Direct application to fallow soils by either spray or ridge furrow distribution
3 Similar application to plots of land with suitable ground cover growth
4 Controlled effluent applications designed to produce crops by the use of suitable irrigation practices. Trees should be considered as a potential crop for such land application

The benefits of crop cover are discussed in the literature (61, 62). Recent studies have shown a 15% increase in the volume growth of pine resulting from the land application of municipal effluent (63). One of the principal benefits of land application is that no NPDES permit is required if there is no runoff.

Most mills choose application by spraying rather than seepage because of the general scarcity of soils suitable for high-rate seepage installations. In spraying, 20% of the effluent volume evaporates.

The disposal site must be evaluated for soil type, direction of groundwater movement, and terrain. Hill terrain has the potential for runoff at high hydraulic loadings. Isolated cases of groundwater contamination have occurred after application of pulping liquors to soil. In these cases, the soil was highly porous sand that allowed rapid transit of the effluents to the water table, creating a high concentration of effluent in a localized area of the groundwater.

For most soils, 40–50 acres are required per million gallons, and the BOD loadings generally must be below 200 lb/acre/day to maintain the soil in an aerobic condition. The pH should be below 6 or above 9, and the sodium absorption ratio should be less than 8.0 (64). Land disposal is effective in removing suspended solids and BOD but has variable performance on color, depending on the soil type. The color bodies may be leached from the soil by rain water.

Land disposal is best suited to small mills, which can use this technique to compensate for low stream flows during warm months. It is not suitable for large mills due to the large land requirements. The operating costs of land disposal compare favorably with those of other methods of secondary treatment. Several small mills that practice the technique have been described in the literature (61, 65–72).

The failures of land disposal that have occurred may have been due to overloading of the land. Calculation of the assimilative capacity of a site is one of the most difficult yet most critical elements of the land application system. There is a high degree of variation among both wastes and soils, and a careful match must be made. Each effluent will react to a specific site in a different manner. In characterizing wastes, the constituents can be classified into three categories:

1 Biodegradable materials
2 Immobile constituents that accumulate, such as metals
3 Mobile constituents that migrate through the soil, such as chlorides and other anions

The fate of each type of constituent in a particular site and the resulting consequences must be assessed.

The advantages of land disposal are low cost, the high purification achieved, and simplicity of operation. The disadvantages are the large amount of land required, freezing during cold weather, soil sealing, and odor. These disadvantages have caused many mills to abandon the land disposal technique, and very few mills have chosen to use this system of secondary waste treatment. At present, however, there seems to be some resurgence of interest in land disposal.

SUMMARY OF BIOTREATMENT

Eight criteria should be used to select the best biotreatment option for a particular mill from among the many available:

1 Efficiency of removal of BOD_5 and suspended solids
2 Amount of operator attention required
3 Sensitivity to climatic conditions
4 Resistance to shock loading conditions
5 Power required
6 Nutrient demand
7 Sludge yield and characteristics
8 Space requirements

The system that ranks highest on these criteria as determined by a particular mill's unique circumstances is the best choice for that mill. All of the systems discussed, except trickling filters, are capable of the BOD_5 removal required; their ranking on the other criteria determines system choice. Actual operating

Table 16-5 Design Parameters for Biotreatment Systems

System	Aeration Time	Organic Loading	F/M	Efficiency (%)
ASB	5–10 days	50#/acre-ft/day	0.1	80–90
Activated sludge	3–8 hours	50#/BOD_5/1,000 ft	0.3–0.5	80–85
Extended aeration	1–3 days	15–20#BOD_5/ 1,000 ft	0.1	80
Oxygen-activated Sludge	1.5–2 hours	100–150#BOD_5/ 1,000 ft^3	0.7	90
Rotating biological surface	2–3 hours	1.5 gal/day/ft^2	1.0	85

Source: Copyright © 1981. TAPPI. Reprinted from *Introduction to Environmental Management for the Pulp and Paper Industry*, Study Guide, Volume 6, with permission.

Table 16-6 Alternative Treatment Process Comparisons (Costs Developed from Generalized Data in the EPA's Assessment Procedures Manual[a] for Municipal Wastewater Treatment Plants)

	Process Alternatives			
Typical Municipal Wastewater	Activated Air	Sludge Oxygen	Rotating Biological Contactors	Lagoon-Aerated/ Facultative
Influent characteristics				
Flow, million gal/day	25	25	25	25
BOD_5, mg/l	210	210	210	210
TSS, mg/l	230	230	230	230
Effluent characteristics[b]				
BOD, mg/l	20	20	25	45
TSS, mg/l	20	20	25	60
Site area required, acres	8.1	7.6	9.9	47
Operating staff[c]	36	36	36	7
On-site energy consumed,[d] BTU/year × 10^9	7.2	7.2	7.2	0
Off-site energy consumed,[e] BTU/year × 10^9	140	170	150	30
Chemicals consumed,[e] tpy	570	570	520	0
Solids handling, dry tpd	14	14	14	Note[g,h]
Project cost,[f] $ × 10^6	20	22	41	5
Annual cost, solids, handling, $ × 10^6, operation and maintenance (O & M) and amortization of capital	1.4	1.6	1.6	Note[i]
Total annual cost, $ × 10^f operation and maintenance (O & M) and amortization of capital	4.0	4.6	6.9	0.9[j]

[a] *Areawide assessment procedures manual*, Appendix H, Point Source control alternatives, performance and cost, EPA 600/9-76-014, July 1976. Cost data adopted have a rated reliability of B ± 30%.

[b] Thirty-day average.

[c] Manpower projections derived from Appendix H appear conservative. The figures shown are representative for a plant of this size and complexity.

[d] Includes electrical energy and incineration auxiliary level.

[e] Chemicals required for solids dewatering (lime, $FeCl_3$).

[f] Land acquisition excluded.

[g] Solids generation in the lagoon is estimated at 10 dry tpd; however, continual degradation will occur. Typically, lagoons have a 10- to 20-year life before solids buildup inhibits treatment performance.

[h] Proposed project with dike construction costs $9.5 × 10^6 (excludes outfall).

[i] Costs vary with sludge buildup rate.

[j] Excludes cost for sludge removal.

experience on the particular waste of interest should weigh heavily in choosing the best system. The ASB is the system used most often in the pulp and paper industry.

For summary and comparison, the design criteria for some of the systems discussed are listed in Table 16-5.

It is difficult to obtain cost comparison data from the industry for different systems operating in specific mills. However, a generalized comparison based on cost data from an EPA manual on municipal treatment plants was presented in reference 73 and is reproduced in Table 16-6 for information. These data clearly show the advantage of the aerated lagoon type of treatment, provided land is inexpensive. The air-activated sludge method is the next most desirable alternative on the basis of these cost data.

REFERENCES

1 J. M. Leach, A. N. Thakore, and C. C. Walden, *Isolation of toxic constituents of Kraft pulp mill effluents*, CPAR Report No. 11-4, Canadian Forest Service, Ottawa, Ontario, 1974.

2 J. M. Leach and A. N. Thakore, *Identification and treatment of the toxic materials in mechanical pulping effluents*, CPAR Report No. 149-2, Canadian Forest Service, Ottawa, Ontario, 1974.

3 H. S. Dugal, J. W. Swanson, E. E. Dickey, and M. A. Buchanan, *Tappi*, 58 (7):132 (1975).

4 H. Gehm, *State-of-the-art review of pulp and paper waste treatment*, Environmental Protection Technology Series EPA-R2-73-184, U.S. Government Printing Office, Washington, D.C., April 1973, p. 32.

5 C. E. Warren, *Biology of Water Pollution Control*, Saunders, Philadelphia, 1971.

6 J. C. Mueller, J. M. Leach, and C. C. Walden, "Detoxification of bleached Kraft mill effluents—a manageable problem," in *1977 TAPPI Environmental Conference*, Technical Association of the Pulp and Paper Industry, Atlanta, 1977, p. 77.

7 C. C. Walden and T. E. Howard, "Toxicity of pulp and paper mill effluents—a review," *Pulp & Paper Magazine of Canada*, 82 (4):115–121 (1981).

8 W. L. Carpenter, *Foaming characteristics of pulping wastes during biological treatment*, Technical Bulletin No. 195, National Council of the Paper Industry for Air and Stream Improvement, New York, 1966.

9 H. Edde, *A critical review of the literature on slime infestation*, Technical Bulletin No. 232, National Council of the Paper Industry for Air and Stream Improvement, New York, 1969.

10 R. A. Conway and R. D. Ross, *Handbook of Industrial Waste Disposal*, Van Nostrand Reinhold, New York, 1980, p. 75.

11 P. L. McCarty and A. W. Lawrence, "1970 unified basis for biological treatment design and operation," *Jour. SED*, 96 (SA3):757 (1970).

12 T. D. Reynolds, *Unit Operations and Processes in Environmental Engineering*, Brooks/Cole Engineering Division, Monetary, Calif., 1982, p. 277.

13 D. Therumirthi, "Design criteria for aerobic aerated lagoons," *Journal of the Environmental Engineering Division, Proceedings of the American Society of Civil Engineering*, 105:135 (1979).

14 D. Barton, *Selection and development of biological wastewater treatment simulation model for use with pulp and paper industry effluents*, Technical Bulletin No. 384, National Council of the Paper Industry for Air and Stream Improvement, New York, 1982.

15 J. McKeown, *The effect of temperature on treatment plant performance and related temperature*

studies, Technical Bulletin No. 312, National Council of the Paper Industry for Air and Stream Improvement, New York, 1978.

16 Metcalf and Eddy, Inc., *Wastewater Engineering: Treatment/Disposal/Reuse*, 2d ed., McGraw-Hill, New York, 1979.

17 W. W. Eckenfelder, Jr., *Industrial Water Pollution Control*, McGraw-Hill, New York, 1966.

18 W. K. Lewis and W. C. Whiteman, "Principles of gas adsorption," *Industrial Engineering and Chemistry*, 16:1215, (1924).

19 *Environmental management in the pulp and paper industry*, UNEP Industry and Environment Manual Series No. 1, Vol. 2, United Nations, New York, 1980, p. VIII-66.

20 W. W. Eckenfelder, Jr., and D. L. Ford, "1966 engineering aspects of surface aeration design," *Proceedings of the XXII Industrial Waste Conference*, Purdue University, 1967.

21 D. Uhlmann et al., "A new design procedure for waste stabilization ponds," *Journal of the Water Pollution Control Federation*, 55 (10):1252–1255.

22 J. Nowacki, *Fortschritte der Wasserchemie und Ihrer Grenzgebiete* (no. 2):135 (1965).

23 J. Nowacki, *Przeglad Papierniczy*, 25:211 (June 1969).

24 J. C. Tracy, *Southern Pulp and Paper Manufacturer*, 33 (2):46 (1970).

25 R. L. Lows and O. B. Burns, Jr., "Oxygen transfer and power cost with turbine type equipment," in *Proceedings of the XIV Industrial Waste Conference*, Purdue University, May 5–7, 1959, *Engineering Extension Series* No. 104.

26 D. E. Druer, "Theory and development of aeration equipment," in *Biological Treatment of Sewage and Industrial Wastes*, Vol. 1, Reinhold, New York, 1956.

27 R. C. Gloppen and J. A. Roeber, *Tappi*, 48 (12):103A (1965).

28 D. B. Buckley, *A study of mixing characteristics of aerated stabilization basins*, Technical Bulletin No. 245, National Council of the Paper Industry for Air and Stream Improvement, New York, 1971.

29 H. Amberg and D. Bochman, "Seasonal effects on the aerated stabilization basin," (summary paper, 1981 TAPPI Environmental Conference), *Tappi*, 64 (4):22 (1981).

30 H. Chen, "Tertiary effluent treatments are costly" (summary paper, 1981 TAPPI Environmental Conference), *Tappi*, 64 (4):22 (1981).

31 J. Vamvakias, *Temperature relationships in aerobic treatment and disposal of pulp and paper wastes*, Technical Bulletin No. 191, National Council of the Paper Industry for Air and Stream Improvement, New York, 1966.

32 C. B. Welch, "Biological treatment of pulp and paper mill effluent," *Tappi*, 62 (12):67–70 (1979).

33 H. Edde, "Field research studies of hydraulic mixing patterns in mechanically aerated stabilization basins," in *Proceedings of the International Congress on Industrial Waste Waters*, Stockholm, Sweden, November 2–6, 1970.

34 I. Gellman, *Practice and research in biological oxidation of pulp and paper mill waste*, Technical Bulletin No. 162, National Council of the Paper Industry for Air and Stream Improvement, New York, 1963.

35 I. Gellman, *Tappi*, 48 (6):106A (1965).

36 H. W. Gehm and I. Gellman, *Journal—Water Pollution Control Federation*, 37:1392 (1965).

37 J. C. Palm, D. Jerkins, and D. Parker, *Journal—Water Pollution Control Federation*, 50 (10):2484–2505 (1980).

38 T. L. Moore and E. A. Kass, "Design of waste treatment plants for pulp and paper industry," *Biological Treatment of Sewage and Industrial Wastes*, Reinhold, New York, 1955, p. 347.

39 B. V. Pearman and O.B. Burns, Jr., "Saturated sludge treatment of wastes from Kraft and neutral sulfite mills," in *Proceedings of the XII Industrial Waste Conference*, Purdue University, May 13–15, 1957. *Engineering Extension Series*, No. 94, p. 48.

40 O. P. Morgan, *Biological waste treatment case histories in the pulp and paper industry*, Technical Bulletin No. 220, National Council of the Paper Industry for Air and Stream Improvement, New York, 1968.

41 F. P. Coughlan, Jr., *Tappi*, 46 (5):191A (1963).

42 R. M. Billings and Q. A. Narum, *Tappi*, 49 (8):70A (1966).

43 J. Butler, *Tappi*, 47 (11):82A (1964).

44 E. L. Bailey, *Pulp & Paper Magazine of Canada*, 67 (7):85 (1966).

45 D. Czappa, *Tappi*, 62 (10):59 (1979).

46 A. J. Condren and P. H. Welch, *Tappi*, 62 (1):69 (1979).

47 D. J. Czappa, *Tappi*, 62 (10):57–61 (1979).

48 J. C. Davis, *Chemical Engineering* (New York), 79:88 (October 1972).

49 D. Buckley, *A comparison of air and oxygen activated sludge process performance in pulp and paper industry*, Technical Bulletin No. 359, National Council of the Paper Industry for Air and Stream Improvement, New York, 1981.

50 R. J. Grader, W. D. South, and B. Djordjevic, *Tappi*, 56 (4):103 (1973).

51 R. I. Dick and K. W. Young, "Analysis of thickening performance of final settling tanks," in *Proceedings of the XXVII Industrial Waste Conference*, Part 1, Purdue University, May 2–4, 1972. *Engineering Extension Series* No. 141, p. 35.

52 K. E. Lowe, *Pulp & Paper*, 49 (4):54 (1975).

53 E. A. Wilcox and A. Thomas, *Oxygen activated sludge wastewater treatment system, design criteria and operating experience*, EPA Technology Transfer Seminar Publication, U.S. Government Printing Office, Washington, D.C., August 1973.

54 M. V. Nelson, *Tappi*, 63 (3):61 (1980).

55 D. Kuenzi and D. Thiel, *Tappi*, 63 (8):60–62 (1980).

56 M. E. Brown, *The physiology of fishes*, Vol. 1, Academic Press, New York, 1957, p. 447.

57 B. Rice and R. van Soest, *Tappi*, 58 (10):104 (1975).

58 A. Pasveer, "New developments in the application of Kessener brushes (aeration rotors) in the activated sludge treatment of trade-waste waters," in *Proceedings of the Second Symposium on the Treatment of Waste Waters*, organized by the Public Health Engineering Section, Department of Civil Engineering, University of Durham, King's College, Newcastle upon Tyne, England, September 14–19, 1959. Pergamon Press, New York, 1960.

59 A. R. LeCompte and M. G. Mandt, "A case study of effluent treatment by channel aeration propelled and oxygenated with ejectors," paper presented at the 44th Annual Conference of the Water Pollution Control Federation, October 1971.

60 H. Edde, *Manual of practice for sludge handling in the pulp and paper industry*, Technical Bulletin No. 190, National Council of the Paper Industry for Air and Stream Improvement, New York, 1966.

61 R. O. Blosser, *Disposal of pulping effluents by land disposal and irrigation*, Technical Bulletin No. 164, National Council of the Paper Industry for Air and Stream Improvement, New York, 1963.

62 B. D. Vercher et al., *Paper mill waste water for crop irrigation and its effects on the soil*, Louisiana State University Agricultural Experiment Station Bulletin No. 604, 1965.

63 G. Taylor, L. Neal, and D. Norbert, "Land application: A new approach to plant effluent treatment," *Pulp and Paper*, 57 (4):81–84 (1983).

64 R. O. Blosser and A. L. Caron, *Tappi*, 48 (5): 43A (1965).

65 D. Voights, "Lagooning and spraying disposal of NSSC pulp mill liquors," in *Proceedings of the X Industrial Waste Conference*, Purdue University, May 9–11, 1955. *Engineering Extension Series* No. 89, p. 497.

66 R. Westenhouse, *Tappi*, 46 (8):160A (1963).

67 W. C. Parsons, "Spray irrigation of wastes from the manufacture of hardboard," in *Proceedings of the XXII Industrial Waste Conference*, Part 2, Purdue University, May 2–4, 1967. *Engineering Extension Series* No. 129, p. 602.

68 W. A. Flower, "Spray irrigation: A positive approach to a perplexing problem," in *Proceedings of the XX Industrial Waste Conference*, Purdue University, May 4–6, 1965. *Engineering Extension Series* No. 118, p. 679.

69 R. O. Blosser, *Pulp and papermill waste disposal by irrigation and land application*, Technical Bulletin No. 124, National Council of the Paper Industry for Air and Stream Improvement, New York, 1959.

70 R. O. Blosser and A. L. Caron, *Recent progress in land disposal of mill effluents*, Technolical Bulletin No. 185, National Council of the Paper Industry for Air and Stream Improvement, New York, 1965.

71 I. Gellman, *Pulp & Paper Magazine of Canada*, 61 (3):T-221 (1960).

72 E. A. Guerri, *Tappi*, 56 (4):95 (1973).

73 H. Cady, "New secondary treatment system brought on line at Georgia Pacific," *Pulp & Paper*, 54 (7):180–182 (July 1980).

17
Emerging Biological Treatment Technologies

In the last chapter, we reviewed existing proven biotreatment technology that is being used in the pulp and paper industry. This chapter will briefly introduce some of the more innovative technologies that have been used or are now emerging from research work that might indicate the direction of biotreatment in the future. I define emerging technologies as those with one or fewer known commercial installations in the paper industry; those with more than one known commercial installation have been treated as established technologies.

The driving forces for new waste treatment technology are:

1 Lower capital and operating costs
2 Lower energy requirements
3 Smaller space requirements
4 Ability to remove toxics consistently
5 Minimal operator attention
6 Insensitivity to shock loads
7 Lower sludge production

These goals are to be achieved while keeping high BOD removal efficiency, producing sludge that settles well, retaining insensitivity to climatic conditions, and retaining a reasonable nutrient requirement.

The present trend seems to be toward developing higher rate systems with smaller plant sizes. The oxygen-activated sludge systems are a step in this direction. But limits to solids levels are soon reached in dispersed-growth systems. Fixed-film systems have been used to overcome this problem. These films can be attached to media, such as in the rotating biological surface system or in plastic media trickling filters, or attached to heavy particles in suspension, such as in the oxitron process.

ROTATING BIOLOGICAL SURFACE (RBS)

The RBS waste treatment system can be thought of as a dynamic trickling filter. Each RBS unit consists of a series of discs mounted on a shaft. A biological growth develops on these discs and is alternately submersed in the waste and exposed to the air as the shaft rotates. Generally, three stages are employed; staging is accomplished by mounting the units in series either in separate tanks or in a single baffled tank.

Pilot studies (1, 2) demonstrated that, for a wide variety of paper mill wastes with BOD strengths below 260 mg/l, 85–90% BOD removal could be achieved at loading rates of 1.5 gal/day/ft^2. Rotation speed, submerged depth, and the separation between discs all affect performance, but the main parameter is surface loading rate. The optimum surface loading rate for each installation must be determined by pilot plant testing. The larger the volume to be treated, the more surface area is required. Since the cost of an RBS system is directly proportional to the surface area required, the process does not have economy of scale. Therefore it is suitable only for small- to medium-sized pulp and paper installations, for which the costs of RBS are comparable to those of activated sludge.

The RBS system appears to have advantages over other biological treatment systems in simplicity of operation, low maintenance, low power consumption, resistance to influent variations, and good secondary sludge characteristics. Three RBS installations had been made in the paper industry as of 1979, but by 1984 only one was still in operation. All experienced operational difficulties, especially with the structural integrity of the lightweight polyethylene discs. These difficulties have slowed the development of this process for further use in the pulp and paper industry.

The amount of biological growth that can be attached to the RBS surface is limited by the surface area available. In seeking higher surface areas for fixed-film systems, dispersed-growth systems around particles and fluidization of particles in a bed have been considered.

HIGH-RATE, FIXED-FILM ACTIVATED SLUDGE

Laboratory studies have been conducted on the addition of particles to an activated sludge system in order to give higher density of mixed liquor, obtain a better-settling sludge, and enable very long solids retention times (3, 4). The process is similar in design and operation to a conventional complete mix-activated sludge system, except that dense solid particles of small size are added to the aeration tank. Several types of solid materials have been tested, including calcium carbonate, kaolin clay, diatomite, and sand. Calcium carbonate is the best material so far evaluated, because of its buffering ability. An optimum particle mass/biomass ratio of 2–4 was found. Too coarse a particle would not remain suspended in the aeration chamber, and too fine a particle would not incorporate into the sludge. A size range of 5–25 μm was recommended, with no particle larger than 44 μm and none smaller than 1–2 μm (4).

The settling velocity and compaction density increase with particle dosage, but a particle mass/biomass ratio greater than 4 is usually not beneficial. The particles develop a fixed-film growth around themselves and allow a very high level of MLVSS to build up. The reactors were operated in a mode in which no sludge at all was wasted from the system, except that which inadvertently escaped with the effluent. The solids retention time was estimated at 130 days. Thus most of the solids generated were forced to evolve as CO_2 and H_2O through endogenous respiration. In a full-scale plant, it might not be advantageous to pay the cost of supplying the air necessary to allow the sludge to autooxidize.

OXITRON PROCESS

The oxitron process is a logical extension of the above concepts, in which the solids are in a fluidized bed, as shown in the schematic of Figure 17-1. The process is being developed and marketed by Dorr Oliver.

Influent free of suspended solids is fed through a high-pressure pumping system to a fluidized bed of sand. Pure oxygen is injected under pressure to give 75 ppm dissolved oxygen. The top of the unit is operated as a clarifier, and a large proportion of the effluent is recycled to give sufficient flow to fluidize the bed. The bed fluidizes at 3.4 l/sec/m^2 (5 gpm/ft^2), but the unit is usually operated at 6.8 l/sec/m^2 (10 gpm/ft^2), and flow can go as high as 16.4 l/sec/m^2 (25 gpm/ft^2) before sand is lost by washout. The solids retention time is controlled by the use of a hydrocyclone.

The biological growth develops around the sand grains, which are about 0.5 mm in size. As the growth accumulates around the grain, the total particle density decreases and the particle rises in the fluidized bed. The top layer is skimmed off with hydrocyclone. The force applied by the hydrocyclone breaks the biological solids away from the sand grain, and the grain is returned to the

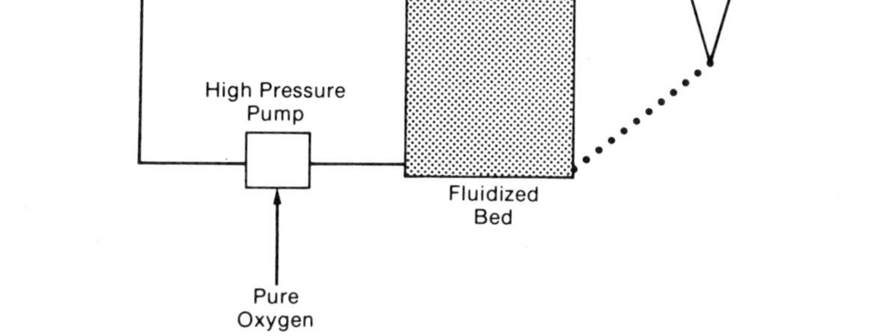

Figure 17-1 Oxitron system

fluidized bed. The system affords very tight control over the solids retention time. Normally, the film thickness is limited to 300 μm. The biological solids sheared off are quite dense. Solids generation rates are on the order of 0.5–1.0 lb of volatile suspended solids per pound of BOD removed.

The system operates at MLVSS levels of 10,000–30,000 mg/l. These solids have a surface area of about 1,000 ft^2 per cubic foot of bed volume. The solids retention time is on the order of 3 days, and the hydraulic retention time is about 20 minutes. More than 90% removal of BOD has been achieved.

The system is efficient and compact, but a major shortcoming is its inability to tolerate any solids in the influent. This would require primary clarification followed by some type of in-line screening device to ensure trouble-free operation. The system also requires pure oxygen and a high-pressure pumping system. The system has economy of scale and apparently will not be cost effective below about 2–3 million gal/day. No commercial installations of the oxitron system have yet been made in the pulp and paper industry, but future new technologies may well contain elements of this system.

AERATED ACTIVATED CARBON FILTER (AACF)

Another interesting system that has been considered in the laboratory is the AACF. Adding powdered or granular activated carbon to a conventional activated sludge system has been tried, but the benefits did not outweigh the costs. In the AACF system, however, the carbon is contained in a fixed-bed design very similar to a downflow granular media filter, except that the grain size is large (4). Large 4 × 12 mesh granular activated carbon is used. It is postulated that the filter operates on the basis that the carbon adsorbs the contaminants and the biological growth regenerates the carbon. In light of the trends we have seen in the three systems previously described, this may not be the actual mode of operation. The filter gives high removal efficiencies, but no AACF system has been operated for a sufficiently long time on pulp and paper effluent to verify that the carbon will actually last a long time. AACF systems have been used successfully with municipal effluents. To keep the system aerobic, air is pumped up through the bed counter to the liquid flow. Surprisingly, the AACF is fairly tolerant of suspended solids, probably due to the large particle size, but it does have to be backwashed periodically. The capital and operating costs of this system would probably be high.

CAPTOR SYSTEM

Another interesting innovation which has been used in a municipal application in England is the Captor system, which resulted from work done at the University of Manchester Institute of Science and Technology (5). In this system, the biological growth is contained in small polyurethane sponges. It is otherwise similar to a conventional diffused-aeration activated sludge system, except that the small (25 × 25 × 12.5 mm) sponges are added to the aeration tank. It is claimed that the entire growth is contained in the sponges, and therefore that the effluent from the aeration tank can be discharged directly through a screen sized to contain the sponges. The effluent is supposed to be clean enough for discharge, thus eliminating the need for a secondary clarifier, but it is unlikely that this effluent would meet U.S. legal requirements for effluent cleanliness.

Because the sponges contain all of the growth, sludge recycling is eliminated and the solids retention time is controlled by removing the sponges from the aeration tank with a conveyer belt. A wringer press removes the solids from the sponges, and the cleaned sponges are returned to the aeration tank. It is claimed that a biomass concentration of up to 6% can be obtained from the sponge wringing process (6). The process gives a very high biomass in the aeration system and is equivalent to an activated sludge system operating at a mixed liquor suspended solids level of 8,000 mg/l. This implies that the plant size could be reduced to about 20% of the size of a conventional activated sludge plant (6). The Captor system should also be useful in upgrading existing overloaded biological treatment systems. Sufficient air must be provided by the diffused aeration system to keep the aeration tank completely mixed, which may require a greater than normal amount of aeration.

DEEP TANK AERATION

A more conservative but still innovative system that has one paper industry commercial installation is the deep tank or deep shaft process (7). As the name implies, this is a modification of the conventional activated sludge system, which employs a deep tank as the aeration chamber; it also replaces the conventional secondary clarifier with a flotation clarifier. The tank depth used in the paper industry commercial application is 21 m, but in municipal applications, oil drilling technology has been used to create shafts 150 m in depth for this process. The deep tank allows a long time for the air supplied by diffused aeration to contact the liquid. The air is injected under pressure; the oxygen transfer efficiency has been estimated to be 2.5–2.8 lb O_2 transferred per horsepower-hour (8), which is high compared to that of a surface aerator. The system operates at a hydraulic retention time of 3.5–7.0 hours and a mean cell retention time of 2–9 days, with removal of 95 + % of soluble BOD. The F/M ratio ranges from 0.55 to 2.0, implying a high rate system. The MLVSS is on the order of 1,700–2,300 mg/l. Nutrients are required in the normal ratio of BOD:nitrogen:phosphorus of 100:5:1. The effluent suspended solids achieved is satisfactory at about 50 mg/l. No flotation aid is added to the flotation clarifier. Hydraulic loading to the clarifier is about 200 l/min/m^2.

This system is very conservative of space. Because of its low surface/volume ratio, it is relatively insensitive to low ambient temperatures. The aeration system has no moving parts in the tank and very efficient oxygen transfer rates. The system is claimed to be good at handling shock loads. It should be free from the normal problems with filamentous growth because such growth would easily float and could then be removed from the system. The soil conditions for supporting the tank are critical, for a high weight is concentrated in a small area. Energy costs for the system are comparable to those of conventional activated sludge. This system would appear to be a potential solution for those mills that have such a variable waste load that they continually have bulking problems with conventional activated sludge.

ANAEROBIC TREATMENT

History

As water usage has been reduced and the strength of pulp and paper mill effluent streams has therefore increased, anaerobic treatment has been reconsidered as an alternative. The first U.S. paper industry application of anaerobic treatment came about for just such reasons (9). An anaerobic pond was installed ahead of an aerobic biological treatment system to reduce the load to the aerobic system when the effluent strength reached 2,000 mg/l.

A second paper industry installation went online in late 1983 at the Roermand paper mill in the Netherlands. It consists of an upflow anaerobic sludge blanket system (UASB) ahead of a conventional aerobic biotreatment system. The anaerobic reactor volume is 1,000 m^3, and it treats 2,400 m^3 of wastewater containing 10,000 kg BOD per day. It generates 3,500 m^3 of gas per day, which is used to produce 36 tons of steam. The UASB system removes 80% of the COD in the wastewater before it enters the aerobic system.

In the past, anaerobic treatment was restricted to high-strength wastes, but with the development of anaerobic attached-film, expanded-bed reactors, it has become possible to treat low-strength waste anaerobically (10).

Anaerobic Sludge Digestion

Anaerobic decomposition has had the most application in sludge digestion. Good discussions of anaerobic digestion of sludges are provided in other works (11, 12), so few details will be given here. Rather, we will mention some of the overall observations that may have relevance to future applications of the system.

For sludge digestion, two popular systems are a low-rate process in which the waste is stored unagitated in a tank, except when feeding the reactor, and a high-rate process in which the reactor is agitated. The low-rate process takes 30–60 days, is loaded at 0.5–1.6 kg volatile solids/m^3/day (0.03–0.1 lb/ft^3/day), and has a detention time of 30–90 days (11). This is a very slow process. The high-rate process has a loading rate of 1.6–6.4 kg volatile solids/m^3/day (0.1–0.4 lb/ft^3/day) and a detention time of 10–20 days, thus demonstrating the effect of agitation. To allow even greater solids levels to build up and to separate the solids retention time from the hydraulic detention time, the anaerobic contact process was developed. In this process, the agitated digester is followed by a clarifier, from which the sludge is returned to the digester.

Studies conducted on high-rate digesters to determine the optimum temperature of digestion have found that there are two optimum temperatures, one for a mesophilic population of bacteria and one for a thermophilic population. Both populations are mixed cultures. The optimum for the mesophilic population is 35°C (98°F), and that of the thermophilic population is 54.4°C (130°F).

The mean cell residence time required in the reactor is obviously a function of temperature. McCarty (13) has determined how the critical mean cell residence time (the time below which reactor failure would occur) varies with temperature; Table 17-1 lists these values ($\Theta_{c\,min}$) at several temperatures. Reasonable levels of solids retention time can be obtained for system design by multiplying $\Theta_{c\,min}$ by 2.5.

The quantity of sludge digestion obtained depends both on the waste being treated and on the reactor design, but usually the volatile solids reductions are on the order of 50–70%.

Fundamentals

The basic principles underlying the process are the same whether sludge digestion or wastewater treatment is being considered. Anaerobic decomposition is a microbial process, and most of the organisms involved are bacteria. Most of the bacteria are facultative anaerobes except for the methane bacteria, which are obligate anaerobes. This is partly why the methane bacteria are a sensitive group.

The basic reaction involved in the anaerobic decomposition process is:

$$\begin{matrix}\text{Organic}\\\text{matter}\end{matrix} + \begin{matrix}\text{combined}\\\text{oxygen}\end{matrix} \xrightarrow{\text{anaerobic bacteria}} \begin{matrix}\text{new}\\\text{cells}\end{matrix} + \begin{matrix}\text{energy}\\\text{for cells}\end{matrix} + CH_4 + CO_2 + \begin{matrix}\text{other}\\\text{products}\end{matrix}$$

The oxygen sources (oxygen in combined form) are CO_3^{-2}, SO_4^{-2}, NO_3^{-1}, and PO_4^{-3}. The other products can be H_2S, H_2, or N_2, depending on the starting organic matter.

Table 17-1 Minimum Mean Cell Residence Times for Sludge Digestion as a Function of Temperature

Temp. (°C)	$\Phi_{c\,min}$
18	11
24	8
30	6
35	4
40	4

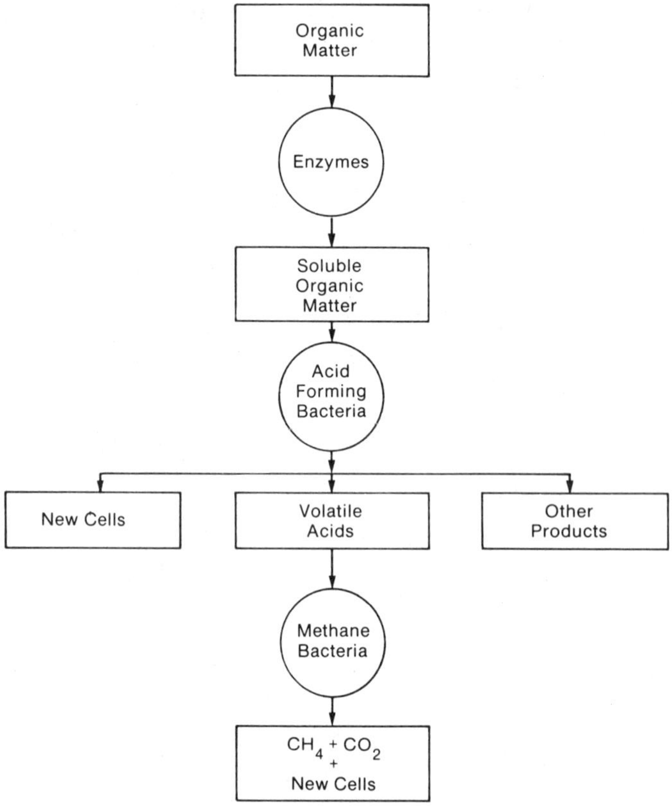

Figure 17-2 Basic anaerobic decomposition reactions

The process can be thought of as consisting of three stages as diagrammed in Figure 17-2. The stages are indicated by the circles on the diagram, which represent agents for accomplishing the various tasks—solubilization, acid formation, and gas formation. In theory any one of the stages could be limiting, but in practice the limiting stage is usually either the first or the third.

Solubilization is accomplished by extracellular enzymes. Some organic molecules are hard to solubilize.

The acid-forming bacteria, a mixed population, are fairly hardy and have a doubling time of about 20 minutes. They convert fats and oils directly into organic acids, but carbohydrates and proteins require intermediate steps.

$$\text{Carbohydrates} \rightarrow \text{simple sugars} \rightarrow \text{alcohol aldehydes} \rightarrow \text{organic acids}$$

$$\text{Proteins} \rightarrow \text{amino acids} \rightarrow \text{organic acids} + NH_3$$

The most common acids formed are acetic and propionic acids. These organic acids are converted to methane, carbon dioxide, and bacterial mass by methane bacteria. The methane bacteria are also a mixed culture with very long generation times, which can be from 2 to 20 days. Since methane bacteria are so slow to generate, they are often the rate-limiting step in the decomposition process. They require a pH of 6.8–7.4; pH variations outside of this range cause them to cease functioning.

If the loading of the anaerobic system is not carefully controlled, it is not uncommon for a "stuck" digester to develop. What probably occurs is that excess feed causes the acid formers to produce excess acids, which decreases the pH and stops the functioning of the methane formers.

Other factors also affect the functioning of methane bacteria. The literature (14) indicates that methane bacteria are sensitive to high levels of alkaline or alkaline earth cations at 1,000–10,000 mg/l, ammonia at 1,500–3,000 mg/l, sulfide ion at 100–200 mg/l, and dissolved volatile acid above about 2,000 mg/l. They are sensitive to ionized heavy metals and to certain organic compounds. Organic compounds of concern in pulp and paper effluents would include formaldehyde above 50–100 mg/l, ethylene dichloride above 150–500 mg/l, and phenol above 300–1,000 mg/l. Despite these sensitivities to a variety of materials and conditions, methane bacteria do function in anaerobic treatment plants.

Anaerobic treatment is a relatively robust process with wide applicability. Because of the two-stage nature of the digestion process, two-stage reactor systems have been tried, but their economics have always been questionable.

Advantages of Anaerobic Treatment

1 The process does not require any aeration equipment with its associated capital and operating costs
2 Less cellular material is produced than from aerobic processes
3 The process will degrade most wastes
4 Oxygen in the form of sulfate and nitrate is used
5 The nutrient requirement is low
6 The mixture of methane gas and CO_2 that is produced can be used as an energy source. Its heating value is typically 22,000–30,000 kJ/m^3

Difficulties with Anaerobic Treatment

1 Above-ambient temperatures are required to achieve the maximum rate of operation
2 Very high biomass levels are required for short hydraulic retention times
3 The generation time for methane bacteria is long (2–11 days at 37°C), so long solids retention times are required
4 Methane bacteria are very sensitive to shock loads, toxic materials, and environmental conditions. Their long generation times mean that the system recovers only slowly from upset conditions
5 Long start-up times—20–30 days for equilibration—are required

6 The potential for odor emission is high
7 It is difficult to produce effluents with good aesthetic properties and with BOD values of 50 mg/l or less

Applications of Anaerobic Treatment

The last difficulty listed relegates anaerobic treatment within the pulp and paper industry to a role as a pretreatment process before conventional aerobic treatment. Of the two commercial installations existing in the paper industry, one is a lagoon operating as a pretreatment, and it performs well (9). The other is an upflow anaerobic sludge blanket, also operating as a pretreatment, whose performance history is not available at this time.

The anaerobic process is also being considered for the treatment of selected strong streams. Pilot plant trials on sulfite mill condensate have been conducted (15), and bench tests have been done with Kraft mill condensate and with total mill effluent (4). In the pilot trials, the greatest interest is in a system with low hydraulic retention time and high solids. Anaerobic fixed-film systems can be operated as fixed-bed reactors, expanded-bed reactors (16), or fluidized-bed reactors. With fixed-bed systems, biomass as high as 35 g/l has been obtained (10).

The treatment system for sulfite mill condensate discussed in reference 15 was a plastic fixed-bed, submerged-media reactor that was loaded at 16 kg/COD/m^3/day and had a hydraulic retention time of 7 hours. The COD reduction achieved varied from 79 to 90%, producing 0.09 g biomass and about 0.33 l CH_4 per gram of COD removed. Surprisingly, toxicity from sulfur compounds, furfurol, and trace organics was not a problem. The sulfite condensate (COD 490 mg/l) was neutralized to pH 7.2 before treatment.

Considerable pilot work has been done at the University of Helsinki on the treatment of pulp and paper effluent (17). Wastewaters from groundwood pulp and paper, thermomechanical pulping, sulfite evaporator condensate, wet debarking, tall oil distillation, and bleached Kraft pulping have been tested, and all of these effluents have proved treatable. The loading rate possible depended on the strength of the waste. For high-strength wastes of COD 7,000–8,000 mg/l, 70% COD removal was obtained at a loading of 10–15 kg COD/m^3/day. For low-strength wastes, 70% COD removal required much lower loadings of 1–3 kg COD/m^3/day.

The biggest problem in the anaerobic treatment of sulfite condensates from chemical pulping is their high sulfur content. Sulfite ion slows methane production at 25 mg/l and completely stops it at 100 mg/l (18). To prevent this from happening, the calcium-based sulfite condensate is stripped with desulfurized gas.

REFERENCES

1 W. J. Gillespie, D. W. Marshall, and A. M. Springer, *Tappi*, 57 (9):112 (1974).
2 D. J. Bennett, T. Needham, and R. E. Summer, *Tappi*, 56 (12):182 (1973).

3 J. V. Maxham and G. A. Dubey, "Effect of fine particle and nutrient addition on the activated sludge process," *Proceedings of the TAPPI Environmental Conference*, Atlanta, Georgia, 1983, pp. 1–15.

4 J. V. Maxham et al., "Innovative biological wastewater treatment processes applied to the treatment of unbleached Kraft mill effluent," *Proceedings of the TAPPI Environmental Conference*, Minnapolis, Minnesota, 1982, pp. 265–285.

5 B. Atkinson et al., *Biotechnology and Bioengineering*, 21:193 (1979).

6 Personal communication, Simon-Hartley, Ltd., Slope on Trent, Staffordshire, England ST4 7BH.

7 D. E. Severeid, *Tappi*, 62 (2):61–63 (1979).

8 M. L. Jackson, *Industrial Wastes*, 31–36 (November/December 1978).

9 C. J. Priest, "A change to anaerobic-aerobic treatment made expanded production possible without expansion of wastewater treatment facilities," *Proceedings of the XXXV Industrial Waste Conference*, Purdue University, May 13–15, 1980.

10 W. J. Jewell, M. S. Switzenbaum, and J. W. Morris, *Journal—Water Pollution Control Federation*, 53 (4):482–490 (April 1981).

11 Metcalf & Eddy, *Wastewater Engineering: Treatment/Disposal/Reuse*, 2d ed., McGraw-Hill, New York, 1979, pp. 455–63, 609–627.

12 T. D. Reynolds, *Unit Operations and Processes in Environmental Engineering*, Brooks/Cole Engineering Division, Monterey, Calif., 1982, pp. 403–435.

13 P. L. McCarty, "Anaerobic waste treatment fundamentals," *Public Works*, 95 (9):107; 95 (10):123; 95 (11):91; 95 (12):95 (1964).

14 R. A. Conway and R. D. Ross, *Handbook of Industrial Waste Disposal*, Van Nostrand Reinhold, New York, 1980.

15 M. M. Benjamin, J. F. Ferguson, and M. E. Buggins, "Treatment of sulfite evaporator condensates with an anaerobic reactor," *Tappi*, 65 (8):86–102 (1982).

16 L. van den Berg and C. P. Lentz, "Comparison between up and downflow anaerobic fixed film reactors of varying surface to volume ratios for treatment of bean blanching waste," *Proceedings of the XXXIV Industrial Waste Conference*, Purdue University, 1979.

17 M. Salkinoja-Salonen et al., "Treatment of pulp and paper industry wastewaters in an anaerobic reactor: From bench size units to full scale operation," Personal Communication, University of Helsinki, Department of General Microbiology, Finland.

18 W. Maaskant and Y. Hobma, "The influence of sulfite on anaerobic effluent treatment," *H_2O*, 14 (25):S96–S98 (1981).

18
Solid Waste Management and Disposal

The paper industry generates a wide variety of solid wastes that need final disposition. In 1974 the U.S. Department of Commerce estimated that 560–630 lb solid waste were generated per ton of production. The largest single source of solid waste is the debarker, which accounts for about half of the total. In the past, barking waste was disposed of either by landfill or by burning in a tepee burner. With the escalating costs of energy, bark has become useful as a feed for hogged fuel boilers, eliminating barker waste as a solid waste problem.

Wastewater treatment generates two types of sludge, primary sludge from the primary clarifier and secondary sludge from the biological treatment system. Sludge disposal may account for half of the operating cost of a waste treatment system. Primary sludge production in the pulp and paper industry is approximately 20 million lb/day (dry basis), and secondary sludge may contribute an additional 10 million lb/day. Wastewater treatment, therefore, generates a total of about 30 million lb solids per day to be disposed of (1, 2). In 1983 an NCASI study (3) reported that the industry generated 2,870 tons/day of primary clarifier sludge, 190 tons/day of secondary clarifier sludge, and 1,800 tons/day of combined primary and secondary sludges from 180 mills responding to the survey. These figures are much lower than those of the Department of Commerce, but they do not represent the entire industry.

If the incoming water is treated for sediment by settling with polymer or alum, a third sludge is generated.

Kraft pulping generates some unique solid wastes, such as slaker grits, unburned lime rejects from the kiln, and dregs from the green liquor clarifiers. These wastes amount to about 30–40 lb per ton of production. Groundwood pulping generates 5–10 lb screening rejects per ton of production. These solids usually end up as primary clarifier sludge.

NATURE OF SLUDGES

The nature of these various sludges differs greatly with the type of pulp and paper manufactured. A bar graph (Figure 18-1) can be used to rank the various sludges on their resistence to dewatering (4). The primary sludges that are high in fiber

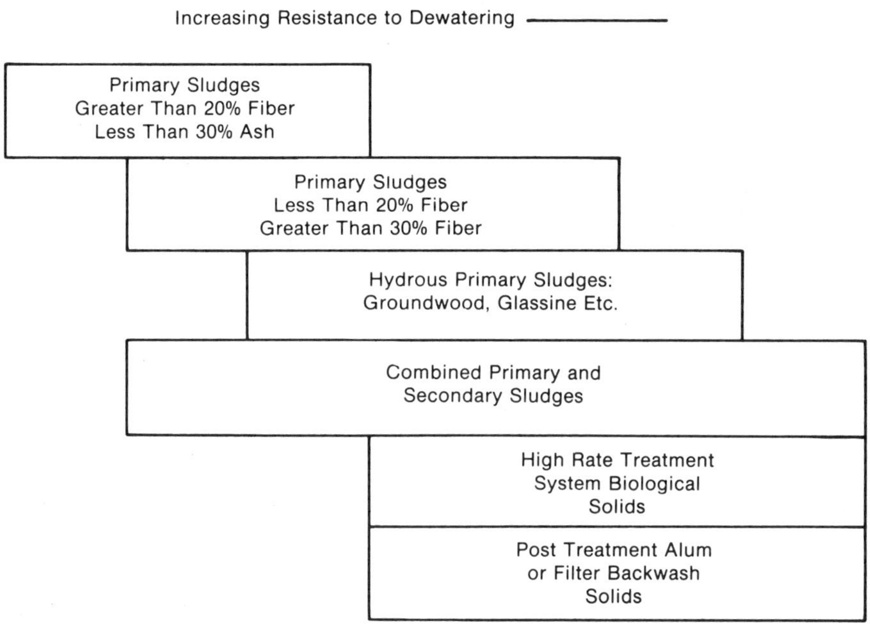

Figure 18-1 Relative dewaterability of pulp and paper mill sludge

and low in ash are the easiest to dewater. The most difficult are the solids from high-rate biological treatment systems and the alum sludges from water treatment. The primary sludge most difficult to dewater is that containing groundwood fines. An in-depth analysis of the nature and composition of primary sludges is provided in reference 5.

Dewaterability is very important because it determines the volume of waste that has to be handled. If a sludge is dewatered from 1 to 5% solids, four-fifths of the volume of material is eliminated; 10% solids represents a nine-tenths volume reduction. The final waste volume is directly related to the cost of disposal.

The usual way to handle the difficult-to-dewater sludges is to blend them with primary sludge. Few problems result if the ratio of primary to secondary sludge is high. However, the more secondary sludge in the mixture, the poorer the dewatering characteristics. The dewatering properties of a 50:50 mixture are very poor. Some mills are approaching this 50:50 ratio and are therefore anticipating difficulties.

In some instances, it may be desirable to dewater the primary sludge separately from the secondary sludge. One example is a situation in which the secondary sludge can be disposed of through land application. Blended sludges are not suitable for such disposal. Another example is a situation in which the primary sludge can be used to produce a by-product or can be reused within the production process, but the blended sludge cannot be so used.

The various alternatives available in sludge dewatering and disposal are outlined schematically in Figure 18-2. The usual treatment is to send sludge to a

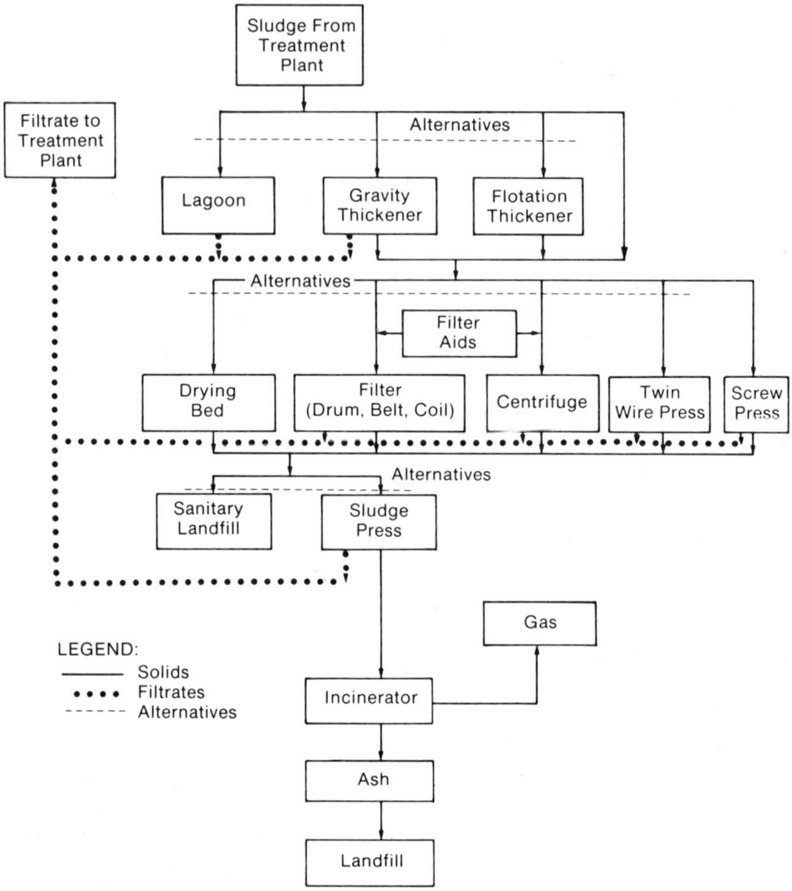

Figure 18-2 Sludge dewatering and disposal alternatives

thickener before it goes to the dewatering equipment. The reason for thickening is that, with rare exceptions, the higher the feed content, the more efficient the dewatering operation in terms of performance, operation, and economy.

Two types of thickeners are common: gravity thickeners and flotation thickeners, which operate on the same principles as gravity clarifiers and flotation clarifiers, previously discussed. The flotation thickener used on secondary sludge can achieve approximately 4% solids, giving it an advantage over the gravity thickener, which can achieve only about 2%.

CONDITIONING AGENTS

In general, the three chemicals used to condition sludges before dewatering, regardless of the type of equipment used, are lime, ferric chloride, and polymers.

The three can be used separately or in conjunction. Ferric chloride has the disadvantage of being highly corrosive but is a very effective conditioning agent. The cost of chemical conditioning can range from $1 to $30 per ton of dry solids treated (6). Sludges that are very difficult to dewater and require high polymer addition rates are the most costly.

Wet air oxidation has also been used as a conditioning process to aid sludge dewatering and has been commercially applied in the paper industry (7) where filler recovery was a side benefit. The installation experienced considerable down time and high maintenance costs. The brightness of the filler recovered was lower than that of filler-grade clay. Severe scaling problems were experienced from calcium sulfate and calcium oxalate. Surprisingly, the report (7) claimed that the process was an economic success, despite its problems. The supernatant produces an additional load to the wastewater treatment plant with its BOD of about 5,000 mg/l.

DEWATERING DEVICES

Six sludge dewatering devices commonly used in the paper industry are rotary vacuum filters, centrifuges, V presses, filter presses, twin-wire presses, and screw presses. Tables 18-1 and 18-2 show the distribution within the industry of the various dewatering devices for primary sludges and mixed primary and secondary sludges, based on data from a 1983 NCASI industry survey (3). In some situations, the primary dewatering device is followed by a press to further increase the solids content.

Table 18-1 Approaches to Primary Sludge Dewatering

	Number of Mills Reporting the Use of a Preconditioning Step				
Dewatering Technology	None	Gravity Thickening	Chemical Conditioning	Thickening Plus Chemicals	Total
None	—	19	0	1	20
Centrifuge	8	4	4	2	18
Vacuum filter	21	19	4	1	45
Belt press	9	3	3	5	20
Pressure filter	3	6	1	1	11
Mechanical press (only)	0	2	2	0	4
Basin drying	16	8	2	1	27
Basin disposal	20	7	0	1	28
Total	77	68	16	12	173

Table 18-2 Approaches to Sludge Dewatering (Mixed Sludges)

	Number of Mills Reporting the Use of a Preconditioning Step				
Dewatering Technology	None	Gravity Thickening	Chemical Conditioning	Thickening Plus Chemicals	Total
None	—	1	0	0	1
Centrifuge	2	0	3	0	5
Vacuum filter	10	3	6	5	24
Belt press	3	0	11	7	21
Pressure filter	0	0	4	1	5
Mechanical press only	0	0	0	1	1
Basin drying	7	0	0	0	7
Basin disposal	2	0	0	0	2
Total	24	4	24	14	66

Vacuum Filters

The vacuum filter, pictured in Figure 18-3, appears to be the most popular device. It operates on the same principle as a pulp decker. The key design parameter is the filter solids loading in $lb/ft^2/hr$, which is usually determined in bench-scale testing with a filter leaf test apparatus. For primary sludges, loading rates of 5–10 $lb/ft^2/hr$ can be achieved without chemical conditioning. Cake solids would be

Figure 18-3 Vacuum filter. Copyright © 1981. TAPPI. Reprinted from *Introduction to Environmental Management for the Pulp and Paper Industry*, Study Guide, Volume 6, with permission

20–30%. Vacuum filter loading rates for combined sludges containing 20–30% secondary sludge solids ranges from 2 to 6 lb dry solids/ft^2/hr, with the average being about 4. From 2 to 7 lb conditioning agents per ton are required to achieve these rates. There is 90–95% solids capture, and the cake produced is about 20% solids. In order for the filter cake to discharge properly from the filter, 10–20% long fiber (greater than 100 mesh) must be present in the sludge (4). Landfill volumes for high-ash and low-ash sludges can be estimated as 0.4 or 0.8 m^3/dry tonne, respectively.

Vacuum filter cakes containing combined-sludge solids can be further dewatered on V presses to approximately 35–40% consistency. A V press is just two discs providing a converging nip that applies pressure to the sludge to squeeze out the water.

Vacuum filters can be equipped with either fabric media or steel coils. Fabric media are often used in situations in which the fiber content is low, the ash content is high, or the solids are otherwise difficult to dewater on a coil filter.

The power costs for operating the large vacuum pump required by a vacuum filter are quite high. Vacuum filters are being replaced by belt presses, which seem to perform as well, if not better, at lower operating costs.

Centrifuges

Three types of centrifuges that have been developed are disc, basket, and scroll centrifuges. Disc centrifuges have found little application in the paper industry. They have been tried as thickening devices, but experience has been unsatisfactory (8). Basket centrifuges have been used to a limited extent for sludges that are very difficult to dewater, but they operate in a batch mode rather than continuously. Usually it is desirable to use the continuous-decanter scroll centrifuge. Special scroll units have been developed for secondary sludge, and they are usually preferred over the basket centrifuge. Figure 18-4 shows a schematic of a scroll centrifuge.

Scroll centrifuges dewatering combined paper industry sludges generally produce cakes of 20–40% consistency at solids capture efficiencies of 85–98% from sludges conditioned with 2–7 lb polymer per ton of sludge solids. Feed rates as high as 400 gal/hr have been used. Centrifuges are favored especially for sludges containing over 40% ash.

Centrifuges operate on the basis of density difference separation, so sludges that are much denser than water, such as high-ash sludges, provide the best application of centrifuges. Centrifuges are also advantageous in the dewatering of secondary sludge. Specially designed scroll centrifuges can dewater secondary sludge from 2–11% solids with 99.9% capture efficiency (9). This performance requires 6–8 kg dry polymer per ton of sludge for conditioning. Polymer addition is necessary for centrifugation of secondary sludge and represents a major part of the cost of dewatering such sludge.

Centrifuges have a relatively low capital cost but can be expensive to operate due to the chemical conditioning agents commonly required, their high power

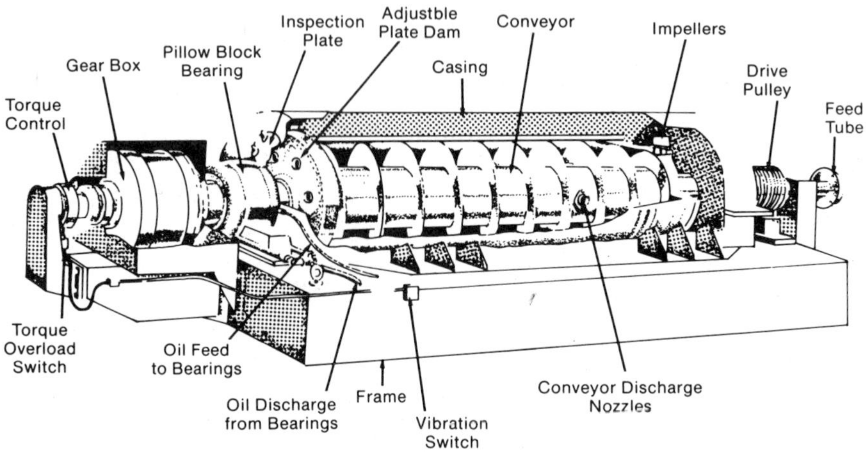

Figure 18-4 Scroll centrifuge. Copyright © 1981. TAPPI. Reprinted from *Introduction to Environmental Management for the Pulp and Paper Industry*, Study Guide, Volume 6, with permission

requirements, and their maintenance costs. Dissatisfaction with centrifugation has been attributed to the following causes:

1 The generation of poor-quality centrates that could cause a buildup of fines in the treatment system (10)
2 The susceptibility of centrifuges to plugging with pieces of bark
3 The severe screw conveyor abrasion experienced at many installations

Integrated mills practicing debarking have an average scroll life of 5 months before resurfacing is required, and both nonintegrated fine paper mills and integrated fine paper mills not practicing debarking have average scroll lives of 7 months. In an attempt to overcome this problem, centrifuge manufacturers have developed a wearplate of ceramic material or of tungsten carbide steel for the end of the conveyor.

V Presses (Mechanical Presses)

V presses have been applied successfully to the dewatering of both centrifuge and vacuum filter cakes containing as much as 30% biological solids. The industry's experience suggests that combined sludges normally encountered require sufficient conditioning for vacuum filtration or centrifugation to render them amenable to V pressing (4). V pressing can be performed to raise the solids content of the sludge high enough for incineration (11, 12). V presses generally produce primary sludge cake consistencies of 30–45%. Either a V press or a screw press would precede most bark boilers burning bark and sludge. The sludge would enter the press at 15–25% solids and be subjected to 690 kPa to raise the solids level to the 30–45% suitable for incineration (6).

Pressure Filters

Pressure filters are the most powerful dewatering devices available, capable of producing the highest cake solids. For combined sludge, cakes of 30–45% consistency can be produced with solids capture efficiency approaching 95–100% (4). To achieve this efficiency level, however, it is necessary to precoat the filter cloth. Precoating is generally thought to be of benefit in aiding cake discharge and in minimizing the frequency of media cleaning. Media cleanliness has been indicated as a crucial parameter in determining the pressure filter cycle time. Various materials, such as diatomaceous earth, fly ash, and cement dust, can be used for precoating. Pressure filtration also requires conditioning of the sludge before filtration.

The main drawback of the pressure filter is that it is a batch operation that requires a lot of operator attention. Continuously operating automatic units have been developed, but they are mechanically complex and therefore subject to many maintenance problems. The relative merits of high- and low-pressure operation remain unresolved.

Twin-Wire Presses

The moving-belt press, as shown in generalized form in Figure 18-5, has received intensive industry interest in the past several years (13, 14). Many paper mills have purchased moving-belt presses. At least five and probably more of these units are being marketed in the United States by various manufacturers. All of the units use essentially the same three dewatering processes—gravity drainage, pressure-induced drainage, and shear-induced drainage. The differences are in the emphasis placed on each of these three processes by a particular unit. Sludge conditioning is required. On primary or combined sludges, moving-belt presses have generated cakes with a consistency comparable to that of two-stage dewatering with V presses, and with similar or somewhat higher conditioning costs and generally lower power consumption.

Conditioning costs range from $0 to $20/tonne for primary sludge and from $30 to $50/tonne for secondary biological sludges. Polymers are commonly used, and some processes use dual-polymer systems (14). Loading rates for presses are usually expressed in l/min/m of belt width. For primary sludge the rates are from 120 to 370 l/min/m and for secondary biological sludge from 62 to 125 l/min/m, based on a survey (14) of industry experience. A third of the installations exceeded these rates, demonstrating that sludge characteristics vary widely from mill to mill. For primary sludge the cake solids are 20–50%, and for secondary sludge 10–20%. Capture efficiency is very high for belt presses, about 95–99% of the solids fed.

Required operator attention is low, only about 1–2 hours per shift. Most of this time is used to tend the auxiliaries such as polymer feed, commutators, and sludge pumps rather than being required by the units themselves. The units require power only to drive the belts; thus they are energy-efficient sludge dewatering

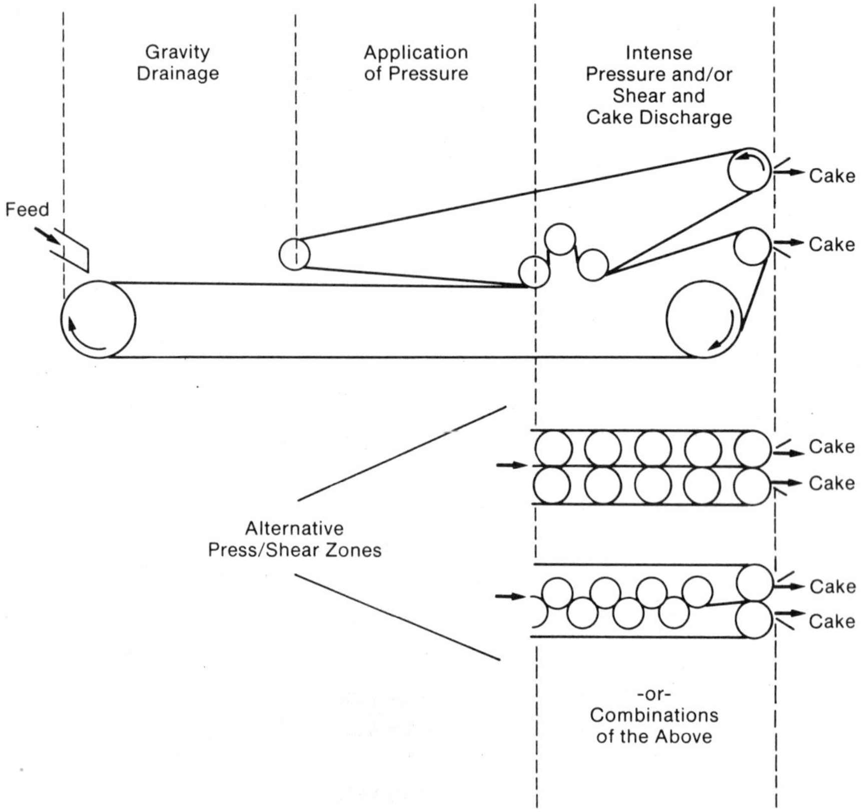

Figure 18-5 Generalized twin-wire press. Copyright © 1981. TAPPI. Reprinted from *Introduction to Environmental Management for the Pulp and Paper Industry*, Study Guide Volume 6, with permission

devices. Another advantage is their ability to operate on secondary biological sludges.

The major operating problem is felt life, which is usually 3 to 4 months. The usual cause of failure is puncture of the felt by incompressible objects in the sludge.

Screw Presses

The latest development in sludge-dewatering equipment is screw presses of new design. Two manufacturers are currently marketing these presses in the United States (15), and several have been sold for primary sludge dewatering. They produce cake solids of 50–55% when operated as the only sludge dewatering device. Solids capture ranges from 70 to 88% with no polymer addition (15). Polymer can be used to improve the capture efficiency, but it has little or no effect on final sludge consistency; therefore it is often not used. These presses appear to be energy efficient.

Screw press designs exist that claim to be able to process secondary biological sludges, but as of May 1984 no U.S. paper industry experience with them had been published. Polymer conditioning would likely be required.

It will be interesting to observe whether screw presses replace twin-wire presses as the dewatering technology of choice for the pulp and paper industry. At this time, insufficient data are available to predict the ultimate success of the new screw press technology.

FINAL DISPOSAL OF DEWATERED SLUDGE

The pulp and paper industry disposes of its dewatered sludge solids primarily by landfilling or by burning for power generation. Table 18-3 summarizes industry experience as of 1979 (3). McKeown reported that 76% of mills dispose of sludge on land, 10% burn it in hogged fuel or power boilers, and an additional 7% incinerate it by other means (6). Since landfill sites are becoming increasingly difficult to obtain, more mills are facing the need to incinerate. Land application of sludge as a soil amendment is also being practiced to a limited extent.

One percent of the total sludge produced is recycled. Sludge reuse is practiced by some waste paperboard mills that reuse the sludge as part of the furnish. Even some secondary biological sludge is reused in this way.

Table 18-3 Combinations of Sludge Dewatering and Disposal Practices Used by Mills in the United States[a,b]

	Incineration		Land Application		Landfill	
	Primary	Mixed	Primary	Mixed	Primary	Mixed
Gravity thickening only	1	0	0	0	9	0
Mechanical dewatering	4	2	3	3	40	34
Dewatering plus supplemental pressing	14	8	1	2	19	9
Basin drying	—	—	3	0	13	7
Total	19	10	7	5	81	50
Percent of total reporting	11	6	4	.3	47	29

[a]Based on responses to the 1979 NCASI Accomplishment Survey,
[b]Units are the number of mills reporting use.

Seven percent of the sludge produced is sent to storage lagoons, where it is left to dry out slowly as the water drains away or evaporates. This practice requires a large land area and is gradually being phased out.

The importance of conducting extensive pilot plant tests before making a final choice of technology for the dewatering and disposal of sludge at a particular mill is reemphasized. Sludge is a highly nonuniform substance whose character can vary from day to day. Extensive evaluation is needed to make certain that the technology chosen will perform as expected.

Incineration

The three types of incineration practiced in the industry are:

1 Burning in an incinerator specifically designed for sludge
2 Burning in the bark boiler
3 Burning in a power boiler that also burns fossil fuel

Burning the sludge in the bark boiler, which is commonly a hogged fuel (combination fuel) boiler, seems to cause few problems except for reduced steam generation and reduced boiler efficiency (16). The burning of sludges containing materials from the causticizing area has in some instances caused unacceptable clinker formation, sludging, or tube fouling. As of 1981, 23 mills burned sludge in combination fuel boilers. The sludge consistency was 20–47% solids (median 33%). Power produced was 6,300–9,500 BTU per pound of volatile sludge solids (median 7,900). The ash content was 2–40% (median 8.8%). The sludge feed rate was 2.5–14 lb sludge solids per 100 lb boiler fuel solids (median 6.9) (16).

Incineration in the bark boiler appears to be acceptable for sludge incineration if such a boiler is available on the mill site and if it can take the increased water load. Dewatering to higher levels—45–50% solids—will make bark boiler incineration an even more attractive alternative and will minimize the effect on boiler operation.

A number of other combustion methods have been tried, including kiln combustion, wet air oxidation, and fluidized bed combustion; only the last appears promising. An advantage of the fluidized bed is that its large mass of hot sand minimizes the effect of variations in moisture content. The Lielahte sulfite mill in Tampere, Finland, practices fluidized bed incineration of 10,000 tons/year of dredged sludge, 1,200 tons/year of coverting waste, 1,000 tons/year of wood yard waste, and 1,000 tons/year of peat (17). The supplementary material provides additional BTU content to the sludge being burned. The Finnish installation automatically feeds oil if the bed temperature falls below 770°C. In the United States, the Packaging Corporation's mill at Filer City, Michigan, practices fluidized bed incineration of combined primary and secondary sludge.

Some requirement for the self-sustaining combustion of sludge are as follows (18):

1 A minimum of 30% dry solids
2 At least 1400 J/g (6,000 BTU/lb)
3 At least 66.5% organic contents

The primary clarifier sludge is dewatered on a vacuum filter and then on a V press before incineration. The auxiliary waste goes through a hammer mill before being fed to the furnace. Figure 18-6 is a schematic diagram of the incineration system. This system is used to generate steam. Its initial cost is $3.5 million. It generates $850,000 of steam per year, giving a return on investment of 22%.

Another disposal technique that has been explored is sludge incineration in the Kraft recovery system (19). An integrated Kraft mill generates about 15 kg of dry secondary sludge per air-dried ton of pulp. This amount represents only 1% of the black liquor solids generated per ton. The solids are not a problem, but the moisture added with secondary sludge at 10–15% solids may cause an increase in evaporator load of up to 20% (19). It is proposed that the sludge be added to the weak black liquor before evaporation. Three potential operating problems can arise from this procedure:

1 Evaporator fouling
2 Accumulation of nonprocess elements
3 Increased dregs and grit loads

Evaporator fouling has proved to be very severe, but it could be eliminated by predigestion of the sludge at pH 13 at 100°C for 20 minutes. The main nonprocess elements that may accumulate are aluminum and chloride. Perhaps the aluminum could be precipitated with the dregs, but this action would further increase the dregs load.

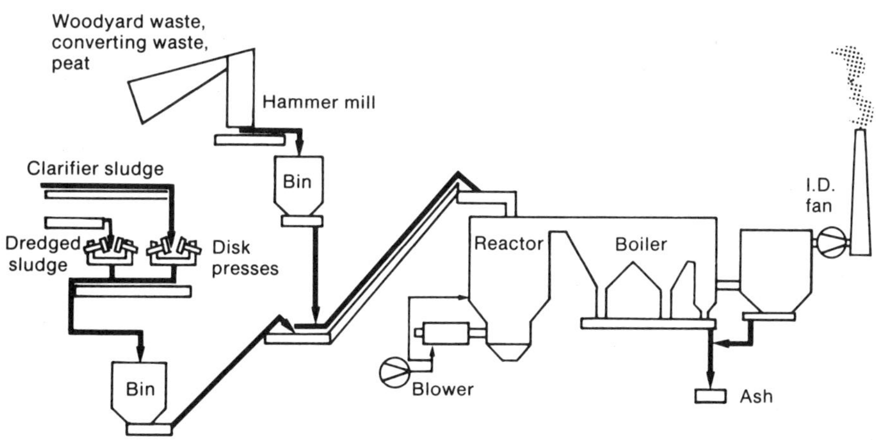

Figure 18-6 Fluidized bed incineration system. Copyright © 1981, reprinted from *Tappi Journal*, 64(9): 150 (1981) Technical Association of the Pulp and Paper Industry, with permission

Adding sludge to the Kraft recovery system has not been tested on a full-scale basis, but it may be worth further consideration if disposal costs are greater than \$0.16 per air-dried ton of sludge solids.

Incineration is not practical for high-ash sludges. Stringent air pollution emissions requirements for combination boilers have diminished the amount of incineration practiced.

Most mills do not produce enough sludge to support fully the operation of even a small dedicated incinerator. However, as landfill regulations become more stringent and fuel prices rise, the point may be reached at which it is economical to install a hogged fuel boiler to generate power through sludge combustion only.

Landfilling

Landfilling or land application of sludge dominates the industry's sludge disposal practices. The average handling cost for land disposal in 1984 was \$10–14/tonne. Normal sanitary landfill practices should be observed in constructing an industrial landfill. Some of the requirements that must be met are as follows (6):

1. The disposal site should be a minimum distance above groundwater
2. All subsurface conduits (such as culverts, gas and water lines) should be removed
3. The site should be above the floodplain and should be protected from flooding
4. The site should be a minimum distance from a public well
5. The site should be a minimum distance from highways
6. The site should be a minimum distance from a watercourse
7. The nearest property line should be a certain distance away

After a site is chosen according to the listed criteria, it should be used in accordance with good operating procedures for sanitary landfills as outlined in reference 20. Studies of the specific requirements for the design of paper mill landfills are found in references 21–23.

Land Application

Numerous trials have been conducted with paper mill sludge as a soil amendment (24). Sludges function only as amendments and not as fertilizers because they do not contain the elemental analysis required of a fertilizer (25). For a soil amendment, the carbon/nitrogen ratio should be 20:1 to 30:1. An average composition of seven different paper mill combined sludges from 10 different mill types was 26:1, so this criterion has been met. The calcium/magnesium ratio should be above 6:1; many combined sludges fail to meet this criterion, but the addition of lime to the sludge fulfills it. Sludges are good soil amendments for sandy soils. Detailed analysis of the seven combined sludges did not indicate a heavy metal problem (18).

Trials have been conducted in which fly ash and either primary or secondary sludge were applied to crop land (24). The fly ash–sludge blends were as effective as commercial fertilizer. In these same trials, lime mud applied to agricultural land performed better than dolomite limestone used for the same purpose (24). Blends of mill solid waste streams may be the answer to the difficulties normally encountered with land application.

The quantity of fly ash is increasing because more stringent air standards allow less particulate emission, stricter water standards do not allow the discharge of fly ash, and oil shortages have promoted the use of more coal and biomass as fuel. Particulate collection produces 80% of the fly ash, and 20% comes from bottom ash (25). The best use of fly ash is to produce Portland cement, but the cement market would soon be saturated. Environmentally, the next best use would be to blend it with primary or secondary sludge and apply it to land. If land application is not feasible, fly ash can be landfilled. Reference 25 gives precautions to follow when fly ash is landfilled.

Criteria have also been proposed for the land spreading of solid waste. The details can be found in reference 26, but briefly the proposed criteria are as follows:

1. The soil–sludge mixture must not have a high content of heavy metal that can be taken up by growing plants
2. The soil waste pH should be 6.5 or higher
3. Excess nitrogen should not be applied beyond that normally taken up by the crop in one season
4. The sludge applied should be free of living pathogenic organisms
5. Solids must be applied in such a manner that they are not available for direct ingestion by domestic animals or humans

LEGISLATION GOVERNING SOLID WASTE DISPOSAL

The basic environmental legislation governing sludge disposal is the Solid Waste Disposal Act as amended by the Resource Conservation and Recovery Act of 1976, otherwise known as PL 94–500. It requires the EPA to establish criteria for sanitary landfills. All landfills that did not comply with these criteria were to be closed.

Some of these criteria are as follows:

1. The landfill should not be located in environmentally sensitive areas such as wetlands, floodplains, or areas of permafrost
2. The landfill should not adversely influence surface waters
3. The site should not harm local groundwater supplies
4. The landfill must not be a place where disease vectors such as rats, birds, or flies can breed

5. The facility must be safe to work at. No toxic or explosive gases can be produced, and no burning of refuse is allowed. The site will probably attract birds, so it must be located such that the birds do not interfere with aircraft
6. Access to the site must be limited by such means as a fence, gate, and guard
7. Noise must be kept to a minimum and in any case cannot exceed 55 dB, the outdoor limit for protection of public health and welfare
8. Covering material must be applied periodically to control odor, dust, and blowing litter
9. Air emissions, including evaporation, sublimination, and oxidation, must be controlled

The landfill technique is the disposal method most widely used in the pulp and paper industry. It is important that the criteria discussed be followed. Given the great concern over improper land disposal of toxic and hazardous substances, our industry's landfills will come under scrutiny even though there is a low probability that they contain toxic or hazardous materials.

REFERENCES

1. "EPA guidelines for thermal processing of solid wastes and for the land disposal of solid wastes," *Environmental Reporter*, 101, 1101; 40 CFR 240 and 241; 39 FR 29328, August 14, 1974; 40 FR 5159, February 4, 1975.
2. *Study of Solid Waste Management in the Pulp and Paper Industry*, U.S. Department of Commerce Report No. PB-234-944, U.S. Government Printing Office, Washington, D.C., 1974.
3. *A compilation of data on the nature and performance of wastewater management systems in the pulp and paper industry*, Special Report No. 83-09, National Council of the Paper Industry for Air and Stream Improvement, New York, 1983.
4. R. A. Miner and D. W. Marshall, *Sludge dewatering practice in the pulp and paper industry*, Technical BuLletin No. 286, National Council of the Paper Industry for Air and Stream Improvement, New York, 1976.
5. T. W. Joyce, A. A. Webb, and H. S. Dugal, *Tappi*, 62 (4):83 (1979).
6. J. J. McKeown, *Tappi*, 62 (8):97 (1979).
7. H. A. Mertz and T. G. Jayne, "Startup and operating experience with Zimpro high pressure wet oxidation system for sludge treatment and clay reclamation," *Proceedings of the TAPPI Environmental Conference*, Savannah, Ga., April 9–11, 1984, p. 75–79.
8. C. A. Barton et al., *Journal—Water Pollution Control Federation*, 40:1471 (1968).
9. W. E. Krepps, *Tappi*, 65 (3):83–85 (1982).
10. A. M. Linsey, J. N. Sullins, and J. R. Fluharty, *Progress in primary and secondary sludge disposal at three integrated southeastern papermills*, Technical Bulletin No. 209, National Council of the Paper Industry for Air and Stream Improvement, New York, 1967.
11. H. W. Gehm, *Pulp & Paper Magazine of Canada*, 61:T-189 (March 1960).
12. J. H. Stovall and D. A. Berry, *Tappi*, 52 (11):2093 (1969).
13. R. A. Miner, *Tappi*, 59 (9):52 (1976).
14. R. Miner, *JWPCF*, 52 (9):2389–2395 (1980).
15. N. K. Toole and J. H. Kirkland, "Pilot studies of screw presses for dewatering primary sludges," *Proceedings of the TAPPI Environmental Conference*, Savannah, Ga., April 9–11, 1984, p. 69–73.

16 R. Miner, *A review of sludge burning practices in combination fuel-fired boilers*, Technical Bulletin No. 360, National Council of the Paper Industry for Air and Stream Improvement, New York, November 1981.

17 R. M. Keppe, *Tappi*, 64 (9):149–151 (1981).

18 J. N. McGovern, J. G. Berbee, and J. G. Bockheim, *Tappi*, 66 (3):115–118 (1983).

19 W. J. Fredrick, T. M. Grace, and T. W. Joyce, *Tappi*, 61 (1):59–62 (1981).

20 R. Tchobanoglous, *Solid Waste Management*, McGraw-Hill, New York, 1975.

21 R. E. Wardwell, S. R. Cooper, and W. A. Charlie, *Tappi*, 61 (12):72 (1978).

22 R. H. Ledbetter, *Design Considerations for Pulp and Papermill Sludge Landfills*, EPA-600/3-76-11, U.S. Government Printing Office, Washington, D.C., December 1976.

23 W. H. Holt, *Tappi*, 66 (9):51–54 (1983).

24 G. G. Simpson, L. D. King, B. L. Corlele, and P. S. Blickensderfer, *Tappi*, 66 (7):71–74 (1983).

25 J. S. Atwell, *Tappi*, 64 (8):67–70 (1981).

26 *Federal Register*, 43 (25):4942–4955 (February 6, 1978).

19
Air Pollution: A Problem Without Boundaries

PROBLEM WITHOUT BOUNDARIES

The water pollution problems previously discussed are of vital concern to humankind because of the important role water plays in sustaining human life. Water follows predictable distribution patterns through rivers, lakes, and oceans, and this distribution system can make water pollution an international problem. But in comparison, air pollution is a problem of faster distribution with no confinement. Air, the medium through which air pollution is conveyed, is more fluid than water and less confined. Air knows no local, state, or national boundaries. Wind action can spread pollutants in air throughout the world in a day or two. The oxygen present in the air is absolutely essential to sustain human and animal life. While a person can survive for several days without water, life without oxygen is possible for only a few minutes. Viewing pollution of the air from this perspective shows us that it has the potential to be a more serious problem than the pollution of water, as well as the potential to change the quality of human life dramatically.

Pollution of the atmosphere affects the lives of millions of people in all parts of the world, but especially those living in large industrialized cities with heavy motor vehicle traffic. Unpleasant odors, fumes, dust, and corrosive gases injure human health and damage crops and property. Most people are willing to accept some environmental deterioration in exchange for a higher standard of living and a greater abundance of consumer goods, but as living standards rise, artificially created air pollution is seen first as a major irritation and later as a threat to the quality of life. The demand for clean, pure air ultimately comes from the public. It comes from people whose quality of life has been affected, as well as from those who are perhaps a little frightened by the scale of modern technology and the possibility that the resulting emissions of pollutants into the atmosphere may lead to undesirable changes in the existing dynamics of nature between competing species or even to catastrophic changes in the climate of the earth. However, this latter concern is of less importance in stimulating people to demand changes than are the effects of air pollution on health and amenity.

HEALTH AND AMENITY

The World Health Organization defines health as "a state of complete physical, mental and social well-being, and not merely the absence of disease or infirmity" (1). This definition implies that in the concept of health, both the usually accepted meaning and amenity must be considered in dealing with air pollution problems. Air might be considered polluted when the concentration of any of its minor constituents is sufficiently high to cause damage to the health of humans or animals or to cause the loss of amenity through the deposition of dust and dirt, emission of unpleasant smells, diminution of sunshine, and so on. A much larger number of people suffer loss of amenity than suffer ill health as a result of air pollution. When actual illness is involved, greater public pressure is brought to bear on the problem. To the general public, a direct correlation is assumed to exist between loss of amenity and illness, but this conclusion may or may not be true.

Air pollution exists in the public mind whenever life is less pleasant than it would otherwise be—loss of amenity. Demonstration of loss of amenity requires no quantitative measurements or establishment of levels of pollutants. A specific nuisance in an area is apparent to any normal person whose senses have not been dulled through fatigue or adaptation. Grit and dust are obvious from visual inspection. They do not travel far in the atmosphere before depositing, so that it is usually easy to identify the source without recourse to elaborate surveys with directional gauges. Malodorous emissions that cause odor nuisances usually do so at such low concentrations that chemical tests for the compounds responsible require elaborate and expensive procedures. An alternative sometimes used is to recruit a panel of observers with "educated noses." This procedure is called "organoleptic testing" and is useful in determining odor thresholds. It can often easily locate the source of the problem.

Loss of sunshine is a different sort of problem because sunshine is normally highly variable. A skilled meteorologist is needed to assess a potential loss accurately. The changes that result may not be easy to notice, or may be perceived only subliminally.

The problem of loss of amenity becomes more important as the quality of life increases and will need to be considered in future legislation regarding air pollution.

HEALTH

Action to reduce air pollution on health grounds can arise in any of three ways:

1. An air pollution incident may occur, followed by an increase in the rate of death or illness that is clearly attributable to the air pollution incident
2. A particular area may contain an industrial site from which dangerous emissions may be expected
3. The terrain of an area where industry is located may favor weather conditions that produce distributed pollution above acceptable levels

The London smog incident of 1952 is a classic example of an event that was followed by wise action. The pollution level in London was high long before this incident, but it served to precipitate the needed action. On December 4, 1952, atmospheric stagnation under a temperature inversion set in over London and prevented the normal natural ventilation of the city. The condition lasted for 4–5 days, and the air pollution increased to about 20 times the normal level. It resulted in about 4,000 deaths, mostly among the elderly and those suffering from bronchial diseases. The deaths caused a political demand for immediate action strong enough to override the fact that the exact cause of the problem was not known. Coal smoke was the likely candidate, but oxides of sulfur or other pollutants could equally well have caused the problem. It was technologically possible to take rapid action to reduce smoke emissions without causing much economic hardship. Successful action to reduce oxides of sulfur would at that time have been technologically and economically impossible. The decision was made to deal with smoke emissions immediately, and this action was spectacularly successful. London became relatively free of serious problems. The lessons to be learned from this case study are that rapid action may be required without waiting for the technical personnel to cross every "t" and dot every "i," and that such action can be completely successful.

In certain instances, there are emission sources that can be considered potentially poisonous through the use of good judgement alone, an example being a lead works. In these cases, preventive measures must be initiated without waiting for the anticipated cases of ill health to develop in the affected community.

The last type of problem is that of an urban area whose geography makes it subject to meteorological conditions that permit the buildup of pollutants to harmful levels. A town located in a deep valley is a simple example of such a situation. An experienced consultant should be able to assess the likelihood of such a problem, based on the industrial and residential structure of the town, its topography, and the local weather conditions. If a potential problem exists, a survey to collect substantiating data should be carried out in sufficient detail to guide the application of an abatement program and to monitor its efficacy, after which positive abatement measures should be taken. Any overly long survey before an abatement program is likely a delaying tactic, however.

LIMITS TO PROTECT HEALTH

In all communities, there is a wide variation in the susceptibility of individuals to the effects of air pollution. Healthy adults can tolerate relatively high concentrations of harmful substances without suffering harm, while the very old, the very young, and the sick are relatively much more sensitive. To handle this variation, a statistical concept must be used. The aim is to protect the great majority of even the susceptible subgroups from significant harm. This purpose is not as definitive as might be desirable, since it is subject to human judgement, but it provides a workable concept.

Evidence indicates that very rough limits can be defined below which the common pollutants appear to cause no harm. These limits vary for different subgroups of individuals and may also be influenced by climatic and other conditions. However, in spite of the uncertainties, the experts consider that available evidence does provide a basis for action. The Clean Air Act provides two bases for action based on the health implications of pollutants. The first is a set of primary standards whose goal is to protect against short-term health problems for the vast majority of the exposed population. Primary standards protect public health. The other is a set of secondary standards designed to prevent adverse effects during the entire lifetime of the exposed population. Secondary standards protect public welfare. The standards are expressed as a maximum allowable concentration for a given time period, and the secondary standards are usually about one-tenth to one-hundredth of the primary standards.

METEOROLOGICAL CONSIDERATIONS

In any area, the level of pollution to which the inhabitants are exposed represents a balance between the rate of emission of the pollutants from their sources and the rate at which these pollutants are blown away by the winds and dispersed harmlessly into the upper air. Adverse conditions for dispersion exist when winds are calm and vertical dispersion processes fail due to the occurrence of a temperature inversion.

In temperature inversions, the temperature of the higher air is greater than that of the air near ground level. Whether or not vertical mixing occurs in the atmosphere depends on the relation of the atmospheric temperature profile to the adiabatic lapse rate, which is the temperature profile that would develop if the temperature of the atmosphere changed with pressure in accordance with the ideal gas law, $PV = RT$. If the atmospheric temperature decrease is greater than the adiabatic lapse rate, vertical mixing occurs and pollutants are dispersed. If the temperature decrease is less than the adiabatic lapse rate or if the temperature actually increases with altitude, an inversion occurs and no mixing takes place (see Fig. 19-1). Essentially, the atmosphere becomes a closed container. Whenever such conditions persist for a day or two, pollutant concentrations may increase to 10–20 times the normal level. For a town surrounded by hills or located in a valley, such a condition can be even more severe.

The weather forecasts of most national meteorological services can predict and observe such situations and can alert local authorities to take appropriate action; most commonly, major emission sources are closed or required to switch to a cleaner-burning fuel source. Installations with very high chimneys, such as those of most modern power stations, may make no contribution to the local air pollution, since the level of the chimney mouth may be above the level of the

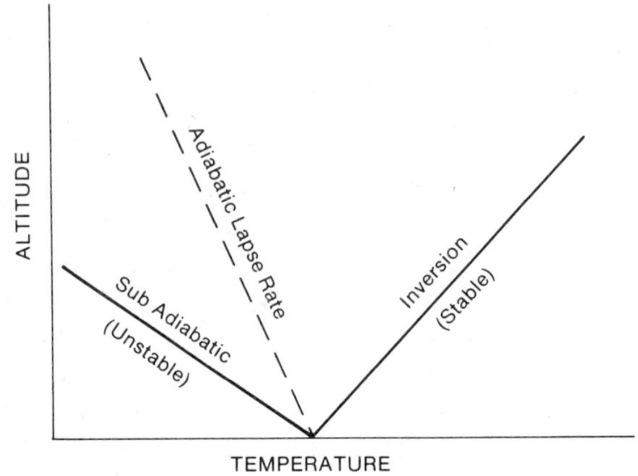

Figure 19-1 Air temperature profiles and their relation to stability. Copyright © 1981. TAPPI. Reprinted from *Introduction to Environmental Management for the Pulp and Paper Industry*, Study Guide, Volume 6, with permission

inversion or the thermal buoyancy of the flue gases may be sufficient to carry them above the inversion level to disperse freely in the upper air.

Shutting down industrial and commercial sources is a serious proposition, and it is advisable to back up the meteorological predictions with reports from ambient air monitoring stations that conclusively show a trend toward a developing incident before such action is taken. Some weather broadcasts have begun giving pollution indexes as part of their normal information. Based on such pollution indexes, a series of shutdown stages can be implemented depending on the severity of the problem. The index information is also useful to respiratory disease sufferers, who can begin to take precautions at pollution levels that are lower than those that have any effect on the majority of the population. A fuller discussion of the air pollution index and how it is determined is given in Chapter 23.

SPECIFIC AIR POLLUTANTS

The six major common air pollutants are particulate matter, carbon monoxide, oxides of sulfur, oxides of nitrogen, photochemical oxidants, and hydrocarbons. National ambient air quality standards for these pollutants will be discussed in Chapter 20. Approximately one-half of the air pollution burden in the United States is associated with emissions from motor vehicles, which is the reason these six pollutants are normally discussed. However, since our concern is with problems specific to the pulp and paper industry, our list of important air pollutants needs to be modified somewhat. Particulate matter and oxides of sulfur are still of concern; in addition, we must consider reduced sulfur compounds and chlorine compounds. Carbon monoxide and oxides of nitrogen are of minor concern, and photochemical oxidants are inconsequential in pulp and paper industry problems. We will now examine the character and potential environmental impact of each of the four principal air pollutants of the pulp and paper industry: particulates, oxides of sulfur, total reduced sulfur compounds, and chlorine compounds.

Particulates

The Kraft pulping process is the major source of particulate emissions, and it emits particles ranging in size from 30 μm for wood char from hogged fuel boilers to less than 1 μm for sodium sulfate, sodium carbonate, and sodium chloride from the recovery furnace.

Material of smaller particle size is of more concern because the health effects of particulates arise from their ability to penetrate into the alveoli of human lungs. Figure 19-2 indicates that the alveoli are located at the ends of the branching air passages and are the organs in which the transfer of oxygen from the inspired air to the bloodstream takes place. Another concern is the ability of phagocytes to take up and transport particulates via the blood into the lymphatic system (2). Entry of particulates into this system is well documented, but the ultimate effect

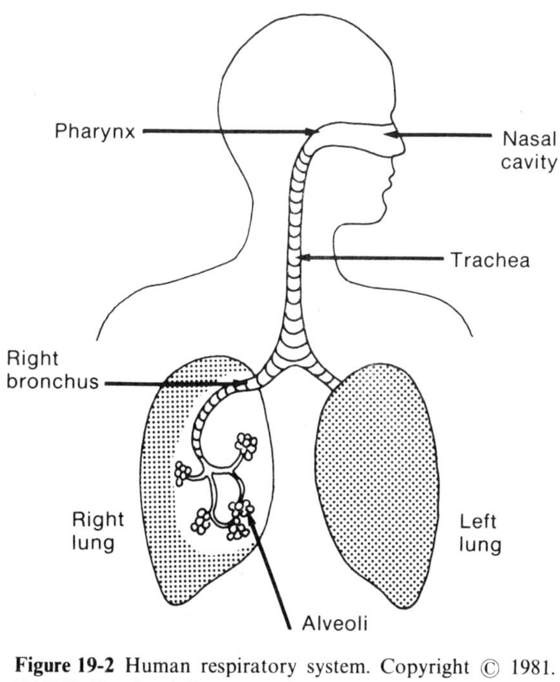

Figure 19-2 Human respiratory system. Copyright © 1981. TAPPI. Reprinted from *Introduction to Environmental Management for the Pulp and Paper Industry*, Study Guide, Volume 6, with permission

upon human health is not known. It has been said that consideration should be given to a size-specific standard for particulates (3).

Particulates can also cause difficulties when they settle out of the air under the influence of gravity. Most settling occurs within 2 miles of the plant site (4). Larger particles can cause soiling and staining, and chemical salts can intensify corrosion problems.

Particles that remain airborne can create an atmospheric haze near the discharge point. Such a haze is often coupled with water vapor transmissions that condense on the particulates to cause a fog-like plume that can persist for several miles.

The lime kiln emits particulates of alkaline calcium oxide and inert calcium carbonate. However, these emissions are at levels that do not raise the concentrations in the vicinity of the kiln to those exceeding the continuous exposure tolerance level of the American Council of Government Industrial Hygienists (ACGIH), which is 5,000 $\mu g/m^3$ for calcium oxide. Studies indicate that pulp mill concentrations are well below this level.

Oxides of Sulfur

The two oxides of sulfur of concern are sulfur dioxide (SO_2) and, to a lesser extent, sulfur trioxide (SO_3). They are emitted primarily from fossil fuel boilers burning high-sulfure fuels and from Kraft pulp mills. Green plants can be severely damaged by continuous exposure to sulfur dioxide concentrations greater than 300 $\mu g/l$. Since the level of SO_2 usually found in Kraft mill stacks is 50–200 $\mu g/l$, some meteorological phenomenon that concentrates the stack emissions must be present for any damage to occur. There is even some evidence that low levels of SO_2 can be beneficial to plants.

Sulfur dioxide is perceived by a sour taste and an irritating odor. The maximum concentration allowed by OSHA is 4 $\mu g/l$, and the maximum time weighted average is 13 $\mu g/l$. SO_2 and SO_3 combine with water in the nasal passages to form sulfurous acid (H_2SO_3) and sulfuric acid (H_2SO_4), which are the cause of the difficulties.

Table 19-1 Odor Threshold Concentrations of TRS Pollutants

Reduced Sulfur Compound		Odor Threshold (ppb)
H_2S	(hydrogen sulfide)	8–20
CH_3-SH	(methyl mercaptan)	2.4
CH_3-S-CH_3	(dimethyl sulfide)	1.2
CH_3-S-S-CH_3	(dimethyl disulfide)	15.5

Source: Copyright © 1981. TAPPI. Reprinted from *Introduction to Environmental Management for the Pulp and Paper Industry*, Study Guide, Volume 6, with permission.

Total Reduced Sulfur Compounds (TRS)

The principle difficulty with TRS emissions is their nauseous odors, which are detectable by the human nose at very low concentrations. Table 19-1 lists the lower detectable limits for hydrogen sulfide, methyl mercaptan, dimethyl sulfide, and dimethyl disulfide, which are the four principle TRS compounds emitted by Kraft mills. These limits were established by organoleptic testing (5).

Chlorine Compounds

Chlorine is used in various forms as a bleaching agent in pulp and paper processes. Bleach plants are the principal source of chlorine compounds, which are extremely corrosive to metals and also hazardous to human health.

REFERENCES

1 S. R. Crarford, "General principles for the abatement and control of air pollution," *Manual on Urban Air Quality Management*, WHO Regional Publications, London, European Series, No. 1, 1979.

2 W. Leithe, *The Analysis of Air Pollutants*, Ann Arbor Science, Ann Arbor, Mich., 1970.

3 C. Cowherd, "Technical basis for a size specific particulate standard," *JAPCA*, 30 (9):971–983 (1980).

4 L. Hastings, R. Freitag, and A. Smith, *Atmosphere and Environmental*, 6 (4):241–246 (1972).

5 M. F. Driesinger and P. C. McGovern, "Monitoring atmospheric sulfur dioxide and correlating its effects on crops and forests in the Sudbury area," *Proceedings of the Conference on the Impact of Air Pollution on Vegetation*, Toronto, Ontario, 1970.

20
Air Pollution Legislation

HISTORY OF AIR REGULATIONS

1881–1955

The only guide we have to what might happen in the future is the examination of similar situations in the past in light of new parameters. During the first century of the existence of the United States, air pollution problems were settled by arbitration among the affected parties. Smoke was the usual complaint, and around 1881 the first legislation was passed declaring emission of smoke to be a public nuisance (1). Gradually people had begun to realize that an attempt should be made to control a problem before it escalates to a more serious level.

The first municipal legislation of this sort was passed in Chicago in 1881, and at about the same time a similar ordinance was enacted in Cincinnati. The first state legislation relating to air pollution was passed in Massachusetts in 1910 to deal with the smoke problem in the Boston area. It took until 1952 for the first comprehensive state regulation to be passed, in Oregon. The first federal air pollution law was passed in 1955. Thus, federal air legislation is a relatively recent development, and the complexity and methodology of federal air legislation are measures of its newness.

All of the early laws were aimed at smoke abatement, and the organization that has evolved into the Air Pollution Control Association began as the Smoke Prevention Association of America.

The most common technique used to assess stack pollution was to monitor visually the opacity of the plume. The Ringelman number chart was developed, each number of which correlated directly with plume opacity. Trained observers were used to determine Ringelman numbers. Many early ordinances prohibited the discharge of stack gases darker than #3 Ringelman, corresponding to 60% opacity. This concern about plume opacity has remained a feature of current regulations, with a gradual shift to lower and lower opacity requirements. Table 20-1 summarizes this trend, reflecting the increasing intolerance of smoke emissions.

As the science of air pollution monitoring developed along with our understanding of health problems related to air pollution, specific pollutants began to be regulated. Smoke became translated into total suspended particulate matter. Such pollutants as oxidants (ozone), carbon monoxide, oxides of

Table 20-1 Percentage of Local Regulations Requiring Visible Stationary Source Emissions to be Less Than the Stated Opacity

Opacity	1940	1950	1960	1965	1975
60	81	69	41	32	3
40	19	31	59	66	33
20	0	0	0	2	56
0	0	0	0	0	8

Source: Reprinted, with permission, from A. C. Stern, "History of air pollution legislation in the United States," *Journal of the Air Pollution Control Association*, Volume 32, number 1, 1982.

nitrogen, hydrogen sulfide, fluorides, beryllium, sulfuric acid, sulfates, lead, lime (CaO), ethylene, and deposited matter were regulated by some state or local agency by 1966 (1).

The federal regulations that evolved mirrored the concerns of the specific time period and promoted the development of state air pollution control agencies to help resolve the problems.

Air pollution problems received national news media attention particularly because of two events—the Donara, Pennsylvania, episode and the Los Angeles smog problem. At Donara between October 27 and 31, 1948, at least 20 persons died and over 6,000 became ill. This news made people aware of their vulnerability to air pollution problems and perhaps willing to allocate money to study them. The City of Los Angeles, local industry, and the State of California spent a considerable amount of money studying the smog problem. This concern prompted T. H. Kuchel, a senator from California, to work toward passage of PL 84-159, the Air Pollution Control Act of 1955, which authorized the federal government to assist state and local air pollution control agencies by providing research, training, and technical assistance.

1955–1963

The basic concept of the 1955 act was extended for 4 years in 1959 by PL 86-365, the Air Pollution Control and Extension act. But undercurrents to bring about significant changes in the legislation were beginning to emerge. In 1958 a Special Subcommittee on Traffic Safety of the House Committee on Interstate and Foreign Commerce held hearings on "unburned hydrocarbons." A bill, HR9368, was introduced to require the Surgeon General to publish standards on the amount of unburned hydrocarbons that are safe for human health. The bill failed to pass the subcommittee because the Public Health Service could not define the concentration of hydrocarbons that should be considered dangerous.

During November 18–20, 1958, the Public Health Service held the first National Conference on Air Pollution. After this conference, Secretary Arthur S.

Flemming recommended that the federal government hold hearings and make recommendations on air pollution problems. The Senate Committee on Public Works and the House Committee on Interstate and Foreign Commerce eventually did so. Much later, in 1977, this Senate committee even changed its name to the Committee on Environment and Public Works (1).

The House Committee used its Subcommittee on Health and Safety to hold hearings on "motor vehicle safety" that included testimony relating to automotive air pollution. After much deliberation and bills that did not pass, these discussions finally led in 1960 to the Motor Vehicle Exhaust Study and to PL 86-493, which required the Secretary of Health, Education, and Welfare HEW to report to Congress in 2 years on "motor vehicles, air pollution, and health," which was done. In 1962 the Air Pollution Control Act was again extended by PL 87-761, but events were underway to promote greater changes that resulted in the Clean Air Act of 1963, PL 88-206.

Four events happened in 1962 that aroused public interest in air pollution and helped prompt congressional action:

1. The book *Silent Spring*, by Rachel Carson, was published (3)
2. There was a smog disaster in London
3. There was an air pollution episode in Birmingham, Alabama
4. The Second National Conference on Air Pollution was held in Washington, D.C.

1963–1969

The Clean Air Act of 1963 was a benchmark piece of legislation, representing the first allocation by the federal government of a significant amount of money for air pollution problems—\$65 million—over a 3-year period (1). The majority of this money was for a grants-in-aid program to develop, establish, and improve state and local air pollution control programs. The law established a conference procedure to resolve interstate air pollution problems. The program of research and technical assistance was not only continued, but time and fiscal limits were removed so that it became a permanently established government activity. The law required semiannual reports to the Congress from the Surgeon General on motor vehicle emissions and established a technical committee to evaluate progress.

These reports on motor vehicle emissions, plus other hearings, resulted in 1965 in PL 89-272, the Motor Vehicle Air Pollution Control Act, which is significant for allowing the Secretary of health, education, and welfare to set emission standards for motor vehicles. Motor vehicles proved to be responsible for about half of the air pollution problem. No deadline was set by the act for setting these emission standards, but they were implemented immediately in the 1965 model year.

The 1963 Clean Air Act was due to expire in 1966, when its life was extended by PL 89-65, the Clean Air Act Amendments. Also in 1966, the Public Health Service created the National Center for Air Pollution Control as a replacement for its

Division of Air Pollution. In December 1966 the Third National Conference on Air Pollution was conducted by the Public Health Service.

Also during 1966, hearings on air pollution were held by the Senate Subcommittee on Air and Water Pollution. President Lyndon B. Johnson strongly supported the development of stronger legislation on air pollution. In an address to Congress titled "Protecting Our National Heritage," he requested legislation incorporating national emission standards for stationary sources, regional air quality commissions to enforce regulations, federally financed and operated regional airsheds, federal assistance to initiate state automobile exhaust inspection, federal regulations on fuel additives, and increased research. All of these features, except national emission standards, were incorporated in PL 90-148, the Air Quality Act of 1967. This law was an attempt to give some national uniformity to the process of regulation of air pollution. The Department of Health, Education, and Welfare (DHEW) was to publish air quality criteria and control technology documents from which the states were to set air quality standards for designated air quality control regions. DHEW could seek federal injunctions in episodic situations. The overall function of this legislation is diagrammed in Figure 20-1.

DHEW was reorganized in 1968, after which the Public Health Service no longer had environmental responsibilities. The National Center for Air Pollution Control was replaced by the National Air Pollution Control Administration, which was a part of the new Consumer Protection and Environmental Health Service.

Further changes were to take place, however. In 1969, President Richard Nixon set up an Advisory Council on Executive Reorganization to determine

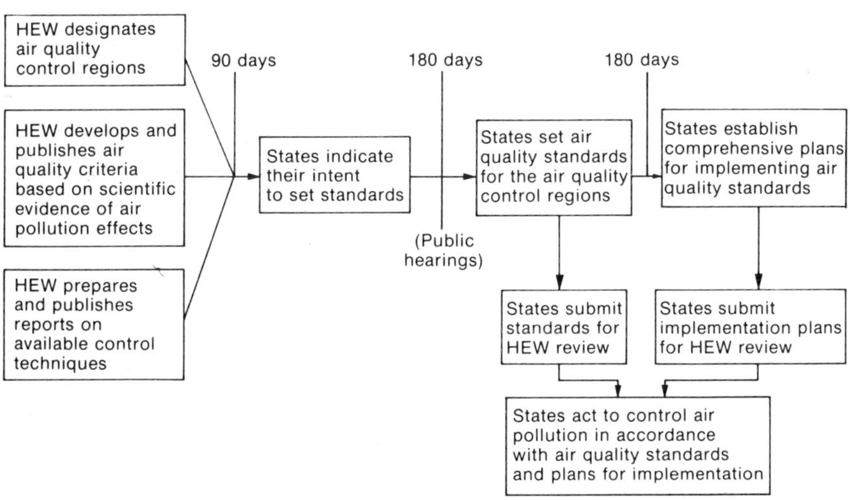

Figure 20-1 Air Quality Act (1967) air pollution abatement procedures. (Reprinted, with permission, from A. C. Stern, "History of air pollution legislation in the United States," *Journal of the Air Pollution Control Association*, volume 32, number 1, 1982.)

whether the federal pollution control program should be reorganized. It recommended the creation of either a new Department of Natural Resources or a special pollution control agency. Nixon decided to form the U.S. Environmental Protection Agency by presidential decree in 1970. The EPA absorbed the National Air Pollution Control Administration as its Air Pollution Control Office. Also, before the creation of the EPA, the President in 1969 set up an advisory committee that evolved into the Council on Environmental Quality.

1970–1977

By 1970 environmental concerns were a dominant issue in American life and the environmental activist organizations held "Earth Day" in Washington, D.C. The political community was not insensitive to environmental concerns and eventually passed PL 91-604, the Clean Air Act Amendments of 1970. This 1970 act covers three basic areas of concern:

1 The attainment and maintenance of national ambient air quality standards
2 The establishment of regulations covering the emission of certain pollutants from mobile and stationary sources
3 The establishment of New Source Performance Standards (NSPS)

The federal government was to establish a set of ambient air quality standards within 6 months and state and interstate agencies were to adopt plans for meeting these standards within 9 months. Each state became an ambient air quality district. Federal standards were to be established based on the technological feasibility of control technology for both new and existing sources, which constituted the beginning of the NSPS program. Regulations on fuel and fuel additives were published. Assembly line testing of automobiles and engines was proposed. Even aircraft engines were to be subject to emissions standards. One clause in the law authorized citizen suits for violation of standards.

The basic scheme proposed was a set of National Ambient Air Quality Standards NAAQS that were to be carefully reviewed and revised for practicality. Then the states were to submit plans for meeting these standards (4), and the plans were to be reviewed by the EPA. The EPA was to issue regulations for the preparation, adoption, and submission of implementation plans.

In a parallel effort, automotive emissions were to be regulated in all new models, and all autos manufactured, rather than only prototypes, were to be tested. The aim was to achieve a 90% reduction in automobile emissions by 1975. The concern with auto emissions was no longer just with hydrocarbons and carbon monoxide, but had expanded to include oxides of nitrogen. After considerable litigation and postponement of effective dates, auto emissions regulation became a fact of life for the automobile industry.

In developing the state implementation plans (SIP), states were to adhere to a criterion of prevention of significant deterioration (PSD) that proved to be very controversial. PSD meant that in areas where current air quality levels were better than the NAAQS, the federal government would not accept an SIP that did not provide, to the maximum extent practical, the continuation of such air quality. This action was to prevent the degradation over time of all air to the level of the NAAQS. Allowing no deterioration anywhere, however, effectively precluded any additional industrial development; therefore some reasonable compromise had to be reached. The way the 1970 law worked is described in the following section.

National Ambient Air Quality Standards. The quantities of various air pollutants emitted must conform to the NAAQS. The EPA established NAAQS for seven pollutants (5): Sulfur dioxide, Total suspended particulates, carbon monoxide, nitrogen oxides, photochemical oxidants, hydrocarbons, and lead. The process involved in setting these standards is diagrammed in Figure 20-2. The NAAQS are to be reviewed every 5 years. The figure shows that the process is complicated enough to take 5 years to complete. Primary standards are health related and are designed to protect the most sensitive groups. Secondary standards are to protect human welfare, including such factors as vegetation, visibility, crops, artificial materials, animals, economic values, and personal comfort. The NAAQS were to be met and maintained by July 1975. The secondary standards were to be attained as soon as possible after that date.

Table 20-2 lists primary and secondary air quality standards by state. A uniform standard was not imposed on all sources of emissions; rather, each state was allowed to apportion the allowable emissions among the existing sources. Each state was given primary responsibility for implementing the act within its boundaries. Each had to prepare an SIP plan for meeting the NAAQS (4) and have the plan approved by the EPA. Most states had developed such a plan by 1972, but many did not meet the 1975 compliance deadline. Therefore a new nationwide deadline of December 31, 1982, for meeting primary NAAQS was set. It was primarily in many urban areas that compliance was not reached by 1975, and many states had to prepare more strict SIPs (6). If such plans were not ready by 1979, federal funding to that area was to be cut off and no new sources of emissions were to be permitted.

New Source Considerations. A different approach was used with new sources, however, for which a procedure termed a "new source review" was instituted. This procedure is basically a permit program that allows a new source to be constructed only if it complies with the Clean Air Act.

Four separate sets of criteria are used in the review process for a new source:

1. It must comply with NAAQS
2. NSPS must be met
3. It must not violate the National Emission Standards for Hazardous Air Pollutants
4. The condition of PSD must be met.

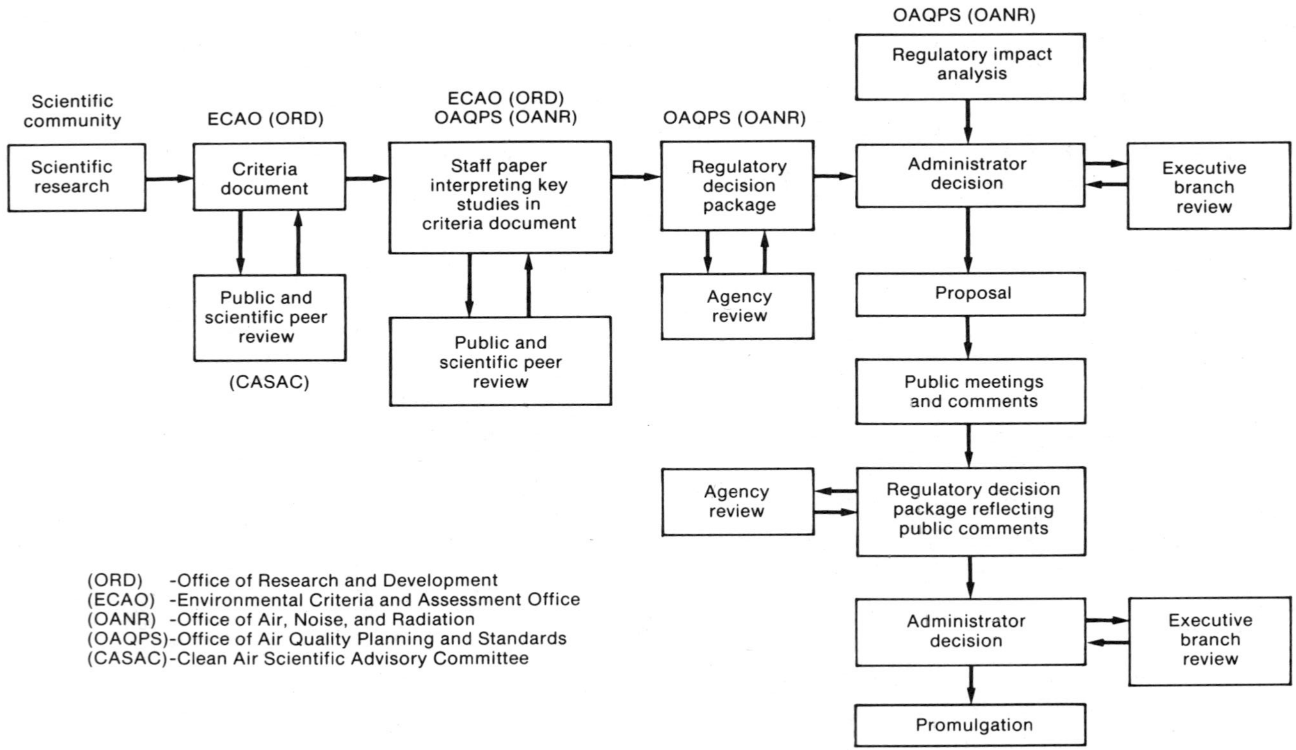

Figure 20-2 NAAQS standards setting process. (Reprinted, with permission, from J. Padgett and H. Richmond, "The process of establishing and revising national ambient air quality standards," *Journal of the Air Pollution Control Association*, volume 33, number 1, 1983.)

Table 20-2 NAAQS By State

State	Sulfur Dioxide		Carbon Monoxide		Nitrogen Dioxide	Photochemical Oxidants ($\mu g/m^3$ max/1 hr)	Nonmethane Hydrocarbons ($\mu g/m^3$ max/3 hr) (6–9 A.M.)	Particulates		Others
	Primary ($\mu g/m^3$)	Secondary ($\mu g/m^3$)	Primary (mg/m^3)	Secondary (mg/m^3)	Annual Arithmetic Mean (AAM) ($\mu g/m^3$)			Primary ($\mu g/m^3$)	Secondary ($\mu g/m^3$)	
Alabama	80—AAM 365—24 hr	1,300—3 hr	10—8 hr 40—1 hr		100	160	160	75 annual geometric mean (AGM) 260—24 hr	60—AGM 150—24 hr	
Alaska	80—AAM 365—24 hr 1,300—3 hr		10—8 hr 40—1 hr		100	160		60—AGM 150—24 hr		Reduced sulfur compounds 50 $\mu g/m^3$—30 min
Arizona	80—AAM 365—24 hr 1,300—3 hr		10—8 hr 40—1 hr		100	160	160	75—AGM 150—24 hr		
Arkansas	80—AAM 365—24 hr	1,300—3 hr	10—8 hr 40—1 hr		100	160	160	75—AGM 260—24 hr	60—AGM 150—24 hr	
California	80—AAM 131—24 hr		10—8 hr 46—1 hr 11—12 hr		100	200	160	60—AGM 100—24 hr	60—AGM 150—24 hr	Hydrogen sulfide 42 $\mu g/m^3$-hr
Colorado	Category I II III AAM— 2 10 15 24-hr max 5 50 100 3-hr max 25 300 700		9 ppm—8 hr 35 ppm—1 hr		0.05 ppm	0.08 ppm	0.24 ppm	Nondesignated areas 150—24 hr Designated areas 180—24 hr		
Connecticut	80—AAM 365—24 hr	60—AAM 260—24 hr	10—8 hr 40—1 hr		100	160	160	75—AGM 260—24 hr	60—AGM 150—24 hr	
Delaware	80—AAM 340—24 hr	60—AAM 260—24 hr 1,300—3 hr	13.8—8 hr 40—1 hr		100	160		70—AGM 200—24 hr 500—1 hr	60—AGM 150—24 hr	Hydrogen sulfide 0.06 ppm—3-min. average 0.03 ppm—1-hr average
Florida	60—AAM 260—24 hr 1,300—3 hr		10—8 hr 40—1 hr		100	160	160	60—AGM 150—24 hr		
Georgia	43—AAM 229—24 hr 715—1 hr		10—8 hr 40—1 hr		100—AAM 300—24 hr	98	98	60—AGM 150—24 hr		

(Contd.)

Table 20-2 *(Contd.)*

State	Sulfur Dioxide		Carbon Monoxide		Nitrogen Dioxide	Photochemical Oxidants ($\mu g/m^3$ max/1 hr)	Nonmethane Hydrocarbons ($\mu g/m^3$ max/3 hr) (6–9 A.M.)	Particulates		Others
	Primary ($\mu g/m^3$)	Secondary ($\mu g/m^3$)	Primary (mg/m^3)	Secondary (mg/m^3)	Annual Arithmetic Mean (AAM) ($\mu g/m^3$)			Primary ($\mu g/m^3$)	Secondary ($\mu g/m^3$)	
Hawaii	20—AAM 80—24 hr		5—8 hr 10—1 hr		70—AAM 150—24 hr	100	100	55—AGM 100—24 hr		
Idaho	80—AAM 365—24 hr	1,300—3 hr	10—8 hr 40—1 hr		100	160	160	75—AGM 260—24 hr	60—AGM 150—24 hr 1,300 hr	Sulfuric acid mist 12 $\mu g/m^3$—24 hr; 30 $\mu g/m^3$—1 hr Beryllium 0.01 $\mu g/m^3$—24 hr
Illinois	0.03 ppm— AAM 0.14 ppm—24 hr	0.50 ppm—3 hr	9 ppm—8 hr 35 ppm—1 hr		0.05 ppm—AAM	0.08 ppm	0.24 ppm	75—AGM 260—24 hr	60—AGM 150—24 hr	
Indiana	80—AAM 365—24 hr	1,300—3 hr	10—8 hr 40—1 hr		100	160	160	75—AGM 260—24 hr	150—24 hr	
Iowa	80—AAM 365—24 hr	1,300—3 hr	9 ppm—8 hr 35 ppm—1 hr		100	160	160	75—AGM 260—24 hr	60—AGM 150—24 hr	
Kansas	None. A standing interpretation by the Kansas attorney general has precluded the adoption of even federal standards on the basis that they would allow deterioration.									
Kentucky	80—AAM 365—24 hr	1,300—3 hr	10—8 hr 40—1 hr		100	160	160	75—AGM 260—24 hr	60—AGM 150—24 hr	
Louisiana	80—AAM 365—24 hr	1,300—3 hr	10—8 hr 40—1 hr		100	160	160	75—AGM 260—24 hr	60—AGM 150—24 hr	
Maine	57 AAM 230—24 hr 1,150—3 hr		10—8 hr 40—1 hr		100	160	160	50—AGM 100—24 hr		
Maryland	80—AAM 365—24 hr	1,300—3 hr	10—8 hr 40—1 hr		100	160	160	75—AGM 260—24 hr	150—24 hr	
Massachusetts	0.03 ppm	0.03 ppm	9 ppm		0.05 ppm	0.08 ppm		75—AGM	60—AGM	
Michigan	80—AAM 365—24 hr	1,300—3 hr	10—8 hr 40—1 hr		100	160	160	75—AGM 260—24 hr	60—AGM 150—24 hr	
Minnesota	0.02 ppm—AAM 0.10 ppm—24 hr 0.25 ppm—3 hr		9 ppm—8 hr 30 ppm—1 hr		0.05 ppm	0.07 ppm	0.24 ppm	75—AGM 260—24 hr	60—AGM 150—24 hr	Hydrogen sulfide 0.05 ppm 1/2 hr, twice per year 0.03 ppm—1/2 hr, twice per year

Mississippi	80—AAM 60—AAM 365—24 hr 260—24 hr 1,300—3 hr	9 ppm—8 hr 35 ppm—1 hr	100	160	160	75—AGM 60—AGM 260—24 hr 150—24 hr	
Missouri	53—AAM 365—24 hr 1,300—3 hr	9 ppm—8 hr 35 ppm—1 hr	100	160		60—AGM 150—24 hr	Sulfuric acid 10 $\mu g/m^3$—24 hr, 30 $\mu g/m^3$—1 hr Hydrogen sulfide 70 $\mu g/m^3$—1/2 hr, twice per year 42 $\mu g/m^3$—1/2 hr, twice every 5 days
Montana	0.02 ppm—AAM 0.10 ppm—24 hr 0.25 ppm—1 hr					75—AGM 200—1% of days per year	Sulfuric acid 4 $\mu g/m^3$—annual average; 12 $\mu g/m^3$—1% of days; 30 $\mu g/m^3$—1 hr, 1% of days; Hydrogen sulfide 0.03 ppm—1/2 hr twice every 5 days 0.05 ppm—1/2 hr twice per year
Nebraska	80—AAM 1,300—3 hr 365—24 hr	9 ppm—8 hr 35 ppm—1 hr	100	160	160	75—AGM 60—AGM 260—24 hr 150—24 hr	
Nevada	60—AAM 260—24 hr 1,300—3 hr	10,000 $\mu g/m^3$—8 hr < 5,000 ft m.s.l. 6,670 $\mu g/m^3$—8 hr > 5,000 ft m.s.l. 40,000 $\mu g/m^3$—1 hr, all elevations	100	160	160	60—AGM 150—24 hr	
New Hampshire	60—AAM	9 ppm—8 hr 35 ppm—1 hr	100	0.08 ppm	160	60—AGM	
New Jersey	80—AAM 60—AAM 365—24 hr 260—24 hr 1,300—3 hr	10—8 hr 40—1 hr	100	160	160	75—AGM 60—AGM 260—24 hr 150—24 hr	
New Mexico	0.02 ppm—AAM 0.10 ppm—24 hr	8.7 ppm—8 hr 13.1 ppm—1 hr	0.05 ppm—AAM 0.10 ppm—24 hr	0.06 ppm	0.19 ppm	60—AGM 150—24 hr	Hydrogen sulfide 0.003 ppm
New York	0.03 ppm—AAM 0.14 ppm—24 hr 0.05 ppm—3 hr	9 ppm—8 hr 35 ppm—1 hr	0.05 ppm	0.08 ppm	0.24 ppm	75—AGM 250—24 hr	
North Carolina	80—AAM	10—8 hr 40—1 hr	100	160	160	60—AGM 150—24 hr	

(*Contd.*)

Table 20-2 (*Contd.*)

State	Sulfur Dioxide		Carbon Monoxide		Nitrogen Dioxide	Photochemical Oxidants ($\mu g/m^3$ max/1 hr)	Nonmethane Hydrocarbons ($\mu g/m^3$ max/3 hr) (6–9 A.M.)	Particulates		Others
	Primary ($\mu g/m^3$)	Secondary ($\mu g/m^3$)	Primary (mg/m^3)	Secondary (mg/m^3)	Annual Arithmetic Mean (AAM) ($\mu g/m^3$)			Primary ($\mu g/m^3$)	Secondary ($\mu g/m^3$)	
North Dakota	60—AAM 260—24 hr 715—1 hr		10—8 hr 40—1 hr		100—AAM 200—1 hr, 1% of 3 month	160 160	160 160	60—AGM 150—24 hr		Hydrogen sulfide 45 $\mu g/m^3$—1/2 hr, twice every 5 days; 75 $\mu g/m^3$—1/2 hr, twice per year Sulfuric acid mist 4 $\mu g/m^3$—AAM; 12 $\mu g/m^3$—24 hr, 1% of time; 30 $\mu g/m^3$—1 hr, 1% of the time
Ohio	60—AAM 260—24 hr		10—8 hr		100	119—1 hr 79—4 hr 40—24 hr	126	60—AGM 150—24 hr		
Oklahoma	80—AAM 365—24 hr	1,300—3 hr	10—8 hr 40—1 hr		100	160	160	75—AGM 260—24 hr	60—AGM 150—24 hr 1,300—3 hr	
Oregon	60—AAM 260—24 hr 1,300—3 hr		10—8 hr 40—1 hr		100	160	160	60—AGM 150—24 hr	15% of samples > 100 $\mu g/m^3$ in 1 month	Lead 3.0 $\mu g/m^3$—1 month Calcium oxide 20 $\mu g/m^3$ as TSP; 0.35 g/month as fallout
Pennsylvania	80—AAM 365—24 hr	1,300—3 hr	10—8 hr 40—1 hr		100	160	160	75—AGM 260—24 hr	60—AGM 150—24 hr 1,300—3 hr	
Rhode Island	80—AAM 365—24 hr	1,300—3 hr	10—8 hr 40—1 hr		100	160	160	75—AGM 260—24 hr	60—AGM 150—24 hr 1,300—3 hr	
South Carolina	80—AAM 365—24 hr	1,300—3 hr	10—8 hr 40—1 hr		100	160	160	75—AGM 250—24 hr	150—24 hr	Gaseous fluorides (HF) ($\mu g/cm^2$) 12 hr avg—3.7 24 hr avg—2.9 1 wk avg—1.6 1 month avg—0.8

South Dakota	BO—AAM 365—24 hr	10—8 hr 40—1 hr	100—AAM 250—24 hr	160	160	60—AGM 150—24 hr	
Tennessee	80—AAM 1,300—3 hr 365—24 hr	10—8 hr 40—1 hr	100	160	160	75—AGM 60—AGM 260—24 hr 150—24 hr	
Texas	80—AAM 1,300—3 hr 365—24 hr	10—8 hr 40—1 hr	100	160	160	75—AGM 60—AGM 260—24 hr 150—24 hr 1,300—3 hr	
Utah	80—AAM 1,300—3 hr 365—24 hr	10—8 hr 40—1 hr	100	160	160	75—AGM 60—AGM 260—24 hr 150—24 hr 1,300—3 hr	
Vermont	0.02 ppm—AAM 0.05 ppm—24 hr 0.10 ppm – 1 hr	8.7 ppm—8 hr 35 ppm—1 hr		0.06 ppm		45—AGM 125—124 hr	
Virginia	80—AAM 1,300—3 hr 365—24 hr	10—8 hr 40—1 hr	100	160	160	75—AGM 60—AGM 260—24 hr 150—24 hr 1,300—3 hr	
Washington	0.02 ppm—AAM 0.10 ppm—24 hr 0.40 ppm—1 hr	10—8 hr 40—1 hr	100	160	160	60—AGM 150—24 hr	
West Virginia	80—AAM 1,300—3 hr 365—24 hr	10—8 hr 40—1 hr	100	160	160	75—AGM 60—AGM 260—24 hr 150—24 hr 1,300—3 hr	
Wisconsin	80—AAM 1,300—3 hr 365—24 hr	10—8 hr 40—1 hr	100	160	160	75—AGM 60—AGM 260—24 hr 150—24 hr 1,300—3 hr	
Wyoming	60—AAM 260—24 hr 1300—3 hr	10—8 hr 40—1 hr	100	160	160	60—AGM 150—24 hr	Hydrogen sulfide 70 μg/m³—1/2 hr, twice per year; 40 μg/m³—1/2 hr, twice every 5 days Fluorides 80 μg/m³ 24 hr
Washington D.C.	75—AAM 57—AAM 285—24 hr 225—24 hr 860—1 hr 715—1 hr	10—8 hr 40—1 hr	100	160	160	75—AGM 60—AGM 160—24 hr 140—24 hr	

Source: Copyright © 1981. TAPPI Reprinted from *Introduction to Environmental Management for the Pulp and Paper Industry*, Study Guide, Volume 6, with permission.

The SIP for meeting NAAQS must include consideration of the new source. If the area is already at its air quality limit, the new source must be equipped with technology allowing it to operate at the lowest achievable emission rate. Arrangements must also be made for other sources to lower their emission rates, perhaps by purchasing additional pollution control equipment for them, before the new source is allowed to operate.

Table 20-3 NSPS for Kraft Pulp Mills

Affected Facility	Pollutant	Emission Level	Monitoring Requirement
Recovery furnace	Particulate	0.044 g/dscf[a] (0.10 g/dscm)[b] corrected to 8% oxygen	No requirement
	Opacity	35%	Continuous
	TRS		
	Straight recovery	5 ppm by volume corrected to 8% oxygen	Continuous
	Cross-recovery	25 ppm by volume corrected to 8% oxygen	
Smelt dissolving tank	Particulate	0.2 lb/ton (0.1 g/kg)	No requirement
	TRS	0.0168 lb/ton (0.0084 g/kg)	No requirement
Lime kiln	Particulate		
	Gaseous fuel	0.067 gr/dscf (0.15 g/dscm) corrected to 8% oxygen	No requirement
	Liquid fuel	0.13 g/dscf (0.30 g/dscm) corrected to 10% oxygen	No requirement
	TRS	8 ppm by volume corrected to 10% oxygen	Continuous
Digester, brown stock washer, evaporator, oxidation, or stripper systems	TRS	5 ppm by volume corrected to 10% oxygen	Continuous Effluent gas incineration temperature; scrubber liquid supply pressure and gas stream pressure loss

[a]dscf is dry standard cubic foot.
[b]dscm is dry standard cubic meter.

The NSPS are emission limitations for specific categories of new stationary sources. Thus a new source must meet these limits by installing appropriate control technology or by the design of the process. Table 20-3 gives the NSPS requirements for Kraft mills. The process of establishing NSPS is discussed in reference 7, and an overall summary of new source regulation is provided in reference 8. There are requirements for continuous monitoring in certain instances.

The emission of hazardous pollutants is regulated under the National Emission Standards for Hazardous Air Pollutants. As of 1984 there were standards for the emission of lead, asbestos, mercury, beryllium, hydrogen sulfide, vinyl chloride, and benzene. If any of these pollutants are likely to be emitted by a new source, a permit must be obtained before construction.

The most restrictive criterion faced by new source is that of PSD. A source is subject to regulation under PSD if it emits any of the pollutants listed in Table 20-4 in amounts greater than those given. PSD has two requirements:

1 That new sources install and operate the best available control technology (BACT) for reducing emissions
2 That all sources locating together in an area not degrade air quality by more than a specified increment

Table 20-4 Pollutant Emission Limits to Qualify for PSD

Pollutant	ton/yr
CO	100
NO_x (as NO_2)	40
SO_2	40
Particulate matter	25
Ozone	40[a]
Lead	0.6
Asbestos	0.007
Beryllium	0.0004
Mercury	0.1
Vinyl chloride	1
Fluorides	3
Sulfuric acid mist	7
Hydrogen sulfide	10
Total reduced sulfur	10
Reduced sulfur compounds	10
Other Clean Air Act pollutants	>0

Source: Reprinted, with permission, from H. M. Ellis and R. Greenway, "Prevention of significant deterioration of air quality," *Journal of the Air Pollution Control Association*, volume 31, number 2, 1981.
[a]Of volatile organic compounds.

The increments of the second requirement vary by the classification of the area, with Class I being the most restrictive and Class III the most liberal (see Table 25-5). Most mills are located in Class II areas. Some difficulties might arise if mills were located near their wood supplies, which in turn were adjacent to park areas.

The area classification system makes PSD a viable concept. Class I areas are those in which practically any air quality deterioration would be considered significant, pristine areas such as national and international parks, national wilderness areas, national memorial parks, national preserves, and national primitive areas.

Class II areas are those of moderate economic growth. The law classified as

Table 20-5 PSD Area Classifications and Allowable Increases for Particulate Matter and Sulfur Dioxide

		NAAQS	
Pollutant	g/m^{3a}	Primary	Secondary
Class I[b]			
Particulate matter			
Annual Geometric Mean (AGM)	5	75	60
24-hour maximum	10	260	150
Sulfur Dioxide			
Annual Arithmetic Mean (AAM)	2	80	—
24-hour maximum	5	365	—
3-hour maximum	25	—	1,300
Class II			
Particulate matter			
AGM	19	75	60
24-hour maximum	37	260	150
Sulfur dioxide			
AAM	20	80	—
24-hour maximum	91	365	—
3-hour maximum	521	—	1,300
Class III			
Particulate matter			
AGM	37	75	60
24-hour maximum	75	260	150
Sulfur dioxide			
AAM	40	80	—
24-hour maximum	182	365	—
3-hour maximum	700	—	1,300

[a]Maximum allowable increase in micrograms per cubic meter (but not to exceed NAAQS),

[b]Initially included most national parks, national wilderness areas, national memorial parks, and international parks.

Class II all the rest of the country not designated Class I and left it to the states to reclassify some areas within their boundaries as Class III.

Class III areas are those of intensive economic growth, major industrial areas in which air quality deterioration would be permitted to rise to some fairly high ceiling. As shown in Table 20-5, the levels are not as high as NAAQS but are considerably more liberal than those of Class II areas.

PSD increments are calculated by the application of an air quality model to the allowable emissions increases from the proposed sources. The allowable emission limit is defined as the most stringent of one of the following:

1 NSPS
2 SIP emissions limits
3 The emission rate specified as a federally enforceable permit condition

The PSD system is retained in the 1977 Clean Air Act Amendments. There are provisions in the PSD system for ambient air compliance monitoring if the anticipated concentrations of pollutants are above those listed in Table 20-6.

1977–1984

The Clean Air Act Amendments of 1970 expired in 1973 and were extended by PL 93-15, the Clean Air Act Extension. This law was replaced by the Clean Air Act Amendments of 1977, PL 95-95, which was still basically in effect in 1984. It is an updated and improved version of the comcepts introduced in the 1970 amendments to the Clean Air Act, with the PSD provisions worked into a viable system. Several recent attempts to modify the law had not succeeded in Congress as of February 1984. The basic structure of the law is sound (9). However,

Table 20-6 Concentration Impacts Above Which Ambient Monitoring is Required

	($\mu g/m^3$)	Average Time
CO	575	8-hour maximum
NO_2	14	24-hour maximum
TSP	10	24-hour maximum
SO_2	13	24-hour maximum
Lead	0.1	24-hour maximum
Mercury	0.25	24-hour maximum
Beryllium	0.0005	24-hour maximum
Fluorides	0.25	24-hour maximum
Vinyl chloride	15	24-hour maximum
Total reduced sulfur	10	1-hour maximum
H_2S	0.04	1-hour maximum
Reduced sulfur compounds	10	1-hour maximum

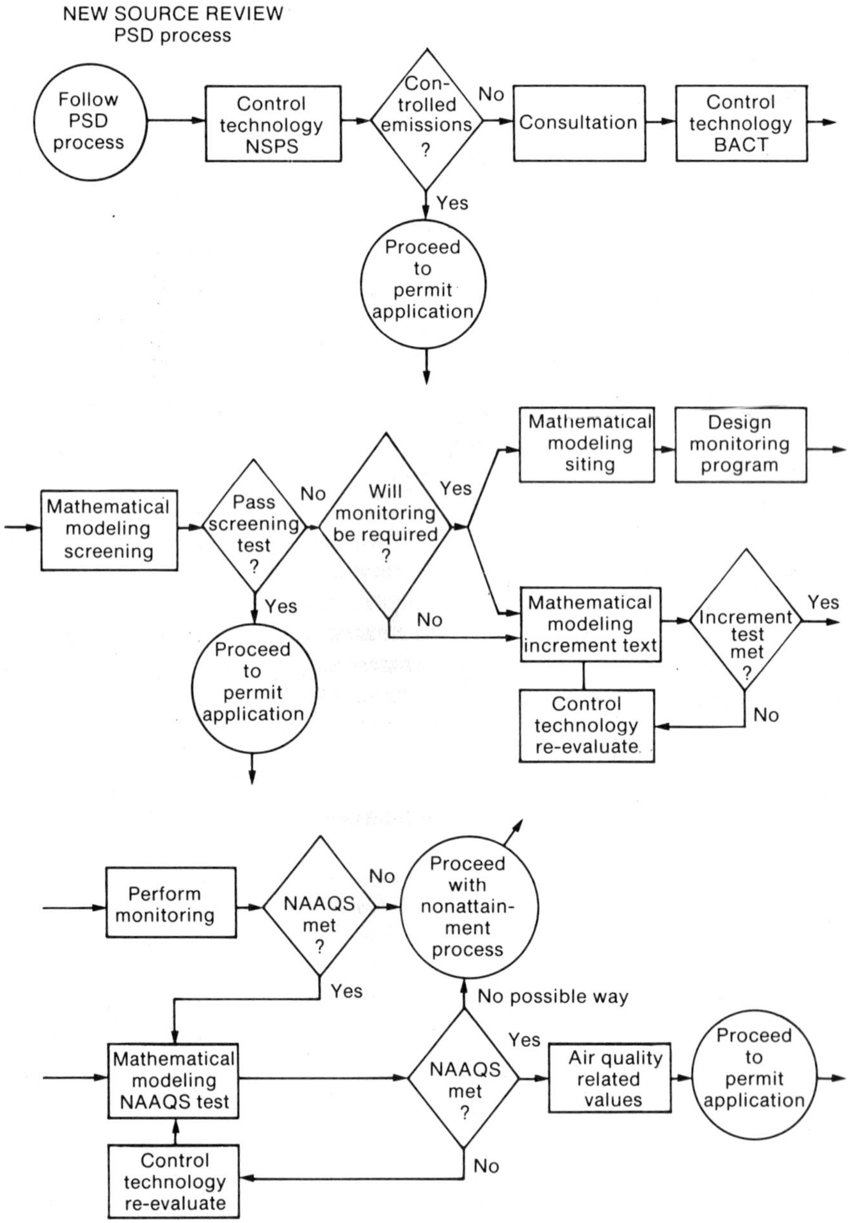

Legend: NSPS - New Source Performance Standards
BACT - Best Available Control Technology
NAAQS - National Ambient Air Quality Standards

Figure 20-3 NSPS procedure. (Reprinted, with permission, from A. C. Stern, "History of air pollution legislation in the United States," *Journal of the Air Pollution Control Association*, volume 32, number 1, 1982.)

complexity could be reduced and the funding of small projects could be made easier.

PSD was explicitly stated in the 1977 amendments, whereas it was only inferred in the 1970 amendments. However, the EPA and the courts have been using the PSD concept since 1970.

The process that must be followed to obtain a permit to construct a new source is diagrammed in Figure 20-3, which shows the roles played by NSPS, BACT, and PSD increments. If the process results in nonattainment, then the process diagrammed in Figure 20-4 must be followed and the Lowest Achievable Emission Rate (LAER) must be implemented. This process reflects the amend-

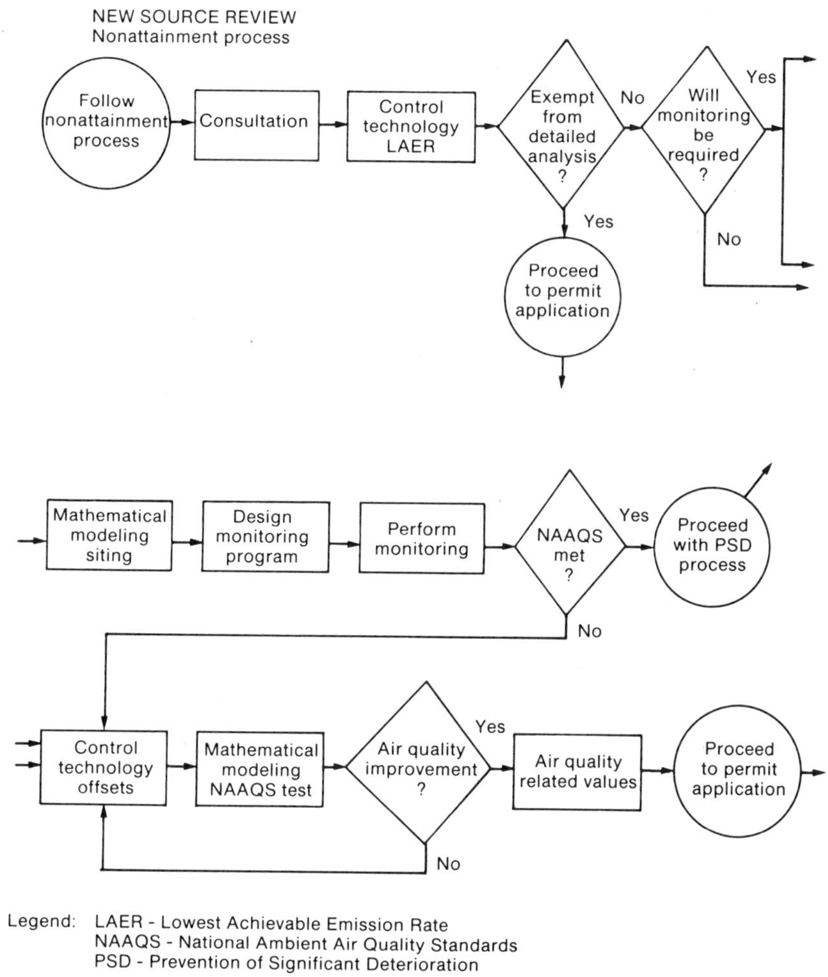

Figure 20-4 Nonattainment process. (Reprinted, with permission, from A. C. Stern, "History of air pollultion legislation in the United States," *Journal of the Air Pollution Control Association*, volume 32, number 1, 1982.)

Table 20-7 Summary of Air Quality Permits in the Forest Products Industry

Case Number	Total Time,[a] Months	Time from Initial Submittal to Final Approval, Months			PSD Permit?	Total Permit Cost, $	Air Quality Modelling st, $	Ambient Monitoring Required?	Cost of Company Monitoring Program, $
		EPA	State	Local					
1	5		3		Yes	5,000		No	
2	6		3		Yes	40,000	10,000	No	
3	6		6		Yes	57,000	*b*	Yes	
4	8	4	3		Yes	83,000	25,000	No	
5	8		6		Yes	10,000		No	
6	9	3	9		Yes	310,000	53,000	Yes	200,000
7	9		7		Yes	24,000	12,000	Yes	
8	10	6	3		Yes	80,000	15,000	No	
9	10	3	3		Yes	75,000	Unknown	No	
10	12	7	5		Yes	35,000		No	
11	12	6		9	Yes	12,000		No	
12	12	7	2		Yes	35,000		No	
13	12	10	4		Yes	75,000	Unknown	Yes	20,000
14	12	9	9		Yes	50,000	30,000	No	
15	12	4	3		Yes	125,000	Unknown	Yes	20,000[c]
16	12	9	6		Yes	40,000	25,000	No	
17	12	8		8	Yes	20,000		No	
18	12		6		Yes	40,000		No	
19	12		4		Yes	20,000		No	
20	13		5		Yes	300,000	6,000	Yes	40,000

21	13	10	1		Yes	12,000	7,000	No	
22	14		14		Yes	65,000	5,000	Yes	30,000
23	16		7		Yes	55,500	35,000	Yes	[d]
24	16	13	5		Yes	190,000	12,000	No[e]	45,000
25	16	12	11		Yes	102,000	60,000	No	
26	17	13	3		Yes	15,000	[b]	No	
27	17	14	14		Yes	70,000	20,000	No	
28	18	12			Yes	500,000	250,000	Yes[f]	50,000
29	18	11	9		Yes	4,000		No	
30	18		11		Yes	320,000	200,000	Yes	65,000
31	18	6		17	Yes	74,000	2,000	Yes	
32	19	9	22	22	Yes	4,500	[b]	No	
33	19		19		Yes	438,000	20,000	Yes	

Source: Reprinted, with permission, from J. E. Pinkerton, "Experience of the forest products industry in obtaining air quality Journal of the Air Pollution Control Association, volume 31, number 11, 1981.

[a] Total time elapsed from the time the company decided to obtain the required air quality permits for the project until all required applications received final approval.

[b] State performed modeling.

[c] Same monitoring program as for Case 13.

[d] The company was required to do postconstruction ambient monitoring.

[e] The final determination by permitting authorities was that no new monitoring was required; however, the company had begun preparations for a new monitoring program.

[f] The monitoring program was initiated to determine the attainment status of the area, but data were later used in the permit application.

ments made to PL 95-95 in August 1980 as a result of a court case that clarified the extent and method for applying the PSD process.

PL 95-95 requires the EPA to apply PSD regulations to both Set I pollutants (particulates and SO_2), which started the process in 1970, and Set II pollutants (hydrocarbons, carbon monoxide, nitrogen oxide, ozone, and lead).

A study of 49 PSD permit applications showed that in only 12% of the cases were NSPS sufficient and therefore the first step in the process (Fig. 20-2) controlled. In 88% of the cases, some other more restrictive aspect of PSD came into play (10).

Much concern has been expressed that the permit process would delay construction. According to this study construction was not delayed, but the permit process took an average of 25 weeks (10).

The experience of the forest products industry with permits for the period July 1, 1976, to April 30, 1981, is summarized in reference 11, which analyzed 65 different permit application processes from a wide variety of sources including Kraft recovery furnaces, lime kilns, fossil fuel boilers, wood residue boilers, solid wood product manufacturing facilities, paper coaters, and printing presses. Unlike the more general experience cited above, 87% of the permits were determined by NSPS. The 13% that were involved in PSD considerations took twice as long to obtain. For a permit involving PSD, the average time was 13 months (range, 5–28 months). Those involving only NSPS averaged 6 months (range, 1–18 months). Air quality modeling was done to provide support information for 35% of the NSPS cases at an average cost of \$8,000. Modeling was conducted in 80% of the PSD cases at a median cost of \$20,000. In general, modeling costs were about 30% of the direct costs of permitting.

In 45% of the cases, the time required to obtain the permit was a major drawback and delayed the start of construction. That the permitting process is a formidable undertaking can be demonstrated by the fact that 16 projects were reported abandoned at the planning stage and that the ultimate locations of 10 other projects were shifted because of perceived difficulties with obtaining air quality permits at the original site.

Table 20-7 summarizes some of the pertinent data on the permitting process. The median total cost for a permit issued under NSPS was about \$30,000 and that for a PSD permit was \$65,000. In 80% of the cases, additional information was requested by the agency before the permit was granted.

Bubble Concept. All decisions on ambient air quality result from either monitoring or modeling. Rather than treating individually each source of emissions in a given plant site, regulatory agencies have allowed a site to be considered under a compliance bubble from which all sources together must achieve a given air quality level. This procedure allows the plant to employ the most cost-effective methods available to achieve the required air quality. It is even possible to voluntarily install equipment that results in air pollutant discharges below those levels permitted under PSD and to "bank" the difference for use at a later time. Permanent source shutdowns or curtailments cannot be banked, however. The EPA has issued an "emissions trading policy statement" (1983), and the State of Oregon (1983) has enacted the first simple generic emissions trading rules that may be adopted by agencies in other states (12).

LEGAL PENALTIES

The Clean Air Act enjoins penalties of up to $25,000 per day and/or imprisonment of up to 1 year for violations. Provision has also been made for eliminating any financial incentives that a company might gain by delaying its purchase of pollution control technology. A penalty equivalent to the cost of installing and operating the necessary equipment for compliance can be levied. Thus, in general, there is no longer any economic justification for not installing pollution control technology, since the cost of the penalties should be the same as the cost of compliance. Delayed compliance would actually cost more due to the resulting increased construction costs.

The above description of PL 95-95 and minor amendments is meant to provide a familiarity with the basic concepts involved in this law rather than an in-depth description of all of its facets. The present law occupies 173 pages and requires a person highly trained in environmental law for its interpretation. The advice of an environmental lawyer should be sought before becoming involved in the permitting system in order to obtain the most advantageous permit.

Congress showed serious concern for compliance with the Clean Air Act when it made federally owned sources subject to the same regulations as private sources. Subjecting these sources to state and local regulation implies suspension of the doctrine of federal supremacy.

It appears that the penalties embodied in the present legislation should have enough force to bring about compliance. The responsibility for implementation lies with state and local agencies, but sufficient tools have been provided to bring about compliance. Ninety percent of all major stationary source air pollution is now in compliance or on a firm schedule leading to compliance (6). For the period 1973–1983, particulate emissions dropped by 20%, carbon monoxide emissions by 40%, and sulfur dioxide emissions by 40% (13).

FUTURE CHANGES NEEDED

Industry is always in favor the liberalizing air pollution regulations, but this is unlikely to happen given the spirit of environmental ferver that still exists in the United States. A fairly objective view of some of the problems with the Clean Air Act as it existed in 1981 is given in a position statement by the Air Pollution Control Association (14). This statement cites the following problems:

1. PSD is worthwhile in concept, but its implementation has become an overly complex regulatory process
2. The Clean Air Act does not provide an effective mechanism and incentives

for the control of interstate and international transportation of air pollutants

3 The use of NAAQS is strongly supported, but insufficient health and welfare effects data and the lack of a clear "threshold" effects level often require changes in approach in setting them

4 The present mechanism (Sec. 112) in the Clean Air Act for addressing the control of hazardous air pollutants is unrealistically restrictive and hence indirectly leads to lack of control

5 The revision process for SIPs is complicated and time-consuming. SIPs have established unnecessary administrative redundancy at the state and federal levels. They provide the means to consider variations in local problems and options available to achieve attainment, but these are difficult to implement

The basic Clean Air Act legislation is sound and has public support. Amendments to correct some or all of the five problems stated will in all probability eventually be adopted.

REFERENCES

1 A. C. Stern, "History of air pollution legislation in the United States," *Journal of the Air Pollution Control Association*, 32 (1):49–61 (1982).

2 A. C. Stern, ed., *Air Pollution*, 3d ed., Vol. V, Academic Press, New York, 1977.

3 R. Carson, *Silent Spring*, Houghton Mifflin, Boston, 1962.

4 *Guidelines for the Development of Air Quality Standards and Implementation Plans*, National Air Pollution Control Administration, Washington, D.C., May 1969.

5 J. Padgett and H. Richmond, "The process of establishing and revising National Ambient Air Quality Standards," *Journal of the Air Pollution Control Association*, 33 (1):13–16 (1983).

6 D. M. Castle, "Reasonable goals for SIP revisions," *Journal of the Air Pollution Control Association*, 29 (1):6–7 (1979).

7 D. Pahl, "EPA program for establishing standards of performance for new stationary sources of air pollution," *Journal of the Air Pollution Control Association*, 33 (5):468–482 (1983).

8 S. K. Tabler, "Federal standards of performance for new stationary sources of air pollution," *Journal of the Air Pollution Control Association*, 29 (8):803–811 (1979).

9 R. S. Stafford, "The exercise of reason," *Journal of the Air Pollution Control Association*, 30 (5):449–453 (1981).

10 R. F. Weiner and D. A. Jaffe, "A study of the PSD permitting process in EPA Region X," *Journal of the Air Pollution Control Association*, 32 (8):797–801 (1983).

11 J. E. Pinkerton, "Experience of the forest products industry in obtaining air quality permits for new and modified sources," *Journal of the Air Pollution Control Association*, 31 (11):1163–1168 (1981).

12 L. Kastow and J. Kawolezyk, "A practical emission trading program," *Journal of the Air Pollution Control Association*, 33 (10):982–984 (1983).

13 K. M. Bennett, "CAA—the challenge for the 80s," *Journal of the Air Pollution Control Association*, 32 (7):691–692 (1982).

14 "APCA issues position statement on CAA amendments," *Tappi*, 64 (10):21–22 (1981).

21
Characteristics of the Air Pollution Problem in the Pulp and Paper Industry

EMISSIONS FROM KRAFT PULPING

At present (1984) the pulping industry in the United States consists of approximately 400 separate chemical and mechanical pulp mills. The Kraft process dominates the pulping industry, and most large plants are Kraft mills. In the Kraft pulping process, the cooking liquor consists of a mixture of sodium hydroxide and sodium sulfide. Due to the presence of sulfur, various sulfide compounds are produced: hydrogen sulfide (H_2S), methyl mercaptan (CH_3SH), dimethyl sulfide $(CH_3)_2S$, and dimethyl disulfide $(CH_3)_2S_2$. In addition, oxides of sulfur (SO_x) and oxides of nitrogen (NO_x) are produced in certain portions of the process. As well as these gaseous emissions, some particulates are emitted, primarily sodium sulfate (Na_2SO_4) and sodium carbonate (Na_2CO_3). Particulates are emitted mainly from the recovery furnace, the lime kiln, the smelt dissolving tank, and power boilers.

Reduced Sulfur Compounds (TRS)

The reduced sulfur compounds, commonly referred to as "total reduced sulfur (TRS)" emissions, are odorous at concentrations of only a few parts per billion; thus elimination or minimization of odor is the major air problem faced by Kraft mills. Many studies have been conducted to understand the mechanism of odor production; Sableski (1) and Hendrickson et al. (2) have reviewed progress in this area.

The major sources of TRS emissions include digester blow and relief gases, vacuum washer hood and seal tank vents, multiple-effect evaporation hot well vents, recovery furnace flue gases, smelt-dissolving tanks, slaker vents, black liquor oxidation tank vents, lime kiln exit vents, and wastewater treatment operations. A summary of representative ranges of TRS emission rates and concentrations by source is given in Tables 21-1 through 21-3 (3, 4). According to Table 21-1, the source of the largest volume of emissions is the recovery furnace, followed closely by the digester blow gases and the washer hood vents. The

Table 21-1 Typical Gas Characteristics of Kraft Pulp Mill Processes

	Process Offgas Characteristic		
Emission Source	Flow Rate[a] m^3/ton (ft^3/ton)	Temperature °C (°F)	Moisture Content, %
Digester, batch			
Blow gases	3–6,000 (96–190,000)	65–100 (150–210)	30–99
Relief gases	0.3–100 (10–3,200)	25–60 (80–140)	3–20
Digester, continuous	0.6–6 (20–200)	75–150 (170–300)	35–70
Washer hood vent	1,500–6,000 (48,000–190,000)	20–45 (70–110)	2–10
Washer seal tank	300–1,000 (9,600–32,000)	55–75 (130–170)	15–35
Evaporator hotwell	0.3–12 (10–400)	80–145 (180–290)	50–90
Black liquor oxidation tower exhaust	500–1,500 (16,000–48,000)	70–80 (160–180)	30–40
Recovery furnace	6,000–12,000 (190,000–380,000)	120–180 (250–360)	25–35
Smelt-dissolving tank	500–1,000 (16,000–32,000)	70–110 (160–230)	35–45
Lime kiln exhaust	1,000–1,600 (32,000–51,000)	65–95 (150–200)	25–35
Lime slaker vent	12–30 (400–1,000)	65–75 (150–170)	20–25

Source: Reference 3.

[a]At standard conditions of dry gas [21.1°C and 760 mmHg (70°F and 29.92 in Hg)]. Flow in cubic meters per tonne and cubic feet per ton.

most concentrated emissions seem to come from the digester blow and relief gases, as can be seen in Table 21-2. Table 21-3 combines information on volume and concentration, reducing it to a common basis of emissions per ton of air-dried pulp. It is apparent that the recovery furnace is the major contributor of TRS emissions. Its large volume of emissions swamps their low concentration to make it the dominant source, but the other sources listed are not insignificant in terms of the total problem.

Oxides of Sulfur and Nitrogen

Oxides of both sulfur and nitrogen can be emitted in varying quantities from specific points in the Kraft system. The dominant source of sulfur dioxide

Table 21-2 Typical Reduced Sulfur Gas Concentrations from Kraft Pulp Mill Sources

	Concentration (ppm by Volume)			
Emission Source	H_2S	CH_3SH	CH_3SCH_3	CH_3SSCH_3
Digester, batch				
Blow gases	0–1,000	0–10,000	100–45,000	10–10,000
Relief gases	0–2,000	10–5,000	100–60,000	100–60,000
Digester, continuous	10–300	500–10,000	1,500–7,500	500–3,000
Washer hood vent	0–5	0–5	0–15	0–3
Washer seal tank	0–2	10–50	10–700	1–150
Evaporator hotwell	600–9,000	300–3,000	500–5,000	500–6,000
Black liquor oxidation tower exhaust	0–10	0–25	10–500	2–95
Recovery furnace (after direct-contact evaporation)	0–1,500	0–200	0–100	2–95
Smelt-dissolving tank	0–75	0–2	0–4	0–3
Lime kiln exhaust	0–250	0–100	0–50	0–20
Lime slaker vent	0–20	0–1	0–1	0–1

Source: Reference 3.

Table 21-3 Typical Reduced Sulfur Gas Emission Rates from Kraft Pulp Mill Sources

	Emission Rate, kg Sulfur per tonne of Air-Dried Pulp			
Emission Source	H_2S	CH_3SH	CH_3SCH_3	CH_3SSCH_3
Digester, batch				
Blow gases	0–0.15	0–1.3	0.05–3.3	0.05–2.0
Relief gases	0–0.05	0–0.3	0.05–0.8	0.05–1.0
Digester, continuous	0–0.1	0.5–1.0	0.05–0.5	0.05–0.4
Washer hood vent	0–0.01	0.05–1.0	0.05–0.5	0.05–0.4
Washer seal tank	0–0.01	0–0.01	0–0.05	0–0.03
Evaporator hotwell	0.05–1.5	0.05–0.8	0.05–1.0	0.05–1.0
Black liquor oxidation tower exhaust	0–0.01	0–0.1	0–0.4	0–0.3
Recovery furnace (after direct-contact evaporation)	0–25	0–2	0–1	0–0.3
Recovery furnace (without direct-contact evaporation)	0–1	0–0.01	0–0.01	0–0.01
Smelt-dissolving tank	0–1	0–0.8	0–0.05	0–0.3
Lime kiln exhaust	0–0.5	0–0.2	0–0.1	0–0.05
Lime slaker vent	0–0.01	0–0.01	0–0.01	0–0.01

Source: Reference 3, 4.

Table 21-4 Typical Emission Concentrations and Rates of SO_x and NO_x from Kraft Pulp Mill Combustion Sources

	Concentration (ppm by Volume)			Emission Rate, kg/t*[a]		
Emission Source	SO_2	SO_3	NO_x (as NO_2)	SO_2	SO_3	NO_x (as NO_2)
Recovery furnace						
No auxiliary fuel	0–1,200	0–100	10–70	0–40	0–4	0.7–5
Auxiliary fuel added	0–1,500	0–150	50–400	0–50	0–6	1.2–10
Lime kiln exhaust	0–200	—	100–260	0–1.4	—	10–25
Smelt-dissolving tank	0–100	—	—	0–0.2	—	—
Power boiler	—	—	161–232	—	—	5–10[b]

Source: References 3, 5.
[a]Kilograms per tonne of air-dried pulp.
[b]Kilograms of NO_x per ton of oil.

emissions is the recovery furnace due to the presence of sulfur in the spent liquor used as fuel. Sulfur trioxide is sometimes emitted when fuel oil is used as an auxiliary fuel. The lime kiln and smelt-dissolving tank also emit some sulfur dioxide. The emission of oxides of nitrogen is more general because nitric acid is formed whenever oxygen and nitrogen, which are both present in air, are exposed to high temperatures. A small part of the nitric oxide formed may further oxidize to nitrogen dioxide. The two compounds, nitric oxide and nitrogen dioxide, are termed "total oxides of nitrogen." Under normal operating conditions, the temperature in the recovery furnace is not high enough to form large quantities of oxides of nitrogen. The main source of NO_x emissions is the lime kiln. Table 21-4 (3, 5) summarizes SO_x and NO_x emission rates from various Kraft mill sources. Large differences in the emission rates are due to the variations in operating conditions at different mills.

The amount of nitrogen oxides formed depends on several factors, the two most important of which are the flame temperature and the oxygen concentration in the flame. Large amounts of NO_x are produced if the flame temperature is above 1,300°C and the oxygen concentration greater than 2%. The major compound resulting from the combustion is nitrogen monoxide; a small quantity (usually less than 5%) of nitrogen dioxide is also formed.

Nonsulfur Organic Compounds

Another type of odorous emission is produced by the hydrocarbons associated with the extractive components of wood, such as terpenes and fatty and rosin acids, as well as those from materials used in processing and converting operations, such as defoamers, pitch control agents, bleach plant chemicals, and so on. These hydrocarbon emissions are small compared to TRS emissions, but they may be odorous, or act as liquid aerosol carriers contaminated with TRS, or undergo photochemical reactions.

Some organic compounds that are not odorous are also emitted by the Kraft process. These compounds include some terpenes, hydrocarbons, alcohols, and other miscellaneous materials released from the wood upon pulping. The significance and emission rates of these compounds are not well characterized. Terpenes and hydrocarbons are photochemical reactants.

The flue gas from the lime kiln and the recovery furnace contain minor amounts of polyaromatic hydrocarbons (PAH), compounds containing several aromatic rings. Some members of this group of compounds are known to be carcinogenic (6, 7).

Particulates

The last type of emission to be introduced is particulates, which are usually classified as primary or secondary. Primary particulates are those produced as a direct result of chemical or physical processes at the emission source. Secondary particulates result from chemical reactions in gaseous emissions that produce liquid or solid materials. Secondary particulates usually range in size from 0.01 to 1.0 μm and have atmospheric residence times of 7–40 days. They are smaller than primary particulates and stay in the atmosphere longer.

Primary particulate emissions are the result of combustion processes; the sources are the recovery boiler, lime kiln, power boiler, and combination boiler burning bark and fossil fuel, of which the recovery boiler is the major source. There is also minor emission from the smelt-dissolving tank, a noncombustion source. The recovery boiler emits mainly sodium sulfate (Na_2SO_4), along with minor quantities of sodium carbonate (Na_2CO_3). If the wood has been soaked

Table 21-5 Typical Concentrations and Emission Rates of Particulate Matter from Kraft Pulp Mill Sources (after Installation of Control Devices)

	Concentration		Emission Rate	
Emission Source	g/scm	(gr/scf)	kg/t	(lb/ton)
Recovery furnace				
After electrostatic precipitation	0.06–1.1	(0.03–0.5)	0.5–12	(1.0–24)
After venturi evaporation	0.9–2.3	(0.4–1.0)	7–25	(14–50)
Lime kiln	0.07–1.1	(0.03–0.5)	0.15–2.5	(0.3–5.0)
Smelt dissolving tank	0.04–2.3	(0.02–1.0)	0.01–0.5	(0.02–1.0)

Source: Reference 3.

in sea water, sodium chloride (NaCl) may also be emitted. The lime kiln and smelt-dissolving tank emissions contain particulates consisting of the calcium and sodium salts of carbonate, hydroxide, sulfate, and chloride.

Table 21-5 gives typical emission rates and concentrations of particulate matter from various sources equipped with appropriate control devices. The particle size distribution for a Kraft mill plume sampled close to its source follows a log normal frequency distribution, with about 50% of the particles being less than 0.4 μm in diameter (8). At long distances downwind from the source, the distribution is no longer log normal but has a smaller average particle size of about 0.15–0.2 μm. Literature evidence (9, 10) indicates that particles greater than 0.08 μm can act as cloud condensation nuclei that may increase rainfall from warm stratus clouds.

Recovery Boiler

The recovery boiler is often considered the heart of the Kraft system, since it is the most expensive piece of equipment in the pulp mill. It is also a major source of air pollution. It serves many functions in the plant: recovery of chemicals, generation of steam, and control of malodorous gases, SO_2, and particulate matter. There are a number of alternatives in the design and operation of recovery boilers, a detailed discussion of which is beyond the scope of this book. A good discussion is contained in the EPA Technology Transfer Series volume *Environmental Pollution Control Pulp and Paper Industry*, Part *I, Air* (3) (Chapter 10).

Very generally, there are basically two types of recovery boilers, those that contain a direct-contact evaporator (which, for the sake of this discussion, will be called the "American approach") and those that do not (the "Scandinavian approach"). The two systems evolved separately due to the differential in fuel costs between the two areas. The American approach is less capital intensive but has higher operating costs than the Scandinavian approach.

The Scandinavian approach produces superior emission characteristics, because a large proportion of the malodorous emissions from the recovery boiler are produced in the direct-contact evaporator. It is possible, however, to achieve satisfactory emission characteristics with the American approach if black liquor oxidation is performed.

In order for black liquor to burn, it must be concentrated to a minimum of 55% solids; usually it is taken to 60–65% dry solids before it is injected into the recovery boiler. Most older recovery boilers in the United States and Canada use a multiple-effect evaporation plant to concentrate the brown stock wash liquor to about 50%. Then one of three alternative types of direct-contact evaporator is used to concentrate the liquor to 65% solids. The three types used are cascade evaporators, cyclone evaporators, and venturi scrubbers. The manufacturer of the boiler usually chooses the evaporator used. A direct contact evaporator of the cascade type is shown in Figure 21-1.

Direct-contact evaporation is responsible for most of the malodorous com-

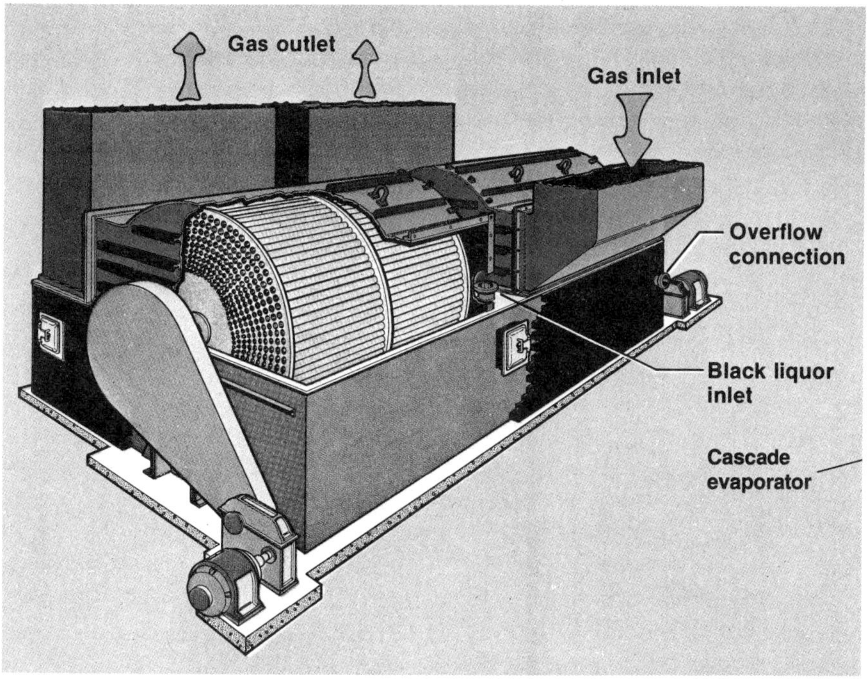

Figure 21-1 Cascade evaporation. (Courtesy C-E Power system.)

pounds produced, the major ones being CH_3SH, CH_3SCH_3, and CH_3SSCH_3. These compounds are absent in a modern recovery boiler of the Scandinavian type. Figure 21-2 presents a schematic of such a boiler. This more modern design has emission problems only with H_2S, SO_2, and particulates. NO_x emissions from a recovery boiler are minimal due to the low temperature of combustion.

The flue gas contains an appreciable quantity of particulate matter, about 40–75 kg per tonne of dry solids fired. This dust is mainly sodium sulfate. Sodium is present as Na_2CO_3 a short distance above the char bed in the furnace. The sodium carbonate reacts with the SO_2 present to form Na_2SO_3, which is then oxidized to Na_2SO_4 by the excess oxygen in the flue gas. The problem of particulate emissions is usually handled by the use of an electrostatic precipitator.

The H_2O and SO_2 emissions are more difficult to control. Their quantity is affected by the operating variables of the furnace, the most significant being the solids firing rate. The conditions in a recovery boiler are such that the furnace can be greatly overloaded and still function, but this overloading has an adverse effect on emissions characteristics, especially on the quantity of H_2S produced. In the past, some U.S. mills commonly operated their recovery furnaces highly overloaded, but emission regulations have pretty well eliminated this practice. For black liquor in the United States, the dry solids load per unit of furnace cross section should be about 14,700 kg/m^2/day (3,000 lb/ft^2/day). Reducing the load further does not improve emissions, so there is a limit to progress by this

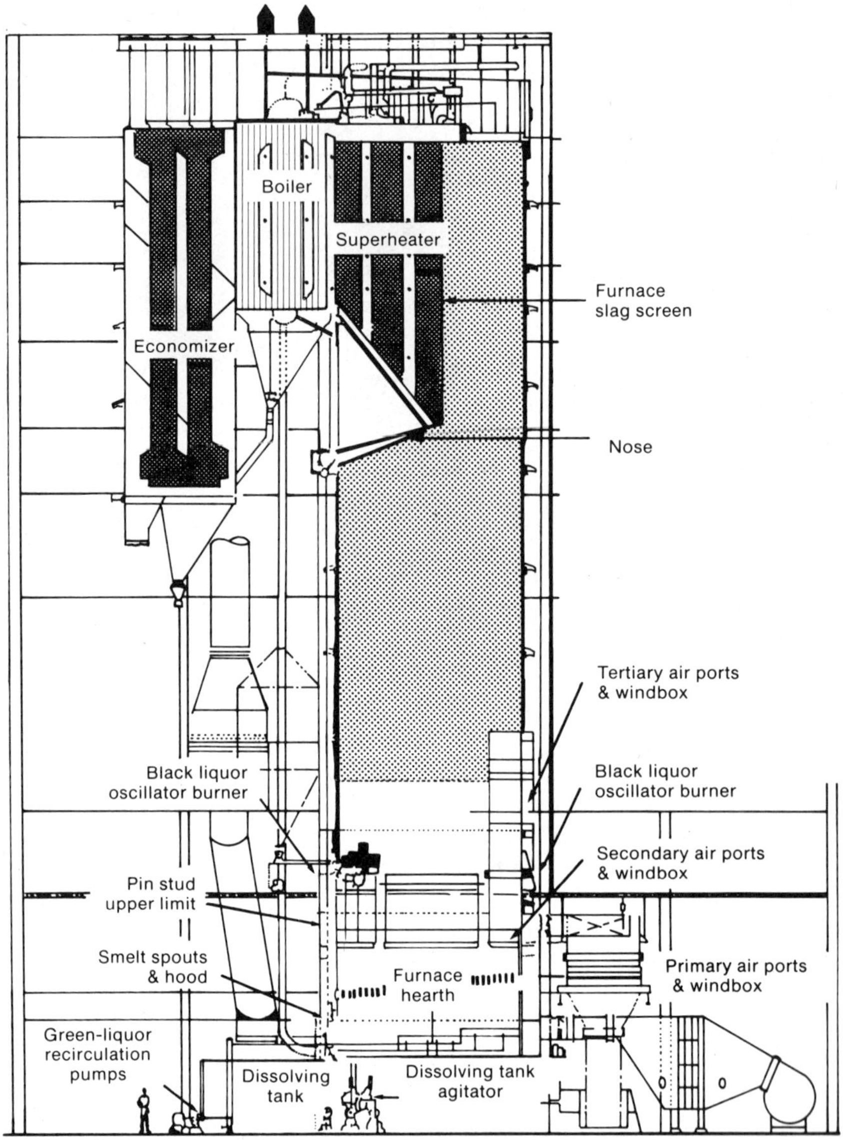

Figure 21-2 Low-odor recovery boiler

route. This would also be a very costly approach, since it would require a larger furnace.

Limited improvement in emission characteristics can be made by the manipulation of air distribution and air temperature. The amount of air required is limited to 10–25% excess. At least 10% excess is required to avoid unburned matter in the flue gas and the formation of H_2S. When more than 25% excess is present, SO_3 is formed in the flue gas. The adsorption of SO_3 by particulates makes them sticky, which must be avoided. However, within these limits, the more excess air present, the better the emission characteristics. Another desirable alternative is to increase the primary air temperature, which gives increased velocity for the same mass of air; this increased velocity, in turn, improves the sodium evaporation from the bed and decreases the release of sulfur. More SO_2 is released and less H_2S. Increasing the primary air temperature decreases the quantity of primary air and increases that of secondary air, which also improves the emission characteristics.

It is also possible to make improvements in emissions by optimizing the distribution pattern of the air in the furnace, but this action is controlled by the design of the furnace itself.

Evaporation

Black liquor evaporation is one of the major sources of gaseous odorous emissions. Two different approaches to evaporation are commonly used, multiple-effect evaporation followed by direct-contact evaporation or direct use of multiple-effect evaporation. The first approach, direct-contact evaporation, produces more emissions of malodorous compounds unless evaporation is preceded by black liquor oxidation.

The dominant odorous compounds in evaporation gases are H_2S and CH_3SH. Scrubbing with white liquor is effective with these compounds as an odor abatement technique, as is incineration. The quantity and composition of evaporator gases depend on the sulfide concentration and pH of the black liquor (11). The type of wood being pulped is also important, with hardwoods usually producing a larger quantity of odorous emissions than softwoods (12). The type of evaporation plant, its age and condition, also affect the quantity of emissions. The variables that affect the quantity of H_2S emitted are pH, temperature, and, most importantly, sulfide concentration. There is an approximately linear relationship between the concentration of Na_2S in the liquor being evaporated and the amount of H_2S in the vapor.

The most common evaporation system is the multiple-effect vacuum evaporation system. There are a wide variety of such systems, which can contain three to seven stages, and falling-film, rising-film, or forced-circulation evaporation can be used. Figure 21-3 illustrates a fairly representative system. Possible odor release points in the system are circled. Circled point 9 represents a major odor source, the output from the condensate gas treatment system from the hotwell. The vapor phase noncondensables are emitted at this point. The noncondensables must be vented from the evaporator system for efficient heat transfer to be maintained, since the vapor from one stage serves to heat the liquor in the next stage.

Venting can be done in a single stage or in two stages; the latter is more efficient. The vapor from the last stage and the vent gases feed to a condenser and then to a vacuum device. Both surface condensers and direct-contact

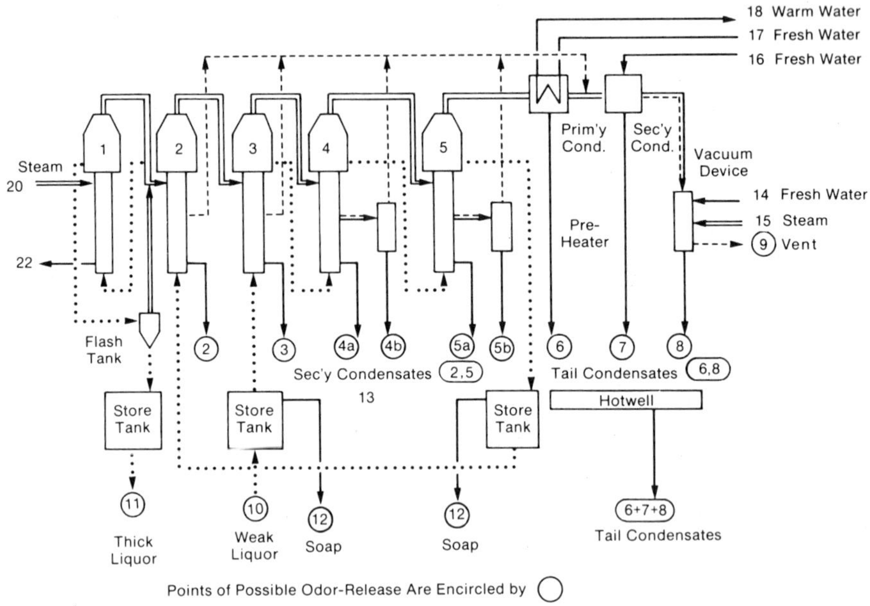

Figure 21-3 Multiple-effect evaporator system

condensers are used. The direct-contact condenser is small and inexpensive. It operates by bringing a large volume of water in contact with the vapor to produce direct condensation. Environmentally, this type is not very desirable because it produces a large volume of contaminated water. The surface condenser, however, produces as condensate only the vapor that condenses. A surface condenser is nothing more than a conventional heat exchange condenser producing condensed vapor on one side and hot water on the other. With this type of condenser, more control over the amount of condensation and subcooling can be exercised. One option of merit is to split the condenser into two parts in which the first condenses 85% of the vapor and the second 15%. This procedure yields two streams of condensate. The first stream, containing 85% of the flow, is a relatively clean stream that is suitable for reuse in the mill. The second stream, 15% of the flow, is highly contaminated and can be sent to treatment. It is more likely that cost-effective treatment can be applied to this small stream rather than to the whole flow. It is certainly more advantageous than having to treat the large volume resulting from direct-contact condensation.

The condensers are followed by a vacuum device that is usually a steam jet ejector or, in some cases, a water ring vacuum pump. Both types allow noncondensable gases to contact fresh cooling water and shift part of the odorous material from the hotwell vent gases to the hotwell condensates.

If a comparison is made only between the quantity of reduced sulfur emissions from evaporation plants containing surface condensers and those form plants with direct-contact condensers, a higher rate of gaseous emissions from surface-condenser plants is found. This is because in direct-contact condensation the odorous gases are absorbed into the condensation water. It would be incorrect, however, to conclude that direct-contact condensation is more desirable environmentally; on the contrary, it creates a more difficult to manage water pollution problem. On balance, most would conclude that a two-stage surface condenser system is more desirable environmentally.

Digester Emissions

The composition and quantity of emissions from digesters differ significantly between batch and continuous digesters. Batch digesters produce more emissions than continuous digesters because of the intermittent nature of their operation. During the blowing of a batch digester, large quantities of vapor are released over a very short time period. Even at their worst, however, digester emissions rank only third as an emissions source in the Kraft process after black liquor combustion and weak black liquor evaporation. The basic approach for minimizing odor in the digester area is to contain the relief and blow gases. These gases consist of turpentine and other organic compounds from the wood and of organic sulfur compounds produced during cooking.

The main sources of odor from a batch digester operation are shown in Figure 21-4. Turpentine forms a significant portion of the odor load. A majority of the turpentine is recovered in the turpentine condensers, but the rest escapes with

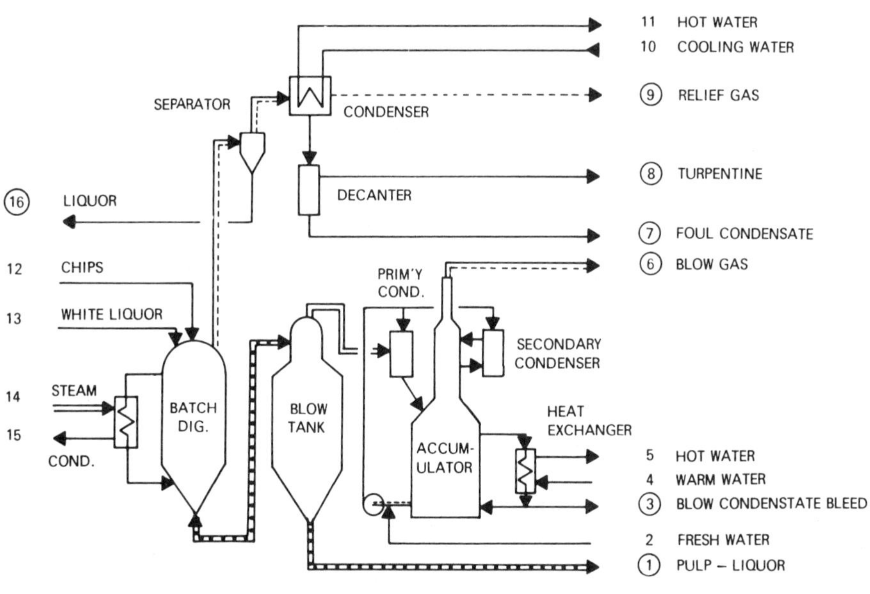

Figure 21-4 Batch digester emissions

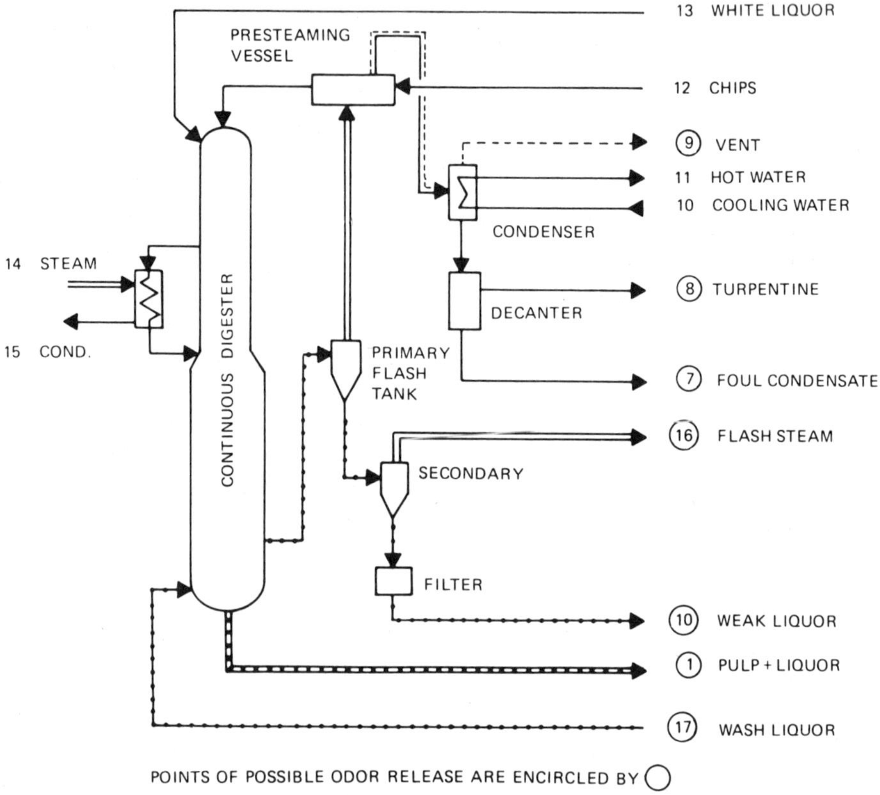

Figure 21-5 Continuous digester emissions

the volatiles and condensed water. In a batch cooking operation it is easy to recover turpentine, but there are large fluctuations in flows.

The sources of odor from a continuous digester are shown in Figure 21-5. In this system, turpentine is mixed with black liquor, making its recovery more difficult and lower in yield, but its odor is easier to control. The volatile sulfur compounds are collected in the blow heat accumulator system. The design of the gas recovery system depends on the type of digester, the design of the blow heat accumulator, and the gas collection system used.

Washer Emissions

Washer emissions are a minor source of air pollution in comparison to combustion, evaporation, and digestion. They are low in concentration and large in volume. The nature of the emissions depends mainly on the type of washing equipment used. However, the process variables—type of wood, sulfur content, pH, and temperature—have minor effects. The two washing processes used are displacement washing and diffusion washing, and several types of equipment

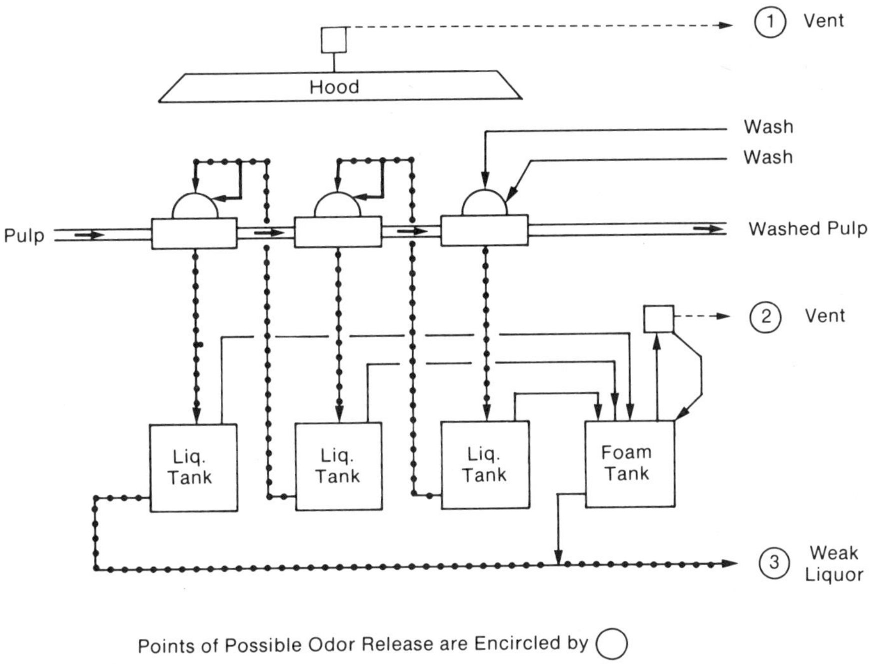

Figure 21-6 Vacuum drum washer

are available for each process. Displacement washing remains more common than diffusion washing, however, and the most common type of washer is the vacuum washer. Figure 21-6 illustrates this system. The two main sources of emissions are the hood vent and the seal tank vents. To decrease mill water usage and effluent volume, condensates are often used as washer water. Such use increases the emissions unless the condensates are carefully stripped of their odorous constituents before use. The only practical way to treat washer gases, especially those from the hood vent, is to incinerate them in an existing boiler. To reduce gas volume, some washer hoods are totally enclosed and pressurized (termed a "pressure washer system"), in which case the volume from both the hood and seal tank vents is small and the emission more concentrated.

Washer emissions can be eliminated entirely by using the more modern diffusion washer systems, but of course, such a switch involves considerable capital cost. Diffusion washing takes place in a closed reactor and ideally involves no air at all. Vents from these washers emit TRS at 3–30 ppm (mean, 17 ppm), which translates into an emission of 0.001 lb/ton compared to 0.1 lb/ton for conventional drum washers (13).

Lime Kiln Emissions

The lime kiln is a source of emissions of particulates, NO_x, and H_2S, and the scrubber used to control its particulates emission may be a source of other organic sulfur compounds. Two types of calciners are used in the Kraft industry, the conventional rotary kiln and the fluidized bed calciner. Rotary kilns are far more common, but some new mills are using fluidized bed systems. The emission characteristics of the fluidized bed calciner are better and its fuel requirements less.

The major design variables include kiln length, kiln diameter, rotation speed, and angle of incline. These variables affect the solids retention time, gas contact area, and temperature, which in turn affect the emission characteristics. Particulate emissions are affected mainly by kiln length and diameter because they determine the gas velocity and gas–solids contact area. The larger contact areas and higher velocities yield increased emission of particulates.

H_2S emission is determined by the amount of excess air and by the Na_2S content of the lime mud. In the presence of sufficient excess air, H_2S is oxidized to SO_2.

The major method used to control particulate emissions from either a rotary kiln or a fluidized bed calciner is liquid scrubbing. The scrubber can be either an impingement type or a venturi type. At present, most new mills are installing venturi scrubbers. Large particles are usually removed beforehand in a mechanical cyclone collector. If digester or evaporator condensate is used as a scrubbing fluid, TRS emissions may be created by a stripping action. Particulate inlet loadings to scrubbing devices on lime kilns can range from 7 to 35 g/m^3 (flow measured at standard conditions of 21.1°C, 1 atm). This amount represents 1–5% of the total dry solids load to the kiln. Particles in wide range of sizes are emitted, with lime particles generally quite large and the sodium chloride

Table 21-6 Particulate Collection Efficiencies of Liquid Scrubbers on Kraft Pulp Mill Lime Kilns

	Impingement Scrubbers		Venturi Scrubbers	
Parameter	Average	Range	Average	Range
Inlet concentration[a]				
g/m^3	27.38	8.00–33.96	18.60	5.85–31.83
g/ft^3	11.94	3.50–14.81	8.11	2.55–13.88
Outlet concentration[a]				
g/m^3	1.78	0.99–3.56	0.73	0.27–2.29
g/ft^3	0.78	0.43–1.56	0.32	0.12–1.00
Removal efficiency, % by wt.	92.2	86.8–96.9	94.8	85.5–99.1
Emission rate[b]				
kg/tonne	1.78	1.14–2.09	1.01	0.33–2.60
lb/ton	3.55	2.28–4.18	2.02	0.66–5.19

Source: Reference 3.

[a] Concentrations are reported at standard conditions of 21.1°C and 760 mmHg (70°F and 29.92 in Hg), dry gas.

[b] Emission rates are based on air-dried ton of pulp basis (i.e., 10% moisture, by weight).

particles small. Table 21-6 gives the collection efficiencies of lime kiln scrubber systems, showing that venturi scrubbers, with 99.1% removal efficiency possible, are more efficient than impingement scrubbers. The scrubbing water should be alkaline and can be recycled to the causticizing system.

The major variables affecting reduced sulfur emissions from lime mud calcining are the level of excess air, the temperature profile in the kiln, and the mud retention time. TRS emissions are minimized at excess oxygen levels greater than 4% by volume (14). Lime kilns with cooler wet end temperatures tend to have relatively high reduced sulfur emissions because the sulfur compounds tend to be volatilized without burning. Oxidation of TRS compounds requires sufficient time at a temperature greater than 760°DC (1,400°F). Short lime kilns tend to have lower TRS emissions than longer kilns because, to achieve the same degree of calcining as in a long kiln, the short kiln must operate at a higher average temperature, thus oxidizing the TRS compounds (15). These high temperatures have the disadvantage, however, of increasing the NO_x emissions. NO_x compounds are formed above 1,300°C (2,200°F) in the presence of excess air. The lime kiln is the only source of significant NO_x emissions in a Kraft mill.

SO_x emissions from the lime kiln are normally minimal because calcium oxide (CaO) can act as an efficient adsorption and reaction medium to form $CaSO_3$ and $CaSO_4$. Long kilns with excess oxygen and high calcination efficiencies have little SO_2 in their stack gas.

Fluidized bed calciners, based on limited testing to date, seem to be relatively free of the TRS, SO_x, and NO_x emission problems of conventional rotary kilns.

Bleaching Emissions

Many Kraft pulps are bleached with chlorine-containing chemicals such as chlorine, chlorine dioxide, or hypochlorite. In the preparation of chlorine dioxide, SO_2 is used. The bleach plant can thus be the source of emissions of Cl_2 (0.1–3.0 kg/ton), ClO_2 (0.1–1.0 kg/ton), and SO_2 (0.1–1.0 kg/ton). The emissions of Cl_2 and SO_2 can be reduced by scrubbing with a caustic solution. Scrubbing with cold water, peroxide, or SO_2 water is effective in reducing ClO_2.

EMISSIONS FROM NEUTRAL SULFITE SEMICHEMICAL (NSSC) PULPING

After the Kraft process, the next most popular pulping process in the United States NSSC. This type of pulping is used mainly to produce corrugating medium. It could be considered a subcategory of sulfite pulping but is probably sufficiently important to be considered by itself. One reason is that conventional sulfite pulping is on the decline in the United States due to its serious environmental problems, whereas NSSC is not so threatened. NSSC pulp is very different from Kraft pulp and is produced in high yield. Wood chips are partially delignified using a combination of either $NaSO_3$ and $NaCO_3$ or NH_4SO_3 and NH_4CO_3

Table 21-7 Representive Air Pollution Emissions from NSSC Pulping and Kraft Pulping[a]

Parameter	Kraft	Soda Semichemical
Reduced sulfur	0.12 (0.25)	0.0002 (0.0004)
SO_2	2.3 (4.6)	0.9 (1.8)
Particulates	2.8 (4.6)	2.7 (5.4)
NO_x	21.9 (43.8)	27.1 (54.2)

Source: Reference 17.
[a]All values in kilograms per tonne, pounds per ton.

at a pH of 6 to 9. Some mills have converted to what is, in principle, a sulfur-free process by using only $NaCO_3$ in the cooking liquor. As of 1977, eight mills in the United States were using the sulfur-free process and more were considering the change (16). As of 1985 the majority of mills have converted to the sulfur-free process or are using a green liquor cook. The absence of sulfur greatly reduces the emission of hydrogen sulfide.

In general, the emissions from NSSC are much less than those from the Kraft process. Both methyl mercaptan and dimethyl sulfide are absent from the gaseous emissions because no Na_2S is present in the pulping liquor. Representative emissions for a semichemical pulping operation are contrasted to those for a representative Kraft operation in Table 21-7 (17). A very low amount of reduced sulfur is emitted. The sulfur emission from the $NaCO_3$ (sulfur-free) process has been traced to sulfur in the fuel oil and process water streams used. The emissions of SO_2, particulates, and NO_x are similar to those of a Kraft mill.

Points of Emission

In general, the points of emission from an NSSC mill are the SO_2 absorption towers, blow pits, spent liquor evaporators, and the fluidized bed reactor or recovery furnace. Evaporator emissions are usually negligible, consisting of a large volume of condensate and a small volume of vent gas.

If the NSSC mill is located next to a Kraft mill, what is termed "cross recovery" is practiced. In cross recovery, the spent NSSC liquor is mixed with Kraft black liquor and burned in the recovery boiler.

If a Kraft mill is not available, the most common way of disposing of NSSC liquor is to concentrate it in evaporators to 30–40% solids and to spray the concentrated liquor into a fluidized bed. The liquor is sprayed from above the bed, so that the flue gases from the bed can complete the concentration of the liquor needed for it to support combustion. Temperatures above the bed are 425–540°C while those in the bed are 650–760°C. The concentrated liquor falls into the bed and burns, giving off energy that heats the bed. Air is supplied to the bed at a temperature 54–93°C cooler than the bed to help dissipate the heat generated by the exothermic reaction. Under normal operating conditions, the emissions from a fluidized bed reactor are minimal, especially if the reactor is equipped with dry or wet mechanical emissions control devices.

Another approach to liquor recovery is to dispose of the liquor in a furnace similar to a Kraft recovery furnace. This is not the cross recovery operation referred to earlier but a separate NSSC recovery system that was developed under the auspices of the Institute of Paper Chemistry. In this reductive furnace system, the highly concentrated liquor is burned in the furnace and the dry inorganics are transformed into a smelt of Na_2CO_3 and Na_2S. The particulate matter generated is mainly Na_2SO_4, as in a Kraft recovery furnace, and is collected by electrostatic precipitation and returned to the furnace. Particulate emissions for this system are approximately 1 kg Na_2SO_4 per tonne of pulp (2 lb/ton). SO_2 emissions vary widely, depending upon the sulfur level and upon which of 10 or more specific recovery schemes is being used. However, in general, SO_2 emissions range from 6 to 20 kg SO_2 per tonne of pulp (12–40 lb/ton) (18).

Galeano and associates (18) did a thorough study of the NSSC recovery furnace system and reported that it is possible to obtain close to zero emission to TRS by using proper distribution and amounts of primary and secondary air flow. However, emission from these furnace systems is generally greater than from a fluidized bed system.

The NO_x emissions from a reductive furnace NSSC recovery system range from 13 to 65 ppm, which is not negligible but is small compared to the 325–535 ppm NO_x emitted by gas-fired power boilers. A certain amount of NO_x emission is difficult to avoid, since its source is the oxidation of nitrogen in the air at the high temperature of the furnace. The amount present in the flue gas from an NSSC recovery furnace seems to be from this source rather than from the presence of other nitrogen compounds.

EMISSIONS FROM SULFITE PULPING

At one time, sulfite pulping was common in the United States, but this situation has changed drastically during the last 25 years for two reasons. First, sulfite pulping produces a lower-yield pulp that is weaker than Kraft pulp. Second, the sulfite process causes appreciable water pollution problems unless chemical recovery is practiced, and this procedure was not common. When faced with the need to install a chemical recovery system in order to meet water quality requirements, most mills chose to shut down their pulp mills and to purchase market pulp for their needs. During the 1970s, no new sulfite mills were built in the United States. Considering its diminishing role, my comments on sulfite

Table 21-8 Main Sulfite Pulping Processes

Cooking Method	pH Range	Cooking Liquor	Base Alternatives
Acid sulfite	1–2	$M(HSO_3)_2$, SO_2	M = Ca, Mg, $(NH_4)_2$, Na_2
Bisulfite	2–6	$M(HSO_3)_2$	M = Mg, $(NH_4)_2$, Na_2
Neutral sulfite	6–9	MSO_3, MCO_3	M = $(NH_4)_2$, Na_2
Alkaline sulfite	10–12	MSO_3, $M(OH)_2$, MS	M = Na_2
Two-stage	6–8 and 1–4	MSO_3, $M(HSO_3)_2$, SO_2	M = Na_2
Two-stage	2 and 5–8	$M(HSO_3)_2$, SO_2, MSO_3	M = Na_2
Three-stage	4, 2, and 8	$M(HSO_3)_2$, SO_2, MSO_3	M = Na_2

Source: Reference 3.

pulping will bc rather brief. This should not be misconstrued to indicate that sulfite pulping is a less technologically complex process than Kraft pulpings; if anything, the reverse is true due to variety of bases and recovery systems available.

Within the sulfite category, there are four main chemicals upon which the cooking process is based: clacium, magnesium, sodium, and ammonia. Recovery of spent liquors ranges from good with magnesium to poor with calcium. For each base, a variety of recovery systems are available; for sodium, for example, there are the Stora, SCA-Bellerud, Tampella, Sevola-Lurgi, and Institute processes, to name only a few modern ones. Thus it would obviously be impractical, within the scope of this book, to go into any detail about sulfite pulping. An overview in more depth can be found in reference 3. Table 21-8 lists the main sulfite pulping processes with their pH ranges, cooking liquor chemicals, and base alternatives, clearly illustrating the diversity of sulfite pulping.

Sulfite pulping processes are relatively free of reduced sulfur compounds that cause odor problems. The main air pollutant from sulfite pulping is SO_2, which has an odor threshold about 1,000 times higher than those of reduced sulfur compounds. As a consequence, sulfite mills do not generally experience complaints about odor from people living nearby. Sulfite mills using a magnesium-based process have a characteristic odor that is principally caused by furfuryl mercaptans and furfuryl methyl sulfides (19). Sulfite mills using an ammonia-based process have a different characteristic odor due to pyrimidines (20).

Besides SO_2, sulfite mills emit particulate matter, oxides of nitrogen, and, in certain instances where reductive furnaces are used in the recovery system, H_2S. For the purposes of this discussion, we will concentrate on the two major pollutants emitted, particulates and SO_2. Many differences in particulate and SO_2 emission exist because of variations in liquor composition and recovery practices. Commonly, SO_2 is scrubbed out of the gas stream with an alkaline solution. Particulates are removed in a variety of ways; cyclones, scrubbers, and electrostatic precipitators are three common alternatives.

Typical emission rates from some representative North American sulfite pulp

Table 21-9 Typical Potential North American Sulfite Pulp Mill Emissions

Parameter (Base Type)	Mill Number 1: Ca Acid Sulfite	2: Mg Acid Sulfite	3: Mg Bisulfite	4: NH_4 Acid Sulfite
Cooking yield, percent of spent sulfite	45	43	50	50
Liquor evaporation spent sulfite	No	Yes	Yes	Yes
Liquor combustion	No	Yes	Yes	Yes
Chemicals recovery	No	Yes	Yes	No
SO_2 emissions, (kg SO_2/ton)				
Digester blow	75	8.5	8.5	1.5[a]
Washers and screens	—	2	2	2
Evaporators	—	2.5	2.5	5
Boiler	—	9.5	10[b]	200
Acid towers	2.5	—	—	0.25
Particulate matter emissions, kg/ton	0	1	1	1.5

Source: Reference 3.
[a]Packed tower absorber with 91% SO_2 removal.
[b]Venturi absorber used after recovery boiler.

mills are given in Table 21-9. The principal sources of emissions are the digester, washers, evaporators, recovery boiler, and acid tower. Note that the first mill listed, which is a calcium-base acid sulfite mill, has no chemical recovery system and would not be allowed to operate under present U.S. water pollution regulations. At the time these data were collected, it was discharging its spent liquor into a watercourse, a practice that is no longer acceptable. This mill would have to change to another base for which a recovery system is possible.

Digester and Blow Pits

Both batch and continuous digesters are used in sulfite pulping. The continuous digesters emits less SO_2 than does the batch type. Blowing the batch digester at the end of the cook is the main source of its SO_2 losses. Three principal methods of discharge are used: the hot blow, the cold blow, and flushing. In the hot blow, the digester contents are blown at 0.14–0.28 mPa (20–40 psig) into the blow pit. In the cold blow, the digester is reduced to atmospheric pressure and its contents are flushed out with spent liquor. In the flush discharge method, the digester is partially filled with spent liquor or hot water, and then its contents are released to the blow pit. The amount of SO_2 released depends on the method of discharge used. The hot blow technique is the least desirable environmentally. The digester top and side relief gases are sent to accumulators, and the gas is used to fortify cooking liquor. The gases released to the blow pit and dump tanks are vented to the atmosphere. Losses as high as 30–75 kg SO_2 per tonne of pulp (60–150 lg/ton) can occur (21). The blowing of batch digesters is a major source of SO_2 emissions and is the largest emission source for most of the mills listed in Table 21-9.

Washer and Evaporator Emissions

Rotary drum washers are the most common type employed. The washing and screening operations release approximately 8 kg SO_2/ton (16 lb/ton) for sulfite cooking and even less for bisulfite and neutral sulfite. Washing losses can be further decreased by using displacement washing in the digester and by changing to diffusion washers.

In the multiple-effect evaporation process, all evaporator effects are ultimately vented to the vacuum system. The vacuum system vent becomes the main emission point for evaporator gases. The point of SO_2 emissions is the hot well of the surface condenser. Acid sulfite evaporator gases contain considerable SO_2 (20–30 kg per tonne of pulp). This SO_2 can be recovered by cycling the gas back to liquor preparation or by scrubbing the SO_2 out with an alkaline solution and returning the solution to liquor preparation.

Recovery Boiler Emissions

The recovery boiler is a significant source of SO_2 and particulate emissions. The best way to control these pollutants is through proper design and operation of the furnace and its auxiliary emissions control devices. The specific requirements depend on which recovery process is being used. In addition to SO_2 and particulates, most furnaces emit a small quantity of NO_x. An exception is the furnace recovering chemicals from an ammonium-based liquor, which may emit major quantities of NO_x. Only limited data on emissions from this type of recovery furnace are available, but one test series showed total nitrogen oxide (as NO_2) concentrations of 200–500 ppm, with peak values of 1,000 ppm (3).

Liquor Preparation

The usual way of producing liquor is to burn either elemental sulfur or pyrites to produce SO_2, which is then absorbed into a carbonate or hydroxide solution of the desired base in an absorption tower. Emissions from the sulfur furnace and absorption tower are usually small except under upset conditions. For calcium-based liquors, SO_2 losses of about 1 kg/tonne (2 lb/ton) are normal. For magnesium-based liquors, losses of 15–25 kg/tonne (30–50 lb/ton) are usual. With ammonium-based cooking liquor, the SO_2 loss is highly pH dependent.

At pH 7.1 the losses of SO_2 and NH_3 are less than 1%, but at pH 6 approximately 10% of the SO_2 is lost. In general, the SO_2 losses in liquor preparation can be kept to a reasonable level if proper control is exercised and the system does not experience upsets.

EMISSIONS FROM POWER BOILERS AND COMBINATION BOILERS

The paper industry is a major consumer of energy, being one of the top five manufacturing industries from the perspective of total energy consumption. A total mill energy balance on an integrated Kraft pulp and paper mill complex shows that such a mill could be energy self-sufficient by burning for power all of its bark, waste wood, and spent liquor. However, no mill has yet achieved this ideal state. As of 1978, the paper industry as a whole was providing 46.6% of its own energy needs—37% from combustion of spent liquor, 4.3% from bark, 4.4% from hogged fuel boilers, 0.6% from hydroelectric sources, and 0.3% from sources classified as "other." By 1983 the industry had become 54.3% energy self-sufficient. Of this amount, 38.2% came from burning spent liquor, 5.5% from bark, 9.5% from hogged fuel, 0.6% from hydroelectric sources, and 0.5% from other sources. The remaining 45.7% of the energy needed by the industry was supplied by natural gas (14.4%), coal (13.8%), residual fuel (oil) (9.9%), electricity (6.4%), steam (0.8%), distillate (0.2%), and liquefied petroleum gas (0.1%).

The industry is trying both to reduce its total energy consumption and to become more energy self-sufficient. It used 36.7% less energy to produce a ton of paper in 1983 than it did in 1972 (22). The present high prices of imported petroleum and the emphasis on energy conservation will serve to move the paper industry closer to the goal of energy self-sufficiency. Mills will seek an economically optimum balance between purchased and self-generated energy.

The concern of this section is the control and minimization of pollution problems arising from self-generated power. The general area of study is that of combustion engineering.

Combustion Engineering

Combustion engineering is the science of burning fuel to generate heat and transforming that heat into useful work with minimum air pollution. The subject is vast, and the student is referred to references 23–25, which are textbooks and handbooks on this topic. In general, a fuel is burned in the presence of air to release heat that is used to generate steam at the appropriate temperature, pressure, and degree of superheat. A by-product of the process is air pollution, the amount of which is a function of fuel characteristics, air/fuel ratio, and boiler type. The burning process itself is affected by fuel chemical composition, heating value, and moisture content, as well as boiler characteristics and operating variables.

Fuel and Boiler Characteristics

The common fuels used by the paper industry are natural gas, fuel oil, coal, waste wood, and bark. Each fuel requires a different type of boiler configuration because of its unique characteristics. The two air emissions of primary importance, because of their quantity, are SO_2 and particulate matter. Other emissions include oxides of nitrogen, carbon monoxide, and trace elements associated with the fuel.

The two critical fuel composition characteristics that affect emissions are the sulfur content and the ash content, the sulfur appearing as SO_2 emissions and the ash as particulate matter. Depending on its fusion temperature, ash may leave the boiler as a liquid or a solid. Approximately 80–95% of the ash material is emitted as fly ash.

Solid and liquid fuels contain traces of heavy metals such as beryllium, mercury, lead, cadmium, zinc, arsenic, and selenium, as well as nonmetals such as chlorine, fluorine, and phosphorus. Since such contaminants contribute to the fuel's emissions, it is easy to understand why gaseous fuels are usually cleaner burning than liquid fuels, which in turn are better than solid fuels.

Carbon monoxide and oxides of nitrogen are toxic gaseous emissions. Carbon monoxide is produced by incomplete combustion, which is undesirable and contrary to the primary aim of energy production. For greater efficiency, proper furnace design and operating procedures try to eliminate carbon monoxide production.

At the temperature conditions of all fossil fuel combustion processes, nitrogen and oxygen combine to form nitric oxide. At lower temperatures, a small amount of the nitric oxide is oxidized to nitrogen dioxide by excess oxygen in the furnace. The reaction for the formation of nitric oxide is very temperature sensitive, so

Table 21-10 Characteristics of Fuel Burned in Power Boilers at Pulp and Paper Mills in the United States

	Heating Value (as Fired)			Percent by Weight Content of:		
Fuel	Units	Average	(Range)	Moisture	Average Ash (Range)	Sulfur
Coal	MJ/kg	31	(24–34)	10 (7–15)	8.1 (3.5–15)	1.9 (0.5–10)
	BTU/lb	13,500	(10–500–14,700)			
Oil	MJ/l	41	(34–43)	0.5 (0–1)	0.1 (0.01–0.2)	1.8 (0.1–3.5)
	BTU/gal	149,000	(122–000–155,000)			
Gas	MJ/m^3	38	(37–40)	0.2 (0–1)	Negligible	Negligible
	BTU/ft^3	1,030	(1,000–1,070)			
Wood	MJ/kg	11	(9.3–13)	25 (10–60)	2.9 (0.1–20)	0.1 (0–0.2)
	BTU/lb	4,600	(4,000–5,500)			

Source: Reference 3.

if high boiler flame temperatures can be avoided, substantial reductions in oxides of nitrogen can be achieved.

Table 21-10 summarizes some representative characteristics of fuels used in pulp and paper mill power boilers. From the ash and sulfur numbers, the expected ranking from cleanest to dirtiest is gas, oil, wood, and coal; this is observed when each fuel is burned in a boiler designed for it at the current state of the art.

The major boiler design parameters that minimize air pollution are:

1 Physical state, chemical composition, and burning characteristics of the fuel
2 Methods used to mix and fire the fuel
3 Combustion chamber volume and configuration
4 Operating parameters such as the firing rate, air level, and air distribution

Combination Boilers

Boilers that can be fired with more than one fuel are referred to as "combination boilers." Combination boilers are one of the industry's techniques for becoming more energy efficient. The usual industry practice is to burn waste wood in combination with another fuel. The wood wastes collectively are called "hogged fuel" and may consist of sawdust, shavings, and slabs from lumber or plywood plants or of bark from a hydraulic or mechanical debarking operation. The design of these furnaces is complicated because of the many alternatives available (more than 40); reference 26 reviews some of these alternatives.

The objectives are to maximize the burning of hogged fuel and to minimize the use of auxiliary fuel. The major parameters affecting the acceptability of hogged fuel are its size and its heating value. The heating value of most wood is 17–21 kg (7,200–9,200 BTU/lb) on a dry basis. The heating value varies with the species, ash content, and moisture of the wood. Moisture is very undesirable due to water's high latent heat of vaporization. The heating value will nearly double when wood is reduced from 63 to 28% moisture, and it is for this reason that dry debarking processes are advantageous if the bark is to be burned.

The major pollutant from hogged fuel boilers is particulate matter. Particulates are strongly related to the ash content, which can arise from the material itself or from grit and dirt adhering to the material. Table 21-11 summarizes the particulate emission characteristics of some combination boilers. If the logs have been stored or transported in sea water, the salt content will cause increased amounts of ash. Most particulates from hogged fuel furnaces are large; for instance, particle sizes from a bark furnace are often 5–10 μm (2–4 10^{-4} in.) (27). These large particles are easy to collect, and an electrostatic precipitator is usually not required. The exception is salt water–soaked material, for which 60–70% of the particulates are less than 1 μm in diameter and difficult to collect.

Besides particulates, wood waste boilers emit NO_x, SO_2, CO, and hydro-

TABLE 21-11 Particulate Emission Characteristics of Selected U.S. Power Boilers

Number of Boilers	Percent of Fuel Supplied, BTU Basis				Collector Equipment		Particulate Concentration, g/m^3(g/ft^3)		Collection Efficiency %	Emission Rate kg/hr(lb/hr)
	Coal	Oil	Gas	B/W[a]	Type[b]	Pressure Drop, In. of Water	Inlet	Outlet		
18	100	0	0	0	C	2.5	4.28 (1.87)	0.85 (0.37)	80	129 (284)
2	100	0	0	0	S	—	—	0.57 (0.25)	—	136 (300)
2	100	0	0	0	P	2.2	11.2 (4.89)	0.98 (0.43)	91	180 (397)
16	0	46	0	54	C	2.7	7.9 (3.47)	1.05 (0.46)	87	140 (309)
2	75	0	0	25	C	3.9	—	0.41 (0.18)	—	73 (160)
2	0	0	62	38	C	2.8	5.3 (2.30)	0.39 (0.17)	93	70 (153)
2	73	16	0	11	C	2.5	—	2.75 (1.20)	—	228 (502)
3	0	25	39	36	C	2.8	4.3 (1.88)	0.71 (0.31)	84	202 (445)
2	0	0	0	100	C	0.2	3.2 (1.40)	0.89 (0.39)	72	44 (96)

source: Copyright © 1981. TAPPI. Reprinted from *International to Environmental Management for the Pulp and Paper Industry*, Study Guide, Volume 6, with permission.

[a] B/W = Bark and wood waste.

[b] C, cyclone; S, liquid scrubber; P, electrostatic precipitator.

carbons volatilized from the wood. NO_x emissions are low because of the relatively low flame temperature in this type of furnace (980–1,200°C, 1,800–2,200°F). SO_2 emissions are usually low because the fuels are low in sulfur. CO emissions are minimized by proper excess air and good combustion conditions in the furnace.

Cogeneration

Another energy-efficient technique is cogeneration. In the pulp and paper industry, "cogeneration" means the burning of fossil fuel to produce high-pressure, superheated steam that is used both for power generation and as process steam. If a given facility uses two separate systems to produce power and process steam, the potential efficiency is about 43%, whereas with cogeneration it is 52% and further improvement may be possible (28). Cogeneration reduces air pollution because more useful energy is produced per unit of fuel input, resulting in less fuel burned and less air emissions generated.

The most common practice uses a back-pressure steam turbine electric generation system in which steam leaving the turbine is at high pressure and can satisfy the process demand. The U.S. paper industry uses this technology extensively and is one of the industrial leaders in the use of cogeneration. Approximately 50% of the process steam used by the industry comes from this process (22).

REFERENCES

1 J. J. Sableski, *Tappi*, 50 (8):35A–39A (1967).

2 E. R. Hendrickson, J. E. Roberson, and J. B. Koogler, *Control of Atmospheric Emissions in the Wood Pulping Industry*, Vols. 1, 2, and 3, Final Report, Contract No. CPA22-69-18 HEW, National Air Pollution Control Administration, Washington, D.C., March 15, 1970.

3 *Environmental pollution control pulp and paper industry, Part 1, Air*, U.S. EPA Technology Transfer Series, U.S. Government Printing Office, Washington, D. C., October 1976.

4 B. Andersson, R. Lövblod, and P. Grennbelt, *Diffuse emissions of odorous sulfur compounds from Kraft pulp mills*, IVLB145, Swedish Water and Air Pollution Research Laboratory, Götenborg, Sweden, 1973.

5 S. F. Galteano and K. M. Leopold, "A survey of emissions of nitrogen oxides in the pulp mill," *Tappi*, 56 (3):74–76 (1973).

6 A. Dipple, "Formation, metabolism, and mechanism of action of polycyclic aromatic hydrocarbons," *Cancer Research*, 43 (suppl 5):2422–2425 (1983).

7 J. Santodonato et al., "Health and ecological assessment of polynuclear aromatic hydrocarbons," *Journal of Environmental Pathology and Toxicology*, 5 (1):1–364 (1981).

8 F. G. Augustine and R. W. Boubel, *Journal of the Air Pollution Control Association*, 25 (6):617–621 (1975).

9 P. F. Fennelly, *Journal of the Air Pollution Control Association*, 25 (7):697–704 (1975).

10 E. E. Hendman, II, P. V. Hobbs, and L. F. Rodke, *Journal of the Air Pollution Control Association*, 27 (3):224–229 (1977).

11 I. B. Douglass, *Tappi*, 52 (9):1738–1741 (1969).

12 K. V. Sarkanen, B. F. Hrulfiorod, L. N. Johanson, and H. S. Gardner, *Tappi*, 53 (5):776–783 (1970).

13 R. P. Fisher, *Emission of reduced sulfur compounds from Kraft process brownstock diffusion washer vents*, Technical Bulletin No. 406, National Council of the Paper Industry for Air and Stream Improvement, New York, 1983.

14 A. L. Caron, *Suggested procedures for the conduct of lime kiln studies to define minimum emissions of reduced sulfur through control of kiln and scrubber operating variables*, Special Report No. 71-01, National Council of the Paper Industry for Air and Stream Improvement, Corvallis, Ore., January 1971.

15 J. E. Walther and H. R. Amberg, *Chemical Engineering Progress*, 66 (3):73–80 (1970).

16 J. P. Hansen, *Pulp and Paper*, 53 (3):116–119 (1978).

17 V. Dallons, *Multimedia assessment of pollution potentials of non-sulfur chemical pulping technology*, EPA-600/2-79-026, U.S. Government Printing Agency, Washington, D. C., January 1979.

18 S. F. Galeano, P. C. Kahn, and R. A. Mack, *Tappi*, 54 (5):741–744 (1971).

19 P. A. Thorn, H. S. Gardner, and B. Hrutiord, "Odorous compounds from magnifite pulping," *Pulp and Paper Canada*, 78 (9):T-215–T-217 (1977).

20 P. Grennfelt, *Investigation of the emission of odorous compounds from a semi-chemical ammonia base pulp mill*, Swedish Water and Air Pollution Research Laboratory, Götenborg, Sweden, 1975.

21 W. D. Johnson and W. L. Sage, *Pulp and Paper*, 45 (13):54–56 (1971).

22 Personal communication, Thomas Grant, American Paper Institute, New York.

23 G. R. Fryling, *Combustion Engineering*, rev. ed., Combustion Engineering Co., New York, 1966.

24 *Steam: Its Generation and Use*, 38th ed., Babcock and Wilcox Co., New York, 1975.

25 R. H. Perry and C. Chilton, "Heat generation, transport and storage," *Chemical Engineers Handbook*, 5th ed., McGraw-Hill, New York, 1975.

26 R. C. Johnson, *Tappi*, 58 (7):102–106 (1975).

27 M. F. Szabo and R. W. Gerstle, *Operation and maintenance of particulate control devices in Kraft pulp mill and crushed stone industries*, EPA-600/2-78-210, U.S. Government Printing Agency, Washington, D. C., October 1978.

28 *Integrated community energy systems planning*, State of the Art Report, Batelle Columbus Laboratories, Columbus, Ohio, November 17, 1977.

22

Alternative Air Emission Control Strategies

This chapter provided an introduction to the common technologies employed to alleviate air pollution problems in the pulp and paper industry. For a more detailed treatment, references 1–3 can be consulted.

The industry's involvement in air pollution control is not a new endeavor; the first attempts to apply electrostatic precipitation were made in 1885, and the process was in commercial use by Cottrell in 1907 (1). Since this early start, the field has been evolving and changing. The National Council of the Paper Industry for Air and Stream Improvement (NCASI) has so far produced 130 technical bulletins summarizing developments in the field of air pollution.

The most stringent requirements that a mill might have to meet are the NSPS or PSD criteria, which may dictate the lowest allowable emission rate (LAER). Table 22-1 gives the NSPS for steam generation, and these stringent requirements will further stimulate the evolution of air pollution control technology. The technologies available to control particulates are well developed. The only problem is to choose the most cost-effective solution for the application under consideration.

PARTICULATE CONTROL EQUIPMENT

The first NSPS requirement is for the control of particulate matter. A wide variety of particulate control equipment exists, with a correspondingly wide range of effectiveness and cost.

Inertial Separators

Inertial separators, also known as "cyclone collectors" or "centrifugal separators," give fair efficiency at low cost. Figure 22-1 illustrates the working principle of this device. The inlet gas velocity is transformed into a double vortex contained within the cyclone. The entering gas spirals down the outside and upward inside the cyclone outlet. Because of their inertia, particulates tend to move toward the outside wall, impacting on it and sliding down to the outlet at the bottom of the cone. The dust is discharged into a collection hopper, while the cleaned

Table 22-1 NSPS for Stream Generation[a]

Particulate matter—A maximum 2-hour average of 0.01 lb particulate per million BTU heat input. Visible emissions may not exceed 20% opacity except for 2 minutes in any 1 hour, when opacity may be up to 40%.

SO_2—A maximum 2-hour average of 0.8 lb SO_2 per million BTU when oil is fired, 1.2 lb SO_2 per million BTU for coal.

NO_x—A maximum 2-hour average of 0.2 lb per million BTU when gas is fired, 0.3 lb per million BTU for oil, and 0.7 per million BTU for coal.

Source: Copyright © 1981. TAPPI. Reprinted from *Introduction to Environmental Management for the Pulp and Paper Industry*, Study Guide, Volume 6, with permission.

[a]Units greater than 250 million BTU per hour of input.

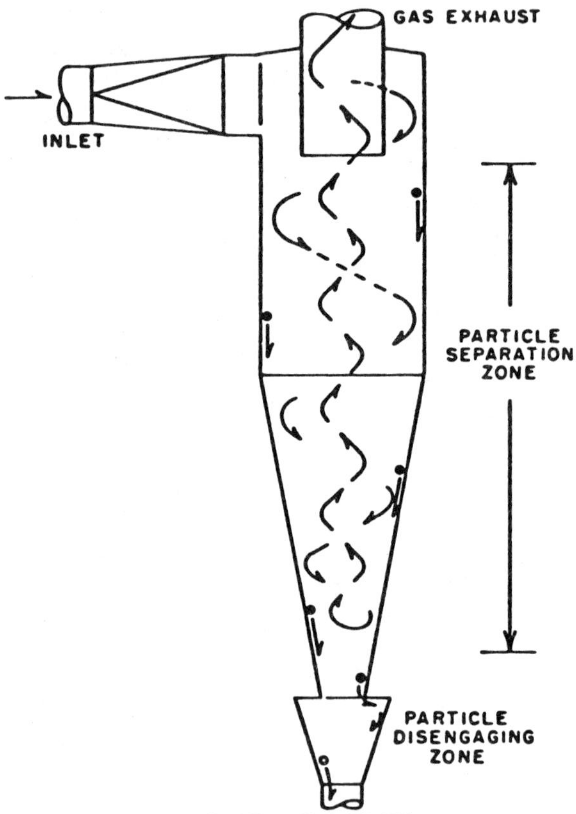

Figure 22-1 Cyclone collector. (From J. P. Casey (Ed.), *Pulp and Paper: Chemistry and Chemical Technology*, 3rd edition, volume 2. Copyright © 1980 by John Wiley & Sons, Inc. Reprinted by permission of John Wiley & Sons, Inc.)

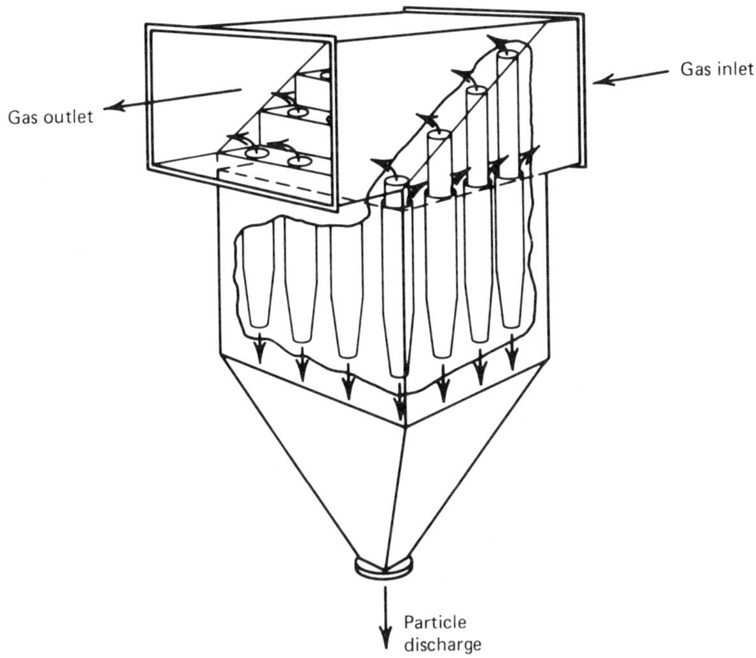

Figure 22-2 Multiple cyclone

gas stream flows upward in vortex fashion to the outlet. These devices have good efficiency for large particles (greater than 30 μm) but poor efficiency for smaller particles. Particles measuring 5–10 μm or less are too fine to be efficiently collected.

Collection efficiency increases not only with larger particle size but also with higher specific gravity of the particle and dust loading. Because of the last factor, fly ash reinjection is often used to improve efficiency (4). Efficiency also increases with increased pressure drop, but this condition is not desirable because it increases operating costs. Another factor affecting efficiency is gas temperature, with temperatures above 370°C (700°F) being detrimental. At any given pressure drop, either increased gas viscosity or increased gas density will increase aerodynamic drag and thereby lower the collection efficiency.

Cyclones that are large in diameter lose efficiency, Thus large gas flows are handled with multiple cyclones, each taking a portion of the flow. Figure 22-2 shows a multiple cyclone, and Figure 22-3 illustrates the increase in collection efficiency that it provides.

The most common application of cyclones is on combination fuel and power boiler installations in which a large portion of the particulate matter has a particle size greater than 30 μm. A survey of 26 boilers fired on wood wastes and other fuels gave collection efficiencies of 75–96%, with an average of 83%. The average inlet particulate loading was 2.74 g/SCF (standard cubic foot) and

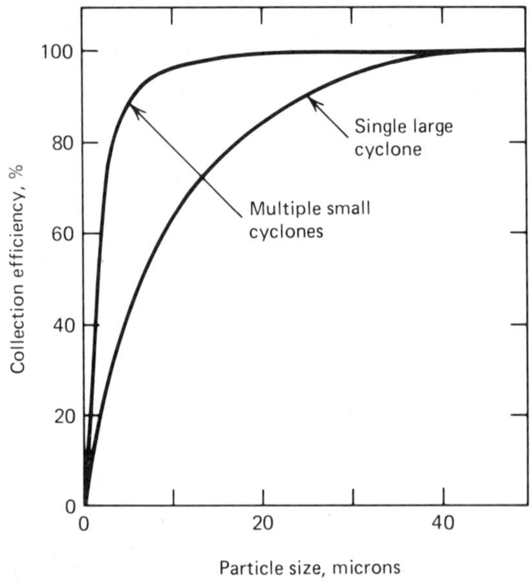

Figure 22-3 Collection efficiency

the outlet load was 0.45 g/SCF. Cyclones are quite cost effective for this type of application.

Scrubbers

Unlike cyclones, scrubbers often perform the dual function of odor control and particulate control, although the latter is their principal application. "Scrubbers" can be defined as air pollution control units in which liquids are used to remove aerosols and soluble gases from exhaust gases. The liquid can actually wet the particle by bringing the particle in contact with a liquid drop, or the wet or dry particle can impinge on a surface and subsequently be washed from it by the liquid. The liquid can be fresh water, a process stream, or a specially prepared solution. Scrubbers come in a large variety of types and sizes, but in the paper industry only four main types are used, although there is a great deal of variation within each type. These four are venturi, cyclone, impingement, and packed-tower scrubbers. The operating principles of each type will be briefly considered.

Figure 22-4 illustrates a venturi scrubber system, which consists of a convergent throat section and a divergent section. As gas enters the throat, it increases in velocity. Liquid is injected into the throat section, becomes fine droplets in the divergent section, and contacts the particles in the turbulence downstream of the throat. The liquid containing the particles is then separated from the gas stream by a cyclone, whose mode of operation was previously discussed. A considerable amount of mechanical energy is dissipated as heat through the turbulence created when the two phases mix. This energy is called

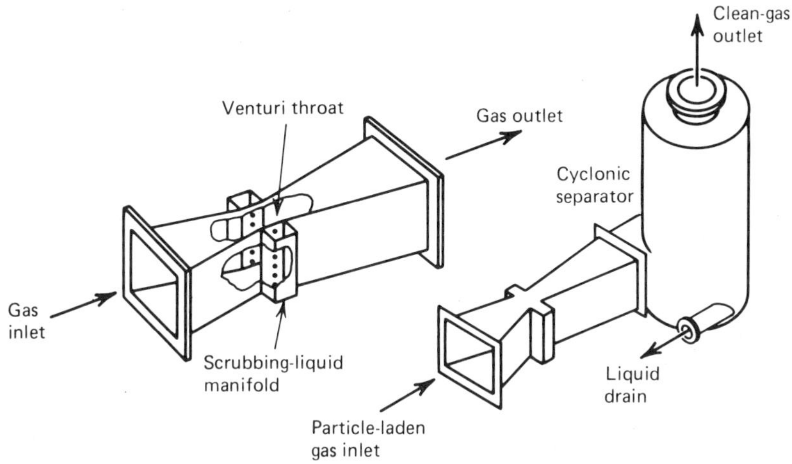

Figure 22-4 Ventury scrubber

"contacting power" and is related to scrubber efficiency (5). The other main determinant of efficiency is the properties of the aerosol. It appears that scrubber geometry and configuration have little to do with efficiency.

In the cyclone scrubber, liquid is sprayed from inside the cyclone onto the gas as it passes through the unit. A cyclone scrubber operates according to the same principles as the dry cyclone previously discussed, but it has a higher collection efficiency. The particles trapped in the liquid droplets strike the outer wall and flow with the liquid out the bottom of the cyclone, while the clean gas exits the top. A cyclone scrubber is also efficient but not quite as good as the venturi scrubber because it has a lower contacting power.

Impingement scrubbers contain two main operations. The gas entering the bottom is sprayed with liquid to remove the large particles. The gas–liquid mixture is then passed through a series of perforated baffles, where the scrubbing liquid is atomized and the particles are trapped in droplets. The liquid droplets fall to the bottom, while the gas passes out the top of the unit.

Packed-tower scrubbers consist of a cylindrical tank filled with packing material that serves to distribute the gas and the liquid flows and to provide a large surface area for contact between them. The liquid enters the top of the tank and flows downward, while the gas enters the bottom and flows up. A variety of materials can be used as packing, ranging from commercially prepared ceramic burl saddles or raschig rings to something as mundane as crushed rock. This type of scrubber creates a lot of turbulence and therefore has a high contacting power.

What is removed besides particulates by any scrubber system depends on the scrubbing liquid used. Water will remove particulates and sulfur dioxides; an alkaline solution will remove particulates, TRS, hydrogen sulfide, sulfur dioxides, and chlorine; a ferric chloride solution will remove particulates and hydrogen sulfide; a dilute black liquor solution will remove particulates and sulfur dioxides; and a solution of hydrogen peroxide is effective in removing chlorine dioxide. The scrubbing solution can be tailored to the job it must accomplish and also to its intended reuse or disposal method. The most versatile solution appears to be caustic. Its scrubbing efficiency can be improved by mixing in some powdered activated carbon.

Mist Pads

Another device that has some of the characteristics of a scrubber is a mist pad. This device consists of a series of fine wire mesh screens about 1 ft thick that serve to demist a gas stream. Droplets in the gas stream condense on the pad and trap particulates and TRS. The pad is periodically backwashed by water sprays.

Both scrubbers and mist pads are wet collection devices. They have the advantage of no secondary dust problem in disposing of the particulates, but the disadvantage of a liquid phase that must be clarified before disposal or reuse. Wet collection devices have been used in the Kraft industry for recovery boiler and lime kiln particulate control for more than 30 years (6, 7). They are also being used on smelt-dissolving tank and lime slaker vents. Wet scrubbers using water as the scrubbing fluid have been employed on hogged fuel boilers with efficiencies of 95–98% by weight. However, they have the disadvantage of producing a visible plume of water vapor. Wet scrubbers have both corrosion and erosion problems (8), which can increase operating costs and decrease efficiency.

Dry Scrubbers

A recent development that may overcome some of the problems of wet scrubbers is the dry scrubber (9). The dry scrubber uses a moving bed of granular material (medium) as the collection material. After scrubbing, the contaminated medium is vibrated to the bottom of the unit to shake off the particulate matter into a storage bin. The cleaned medium is then recycled back to the top of the scrubbing bed. This type of scrubber removes only particulate matter.

The concept of dry scrubbing is not new (some of the original patents were granted to Thomas Edison), but a cost-effective device has only recently been developed. The dry scrubber is being used on furnaces burning wood or bark as part of their fuel (9, 10). It has a low pressure drop of 6–20 cm (3–8 in.) water and does an effective job of removing particulates. The unit does not achieve high removal efficiencies, however. In one application on a fuel oil–bark boiler, the maximum removal efficiency was 86% (10). It may, however, provide a cost-effective method to upgrade a dry cyclone or multiclone unit to allow compliance with regulations without resorting to a wet scrubber system.

Particulate collection efficiencies of 96–98% were achieved with a dry scrubber by using an improved electrostatic granular system (11). An electrostatic grid within the moving bed greatly enhanced collection efficiency.

The efficiency of the electrostatically enhanced dry scrubber results from three collection phenomena—impaction, diffusion, and electrostatic attraction. This new scrubber has been installed on one of the largest hog fuel boilers in the United States. It accepts 11,894 m^3/min actual flow (420,000 ft^3/min) at 174°C (345°F) with 20% moisture. The emissions after scrubbing are 0.176–0.424 grain/standard m^3 (0.005–0.012 grain/dry standard ft^3). The pressure drop is 7–13 cm (2.8–5.2 in.) of water. This unit achieves 99% particulate collection efficiency (11).

Fabric Filters

Probably the most efficient device for the collection of fine particulate matter is the fabric filter, but it is not widely used in the paper industry. Figure 22-5 illustrates a bag house type of fabric filter installation. Cyclindrical cloth bags enclosed in a shelter are used to strain the particulate matter from gas that is

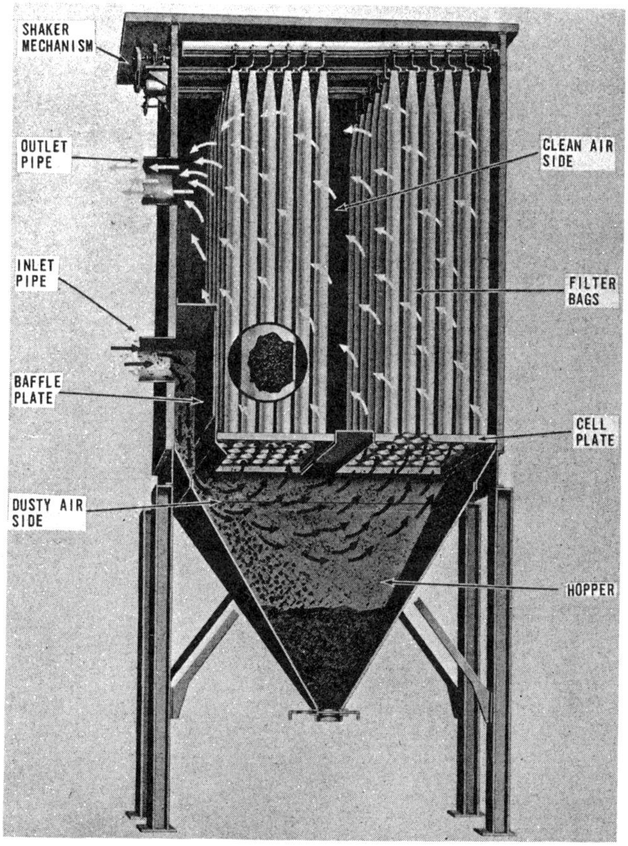

Figure 22-5 Fabric filter. (From J. P. Casey (Ed.), *Pulp and Paper: Chemistry and Chemical Technology*, 3rd edition volume 2. Copyright © 1980 by John Wiley & Sons, Inc. Reprinted by permission of John Wiley & Sons, Inc.)

passed through them. As particles collect on the cloth, they act as an additional sieve for other particles. The collected particles are removed by various means, such as shaking, reversing the gas flow, or impinging a high-velocity air stream, and the particles drop into a collection hopper below the filter. These filters are very efficient event for submicron-sized particles; efficiencies of 99% or better are possible at fairly reasonable pressure drops.

One limitation is the temperature sensitivity of the filter material. For most commercially available fabrics, the temperature must be kept below 315°C (600°F). Sometimes these installations have caught fire and caused extensive damage. Bag life is also limited because of the flexing during particulate removal; 18–24 months is normal bag life.

Despite their disadvantages, fabric filters have been applied in the paper industry on coal-fired power boilers, on wood-fired boilers, and on one lime storage fugitive dust collection system. Reference 12 reviews the limited applications of fabric filtration in the paper industry.

Electrostatic Precipitators (ESP)

An ESP consists of discharging and charging electrodes and a high-voltage rectifier to convert alternating current (AC) to direct current (DC). The discharge electrodes negatively charge the particulates in the gas stream, which are then attracted to the positively charged collecting electrodes. The collecting electrodes are grounded, causing the collected particles to lose their charge and fall to the collecting surface. The particles are removed from the collecting surface by a rapping mechanism. Both a dry-bottom design and a wet-bottom design exist, the latter developed because of the high maintenance costs involved in the former. Reference 13 discusses the proper design of a wet bottom for a precipitator. Figure 22-6 shows a typical ESP of the wet-bottom type.

Electrostatic precipitators have been employed in the paper industry since 1916, and considerable study has been devoted to potential improvements in their design. References 14 and 15 contain many of the design details that might be required, which will not be repeated here. The five essential design parameters that need careful selection are correct contact time, uniform gas distribution throughout the fields, design and arrangement of the electrodes, technique for maintaining clean electrodes, and correct voltage. ESPs are commonly used on the recovery boiler in a Kraft mill. The use of older ESPS, which alone would not meet emission requirements, can be followed by scrubbing.

The collection efficiency of ESPs is generally greater than 99%. In a survey of 87 Kraft recovery boilers equipped with direct-contact evaporators and ESPs, the emissions ranged from 2.1 to 75.2 lb of particulates per ton of air-dried pulp (TADP) (4). Efficiency is affected by a variety of factors, but it usually increases as the plate area is increased and the flue gas temperature is decreased. For example, a unit with a 200 m^2 plate area exhibited an efficiency of 98.9% at a gas temperature of 130°C and only 92.5% at 380°C (16). Another important factor is the gas velocity, which must be limited to avoid reentrainment of

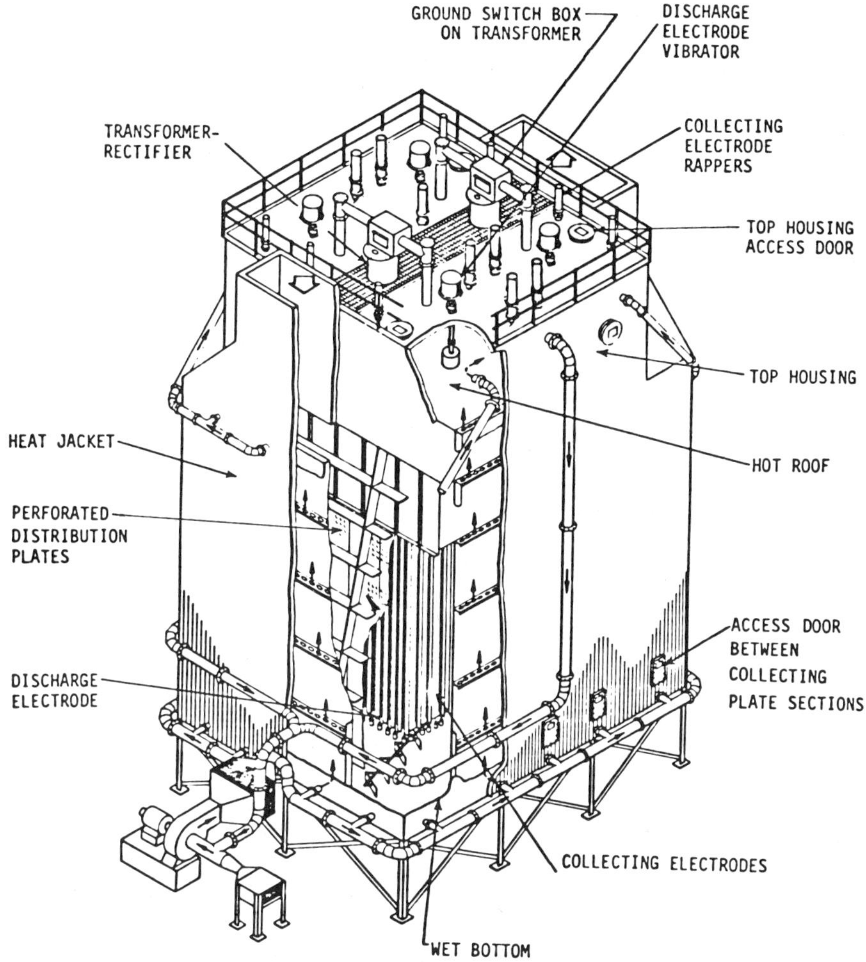

Figure 22-6 Electrostatic precipitator. (From J. P. Casey (Ed.), *Pulp and Paper: Chemistry and Chemical Technology*, 3rd edition, volume 2, Copyright © 1980 by John Wiley & Sons, Inc. Reprinted by permission of John Wiley & Sons, Inc.)

particulates. In normal design practice, gas velocity does not exceed 1 m/sec (3.5 ft/sec). Proper alignment of the emitting electrodes with respect to the collecting plates is also critical.

The paper industry has had a lot of experience in designing ESPs for recovery furnaces, As a result, the manufacturers have been able to guarantee very high removal efficiencies. ESPs are obviously the technology of choice for particulate control on the recovery furnace.

These devices require good maintenance in order to retain their high efficiencies. Problems arise from corrosion, rapper failure, plugging of gas distribution plates, and other causes. The ESP is an expensive piece of air pollution control equipment, but it does an excellent job and saves a considerable amount of environmental damage from particulates. It also saves salt cake, which is an economic inducement to use the device.

The TAPPI Air Committee conducts surveys of experience with ESPs; references 17 and 18 summarize this experience for the two types of recovery furnaces used in the paper industry. For recovery furnaces with direct-contact evaporators, the recent survey showed that ESPs had an average collection efficiency of 99.6% with an average dust outlet of 0.021 grains/actual m^3 (0.009 grains/SCF). The collection velocity averaged 1.2 m/sec. (These data are based on a study of 10 units.) Corrosion still remains a serious problem for one-third of the units surveyed. Other difficulties reported were problems with wires, rappers, and dampers, and with plugging of inlet distribution devices. The wet-bottom units were most popular; they averaged 79 hr/year of down time and maintenance costs of $19,000/year.

For recovery furnaces that are not preceded by direct-contact evaporation, the collection efficiencies were more variable, but a larger number of units were surveyed. Fourteen of the units performed in the efficiency range of 99.6–99.8%, 12 in the range of 99.4–99.6%, 7 in the range of 99.0–99.4%, and 9 in the range of 98.0–99.0%.

For noncontact furnaces equipped with ESPs, the down time for wet-bottom units averaged only 25 hr/year, whereas for dry-bottom units it averaged 241 hr/year. It is easy to see why the wet-bottom units are preferred. The average maintenance cost for furnaces with wet-bottom units was $38,000/year. Corrosion was reported as a large problem in 63% of the units surveyed. Half of the dry-bottom devices reported conveyor problems. Salt cake buildup was a problem in 57% of the units.

The collection efficiency for a given type of furnace, and for propoerly designed ESPs, seemed to vary directly with the throughput rate per specific collection area. However, this relation does not hold between different types of furnaces. For direct-contact evaporators, an average of 84 m^2/actual m^3/sec of specific collection area was provided, whereas furnaces without direct contact evaporators on average provided 118 m^2/actual m^3/sec. The nature of the particles generated in these two types of furnaces must vary, because the furnace with a larger collection area had a lower collection efficiency.

ODOR CONTROL STRATEGIES AND EQUIPMENT

In general, noxious gaseous emissions are handled in one of four ways—condensation, oxidation, incineration, or adsorption—and the type chosen depends on the nature of the source. Low-volume, high-concentration sources such as digester relief and evaporator noncondensables are handled differently from high-volume, low-concentration sources such as gases from brown stock washing and the smelt-dissolving tanks. Because the U.S. pulp and paper industry consists of plants that were constructed at different times to unique plans, no set of all-encompassing solutions exists that can be transferred unmodified from one mill to the next. Each mill needs to be considered as a unique case for which an optimum solution must be found by applying the general principles that will comprise the remainder of this discussion on odor control. The major problems are TRS in the Kraft industry and SO_2 emissions in the sulfite industry. Incineration and black liquor oxidation are the only two control technologies widely used.

Condensation

Condensation is the favored technique, where possible, because it is easy, efficient, and energy conscious. Condensation changes an air pollution problem into a water pollution problem that can easily be handled by biological treatment. Condensation is practiced in a Kraft pulp mill in the digester and evaporator area, where it provides heat recovery and the reduction of gaseous emissions of TRS and organic nonsulfur compounds.

Chlorination

A technique that is effective but expensive is oxidation of TRS compounds by the use of chlorine. Chlorination of various streams is effective in reducing odor but costly because these compounds have a high chlorine demand. Chlorination is impractical unless a source of inexpensive chlorine is readily available.

Incineration

Incineration is also an oxidation process, and a cost-effective one. Often gases that cannot be condensed are accumulated and incinerated. There are several furnaces in a pulp mill that can be used, or a separate one just for this purpose can be constructed. Construction of a separate incinerator is unusual; commonly, a suitable furnace in the existing plant is used. In a Kraft mill, the best choice is probably the lime kiln because of its high temperature and stability of operation. A second choice is the recovery furnace, which is sometimes used during periods when the lime kiln is not in operation. The problem with the recovery boiler is that such use may increase the frequency of recovery boiler explosions, which is obviously undesirable. The cause of such explosions is still poorly understood, and anything that might affect the boiler's operation is suspect. Hogged fuel boilers usually do not operate at a high enough temperature, but oil- or gas-fired power boilers are suitable. Special care must be taken to prevent explosive conditions from developing in the feed lines by installing flame arresters and by maintaining the gas flow velocity above a critical level. The concentration of the gas must be kept either above or below the explosive concentration to avoid problems.

Black Liquor Oxidation

The main source of emissions in the Kraft process is the recovery boiler. One of the main causes of the high emission rate is the stripping of TRS gases during direct-contact evaporation. The next section will discuss elimination of direct-contact emissions. However, to conform to emission standards while using direct-contact evaporation, black liquor oxidation gives emission characteristics comparable to those of a low-odor recovery boiler design. The central idea is to oxidize all reduced sulfur compounds present to their oxidized state before direct contact evaporation so that no TRS will be emitted. The oxidation can be performed on either weak or strong black liquor, and either air or molecular oxygen can be used. From three to five times the stoichiometric oxygen requirement is usually supplied to assure complete reaction. Normally, the air/liquor ratios are greater than 110–190 m^3 air per cubic meter of black liquor (15 25 ft^3 of air per gallon), with retention times of 2.0–2.5 hours and minimum liquor depths of 2.7–3.6 m (9–12 ft).

Black liquor oxidation has been extensively studied, A brief overview of the principles involved and the various alternatives available will be given. For a more detailed treatment, references 16 and 19–22 can be consulted. These references are mainly reviews of the alternatives; their bibliographies, in turn, will lead to the details of specific processes.

Tables 22-2 and 22-3 show some representative benefits that can be gained from black liquor oxidation in emissions from the recovery boiler and non-condensable gases from the evaporators. Significant progress is being made in reducing all TRS emissions, and especially dramatic results have been achieved with hydrogen sulfide.

In general, black liquor oxidation affects the Kraft mill system in ways other than by reducing malodorous emissions. There is an increase in the yield of tall oil, reduced corrosion and scaling of evaporator tubes, improved condensate quality, lower chemical makeup requirements for lime and salt cake, and a

Table 22-2 Effect of Black Liquor Oxidation on Sulfur Gas Emissions during Direct-Contact Evaporation

	Liquor			
	Unoxidized		Oxidized	
Sulfur Compound	kg S/ton	(lb S/ton)	kg S/ton	lb S/ton
H_2S	2.50–15.00	(5.0–30)	0.05–1.00	(0.1–2.0)
CH_3HS	0.15–1.00	(0.30–2.0)	0.02–0.10	(0.04–0.2)
CH_3SCH_3	0.02–0.08	(0.04–0.16)	0.01–0.03	(0.02–0.06)
CH_3SSCH_3	0.05–0.15	(0.1–0.3)	0.01–0.08	(0.02–0.16)

Source: Reference 16.

Table 22-3 Effect of Weak Black Liquor Oxidation on Malodorous Sulfur Gas Emissions from Evaporator Noncondensable Gases

	Liquor			
	Unoxidized		Oxidized	
Sulfur Compound	kg S/ton	lb S/ton	kg S/ton	(lb S/ton)
H_2S	0.05–1.40	(0.10–2.8)	0.00–0.01	(0–0.02)
CH_3SH	0.05–0.50	(0.10–1.0)	0.05–0.10	(0.1–0.2)
CH_3SCH_3	0.01–0.02	(0.02–0.04)	0.01–0.04	(0.02–0.08)
CH_3SSCH_3	0.00–0.01	(0.0–0.02)	0.02–0.05	(0.04–0.10)

Source: Reference 16.

reduction of 2–6% in black liquor heating value. Because TRS emissions are reduced, the sulfur stays in the system and its sulfidity increases. This change in the Na/S ratio needs to be controlled, and caustic is added as part of the makeup chemicals rather than just salt cake. This procedure increases the chemical costs slightly.

Recent research has shown that black liquor oxidation may aid in limiting SO_2 production in the recovery furnace (23). The production of H_2S is about three times higher from sulfide sulfur than from thiosulfate sulfur. The pyrolysis reactions are clearly quite complex, but sulfide and thiosulfate clearly do behave differently. Even sulfate is not completely stable during pyrolysis, and some reduced sulfur species can be formed from sulfate at temperatures below 550°C (23).

There are several disadvantages (24) to the black liquor oxidation process:

1 Methyl sulfides remain volatile and odorous
2 Incomplete oxidation and reversion are possible
3 Sulfur or polysulfide may form, only to revert later as hydrosulfide ion
4 The heating value of the black liquor is reduced
5 The effluent gas from the oxidation process may be contaminated with volatile sulfur compounds

Weak Liquor Oxidation. Anyone who has worked with black liquor knows that it is an unpleasant material that gets nastier as it becomes more concentrated. Logically, therefore, it might be easier to develop an oxidation system for weak liquor than for strong liquor. A number of weak liquor oxidation systems have in fact been developed and are in use in the Pacific Northwest, upper Midwest, and Northeast, where highly resinous pine wood species are not pulped. Figure 22-7 shows a representative weak black liquor oxidation system employing a multiple-sieve tray tower system. Other systems employ porous plate diffusers,

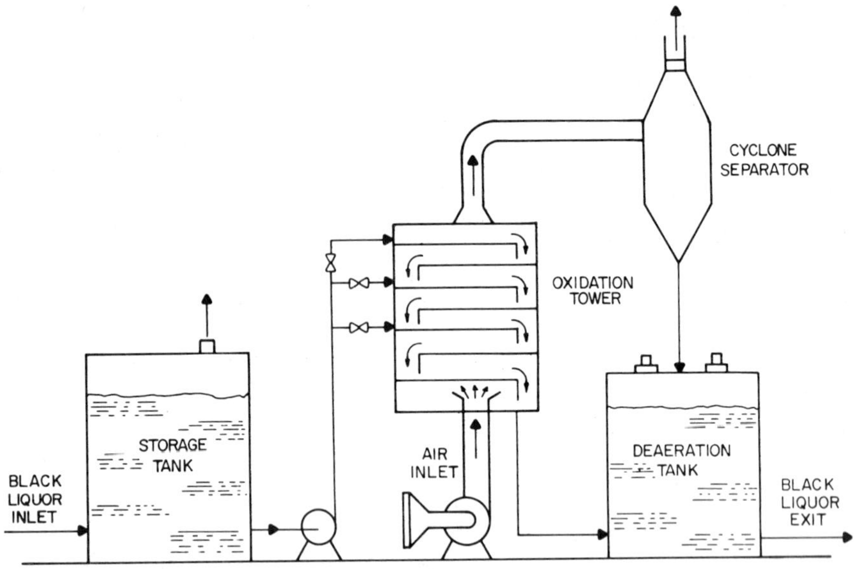

Figure 22-7 Weak liquor oxidation

packed towers, agitated air spargers, and rotating fluid contacters—a variety of options.

However, there are several problems with weak liquor oxidation:

1 Excessive foaming occurs when resinous pine wood species are pulped
2 Incomplete Na_2S oxidation is caused by improperly designed or underdesigned systems
3 Reversion of oxidized species occurs during evaporation
4 Increasing the pulping capacity without expanding the black liquor oxidation facilities causes system overload, resulting in inadequate liquor retention time
5 It is impossible to achieve effective oxygen mass transfer from air to black liquor

These considerations, particularly the foaming and reversion problems, were the principal driving forces in the development of strong liquor oxidation.

Strong Liquor Oxidation. Strong liquor oxidation is widely practiced, especially in the southern United States. Figure 22-8 shows a typical strong liquor oxidation system. It is a completely mixed system in which agitation is provided by the air sparged into it. Na_2S oxidation efficiencies of 97–98% can be achieved at a 30-g/l inlet concentration with a 2.5-hour detention time. Higher oxidation efficiencies, in the 99–100% range, can be achieved with two-stage systems. An efficiency of 99.95% has been reported, with a resulting Na_2S

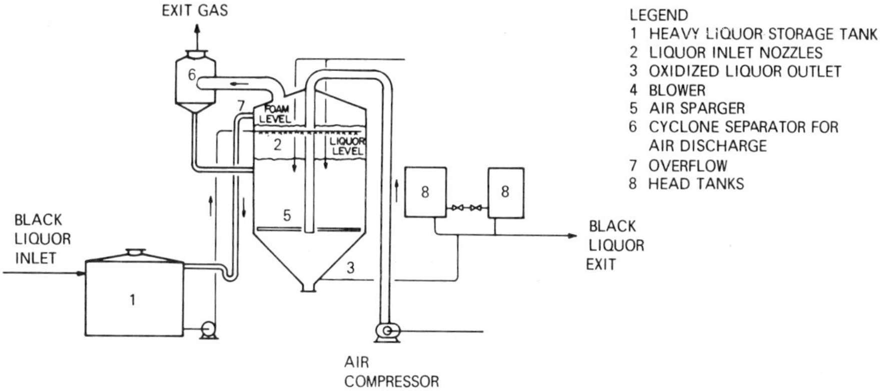

Figure 22-8 Strong liquor oxidation

concentration in the exit liquor of 0.02 g/l. This result yields H_2S emissions from the recovery furnace of about 2 ppm, compared with 50 ppm from the single-stage system previously in operation at the same location (16).

Combined Systems. Another development aimed at improving efficiency is the combination weak and strong liquor oxidation system. The combination system is basically a weak liquor oxidation system plus a strong liquor system to polish the strong liquor after evaporation, mainly to counter any Na_2S reversion that has taken place during evaporation. Nearby 100% efficiencies have been reported with the combination systems (25). Pure oxygen systems have also been developed with the aim of improving efficiency and reducing foaming because less vapor volume is involved.

Low-Odor Recovery Boilers

The disadvantages of black liquor oxidation have convinced many mills to opt for a low-odor recovery boiler instead. In this system, the direct contact evaporator is eliminated and the black liquor is concentrated from 50 to 65% via a forced-circulation, multiple-effect evaporator. It is then fed into a low-odor recovery boiler, which has a different design than one used for firing liquor from a direct-contact evaporator because the gases from the furnace exit up the stack rather than going to the direct-contact evaporator. In order to recover the heat that is normally used in the direct-contact evaporator, an additional economizer is installed inside the furnace. There is obviously a lot more to the design of a low-odor recovery furnace than this, and if details are needed, reference 16 should be consulted.

The Scandinavians were the first to develop and refine this type of furnace. At present, two American manufacturers, Babcock and Wilcox and Combustion Engineering, offer such a furnace. They contend that there are appreciable differences between their designs, and a careful study of both should be made

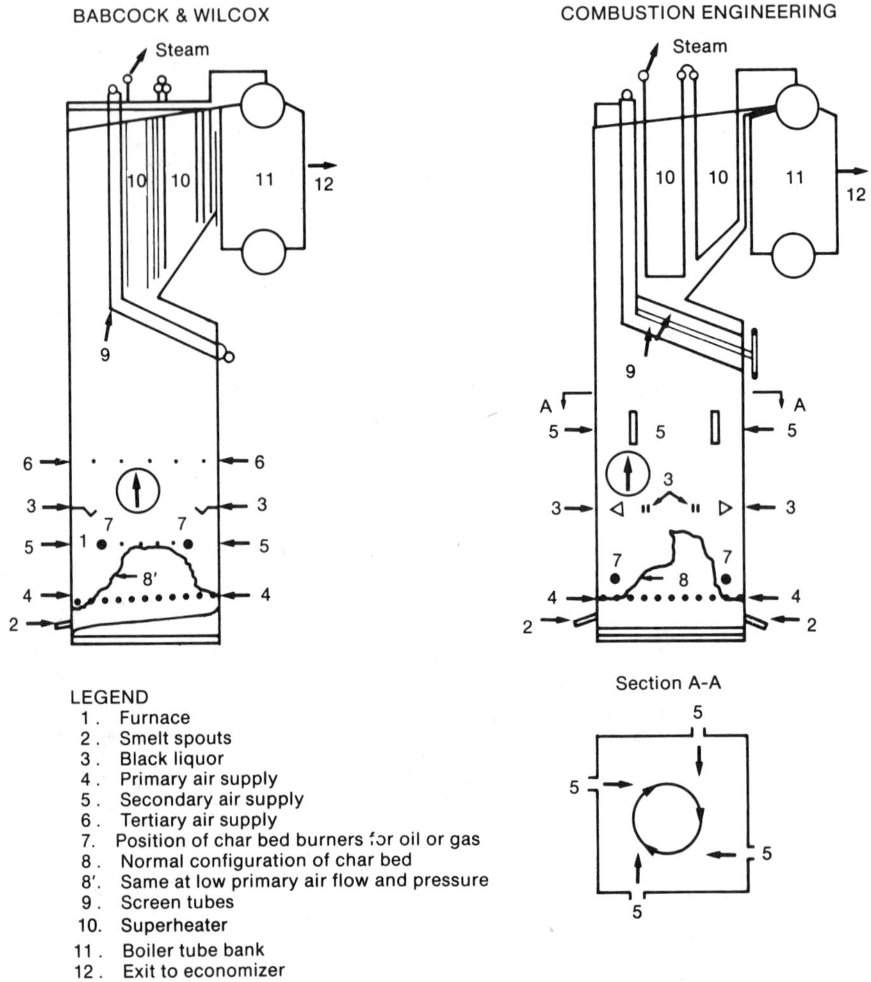

Figure 22-9 Low-order recovery furnaces

before making a choice. Such a detailed examination is beyond the scope of this presentation. Figure 22-9 illustrates these two different designs for low-odor furnaces by American manufacturers.

In-Plant Process Modifications

As was stated previously, the decision on how to control emissions must be made by developing an optimum solution for each mill individually. There are, however, a number of case histories on new mills (26–29) and on the conversion of older mills to low-emission status (30–34) that can be consulted for aid and suggestions in devising the optimum scheme for a particular mill.

A major cause of high emission rates is operation of the system in a state of overload or upset. The best pollution control strategy will be ineffective if proper process control is not maintained. This is especially true of the Kraft mill recovery boiler, whose emission characteristics degenerate significantly in an overloaded, upset state. If the mill was constructed in an era when it was common to undersize the recovery boiler, the overload condition can be relieved by installing a fluidized bed incinerator to burn surplus black liquor (35). A fluidized bed can be installed at significantly less cost than an additional recovery boiler.

Another in-plant process control alternative to help alleviate TRS emissions is to lower the sulfidity of the cooking liquor. Both methyl sulfide and methyl mercaptan are unavoidably produced in Kraft pulping. The release of these compounds can be minimized by keeping the sulfidity below 20% (24), but this is not common practice in U.S. Kraft mills. There may also be energy savings associated with a reduction in sulfidity, as well as possible compromises in pulp quality.

Mills with a batch digester could convert to a cold blow process to conserve energy and reduce TRS emissions. In the United States, the Rader Company has developed this process. It consists of displacing the hot black liquor at the end of the batch cook with cooler liquor from the filtrate tank and then blowing the digester with compressed air. The displaced hot liquor is used to heat incoming white liquor and to preheat chips for the next cook. Since the digester is blown cold, considerably less TRS emissions are produced. The cold blow produces 0.013 kg TRS/ton (0.03 lb TRS/ton), whereas a conventional hot blow might produce 0.091 kg TRS/ton (0.2 lb TRS/ton) (36). This emissions level is not low enough to eliminate the need for incineration, however. This system seems very advantageous. The only disadvantage is that more of the methanol stays with the black liquor to appear in the evaporator condensate. It is estimated that 3.8 kg (8.4 lb) methanol per ton will be produced in the cold blow process versus 2.4 kg/ton (5.2 lb/ton) for a conventional hot blow (36).

Another alternative to reduce odor emissions and conserve water is the deodorization and reuse of condensates. Stripping of contaminated condensates has proved an efficient and economical way of removing odor and BOD. Two principal techniques are air stripping and steam stripping. The details of these stripping processes are given in the literature (37–38). Air stripping is usually done with stack gas from a power boiler; it is important to keep the pH below 9 and the condensate feed as hot as possible. Above pH 9, foaming becomes severe and efficiency declines. Air stripping is cheaper to install and operate than steam stripping, but it is also less effective. Steam stripping, which is usually done in a specially constructed stripping column, can achieve any degree of TRS removal desired, depending on the amount of steam used. Figure 22-10 illustrates the relationship of steam consumption to TRS and BOD removal. Thus the problem is a matter of seeking the economic optimum. The condensates represent approximately one-third of the BOD load from a Kraft mill, so a significant amount of BOD can be removed by this method. The stripped gases

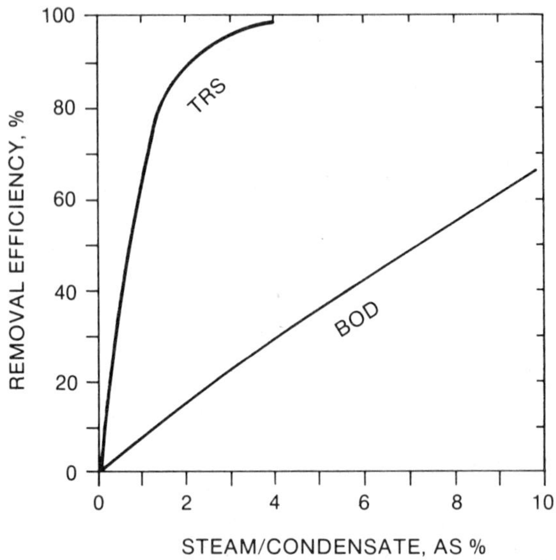

Figure 22-10 TRS and BOD stripping as a function of steam use (10 theoretical plates). Copyright © 1981. TAPPI. Reprinted from *Introduction to Environmental Management for the Pulp and Paper Industry*, Study Guide, Volume 6, with permission

are incinerated. The cleaned condensates can be reused in pulp washing, as scrubbing fluid for lime kilns, or as white liquor makeup water.

Emissions of nitrogen oxides and sulfur oxides should also be discussed under the topic of in-plant controls. As of 1984, the industry had not been required to install special equipment to handle these gases, but extensive literature on possible methods does exist (39–42). Many techniques are based on scrubbing the gas with calcium-based alkalis, lime, and limestone. Insoluble calcium–sulfur compounds are formed that must be removed from the scrubbing solution and disposed of in a landfill. Removal efficiencies range from 72 to 80%, and there are some problems with scaling. A process using magnesium-containing lime has been developed that eliminates scaling and improves efficiency (41).

SORTING OUT THE ALTERNATIVES

In order to decide what alternatives to choose for each emission, the mill environmental engineer should prepare various alternative total mill emission control plans that not only conform to present regulations but can be modified to meet expected future regulations. A cost estimate for each plan should then be prepared to see which is the most cost effective. A probable profit approach, as outlined in reference 43, should be used in this assessment. Table 22-4, which lists all of the sources of emissions in a Kraft mill and the alternative technologies

Table 22-4 Potential Kraft Mill Air Pollution Controls and the Pollutants Removed

	Source and Pollutant																											
	Digester			Evaporator			Washer		Recovery boiler						Lime burning—dust handling				Smelt-dissolving tank		Power Boiler					Bleach plant		
Controls used	TRS	especially H_2S	Organic non-S cmpds.	TRS	Especially H_2S	Organic non-S cmpds.	TRS	Organic non-S cmpds.	TRS	especially H_2S	Organic non-S cmpds	Particulates	NO_x	SO_x	TRS	Particulates	NO_x	SO_x	TRS	Particulates		Organic non-S cmpds.	Particulates	NO_x	SO_x	SO_x	CL_x	CLO_2
Scrubbers (unspecified)				X											X	X	X	X	X							X	X	X
Venturi												X		X		X			X	X			X					
Cyclone																X			X	X								
Impingement																X							X					
Absorption/packed tower	X	X		X	X				X	X		X		X					X	X						X		
ESP									X			X				X							X					
Cyclone separator																X							X					
Fabric filter (bag house)																							X					
Mist pads																				X								
Chlorination	X			X																								
Combination																												
ESP and scrubber												X							X	X								
Cyclone separated and scrubber																X												
Mist pads and scrubber																			X	X								
Black liquor oxidation				X	X																							
Incineration in lime kiln/recovery or power boiler	X		X	X		X	X	X																				
Condensation	X		X																									
Internal controls											X		X								X		X	X				
Venturi and packed tower									X			X	X															

Source: Copyright © 1981. TAPPI. Reprinted from *Introduction to Environmental Management for the Pulp and Paper Industry*, Study Guide, Volume 6, with permission.

Table 22-5 Potential Sulfite Mill Air Pollution Controls and Pollutants Removed

Controls Used	Source and Pollutant									
	Digester	Evaporator	Washer	Recovery boiler				Power boilers		
	SO_x	SO_x	SO_x	SO_x	NO_x	H_2S	Particulates	SO_x	NO_x	Particulates	Organic non-S Cmpds.
Scrubbers (unspecified)	X	X	X	X							
Venturi				X						X	
Absorption/packed tower	X	X				X					
ESP							X				
Cyclone separator							X			X	
Fabric filter										X	
ESP and scrubber							X				
Internal controls	X	X			X			X	X		X

Source: Copyright © 1981. TAPPI. Reprinted from *Introduction to Environmental Management for the Pulp and Paper Industry*, Study Guide, Volume 6, with permission.

available for dealing with each, should serve as an aid in developing various total mill emissions control schemes. The table presents, as simply as possible, most of the alternatives available to the environmental engineer. Table 22-5 presents similar information for the sulfite mill. These tables can serve as the starting point for developing alternative plans that best fit the mill under consideration; additional ideas and special mill factors will also need to be included.

COSTS

In order to choose among the various alternatives, the costs of each one must be determined. Table 22-6 lists the capital, operating, and maintenance costs for various air pollution control devices; comments on the effectiveness of each device are given. Although cost figures must be tailored to the specific location and circumstances of each mill, the figures given in this table can serve as a starting point. Its purpose is mainly to give some appreciation of relative costs; references 2, 12, and 16 can also be helpful in this regard. Inflation and other cost changes must be taken into account, of course.

Cost estimation is always difficult and might best be called cost "guesstimation"; the many different factors involve a lot of uncertainty. A reputable

Table 22-6 Summary of Air Pollution Control Technologies

Device	Percent Efficiency (Pollutants)[a]	Cost[b] Capital	O & M (Date)	Advantages/Disadvantages	Usage[c]
Venturi Scrubber	95–99 +	\$22–27/TPD \$0.6–4/CFM	\$6–10/TPD/ year (10/76)	Requires less ground space than most particulate controllers, has few internal parts to maintain. Cost of operation = 50–60% more than ESP	Used in many lime kilns; limited use in recovery boilers and smelt-dissolving tanks
Cyclone scrubber	65–95	\$0.5–2.1/CFM		—	Normally used in conjunction with mist pads in a lime kiln
Impingement scrubber	92–97	\$29–32/TPD	\$3–6/TPD/ year	Efficient on coarse particulates ($> 10\ \mu m$)	Used in a few lime kilns and in wood-burning power boilers
Absorption/ packed tower	80–90 (TRS) 85–98 (particulates) 98–99 (H_2S) 90–97 (SO_2)	\$2–8/CFM plus installation		Subject to plugging and channeling	Not widely used
ESP	91–99 +	\$0.4–2.3 × 10^6 \$0.9–4.2/CFM	\$0.05–0.7/CFM/ year (1971)	Economical, with simple, low-cost operation High initial cost, large space requirements, cannot be efficiently used on explosive gases; may foul with SO_2 sensitive to flow variation	Widespread usage in recovery and power boilers; sometimes used in lime kilns

(*Contd.*)

Table 22-6 (*Contd.*)

Device	Percent Efficiency (Pollutants)[a]	Cost[b] Capital	Cost[b] O & M (Date)	Advantages/Disadvantages	Usage[c]
Cyclone separator	65–99 +	\$0.59–1.0/CFM plus installation		Efficient on coarse particulates and high concentration Excellent as a primary collection system	Typical in lime kilns; also found in coal- and wood-fired power boilers and in some recovery furnaces
Fabric filter	99 +	\$0.91–4/CFM	0.12–0.32/ CFM/YEAR (1971)	Useful on gases of low humidity, volume, and temperature, and to trap particulates that are economical to recover	Not extensively used in any process
Black liquor oxidation	50–100 (TRS) 95–99 (H_2S) 50–100 (TRS)	\$350–1,000/TPD	\$10–200/ TDP/year (11/73)	Can provide better performance in a recovery boiler for chemical recovery	Frequently used
Mist pads	70–90	—	—	—	Used for some smelt-dissolving tanks
Incineration	—	\$108–162/TPD	\$0.01–0.05/ TPD/year (10/76)	—	Lime kiln incineration relatively widespread

Source: Reference 44.

[a]Unless otherwise specified, efficiency is based on particulates.

[b]Cost figures based on TPD (tons of product per day) and CFM (cubic feet of gas treated per minute). O & M (operation and maintenance) costs do not include any credits (such as any chemical or heat recovery).

engineering consulting firm, preferably one that is familiar with pollution control problems in the industry and has done similar jobs recently, should be consulted for the best, most realistic current estimates.

For a new mill in 1985, the capital expenditure for environmental controls for both air and water amounted to 12.5–15% of the total capital cost of the mill (45, 46). This is not an insignificant figure, and it reflects the increased attention that mills must pay to externalities.

RETROFITTING ENVIRONMENTAL CONTROL

The problems involved in making an old mill reasonable compliant with odor regulations are considerably different from those involved in designing a system as an integral part of a new mill. Table 22-7 compares the guidelines for existing mills with the NSPS. The existing source limits are more liberal for the lime kiln and the recovery furnace, which are the two major sources of emissions. For an existing mill, the goal is to reduce odor to an acceptable level, not to eliminate it. There are no reported cases of retrofit mills achieving elimination of all odors (33).

Table 22-7 Regulatory Limits for Kraft Pulp Mill TRS Control

	NSPS	Existing Mill Emission Guidelines
Recovery furnace, ppmv		
Old design	5[a]	20[a]
New design	5[a]	5[a]
Cross recovery	25[a]	25[a]
Digester system, ppmv	5	5
Multiple-effect evaporator system, ppmv	5	5
Lime kiln, ppmv	8	20[b]
Brown stock washer system, ppmv	5	No control
Black liquor oxidation system, ppmv	5	No control
Condensate stripper system,[c] ppmv	5	5
Smelt-dissolving tank, mg/kg of black liquor solids	8.4	8.4

Source: Reference 33. Copyright © 1980, reprinted from *Tappi Journal*, 63 (8):53 (1980) Technical Association of the Pulp and Paper Industry, with permission.

[a]To be achieved in 99% of contiguous 12-hour periods each quarter.

[b]To be achieved in 98% of contiguous 12-hour periods each quarter.

[c]Condensate stripping is not required, but control of emissions from the stripper is.

Before beginning an odor minimization program, data must be collected about each source to determine the size of the control equipment required. A good rule of thumb is that 5 days of continuous data on all major sources is required to make an intelligent decision on equipment (33).

The usual approach is to collect and incinerate. Problems can be expected in fitting the collection hoods and duct system into the existing space. Incineration at a temperature greater than 816°C (1,500°F) renders gases odor free (34).

One mill chose to collect and incinerate in a coal-fired power boiler. It incinerated digester blow gas, digester relief gas, foam tower vent gas, evaporator hotwell vent gas, condensate air stripper vent gas, strong liquor oxidation vent gas, and brown stock washer vent gas (34). The project cost $2.2 million, or $10,092 per ton of pulping capacity. This effect did not render the mill odor free, but it did bring the odor down to a tolerable level and minimized community complaints.

REFERENCES

1 J. A. Danielson, *Air Pollution Engineering Manual*, U.S. Environmental Protection Agency, Research Triangle Park, N.C., May 1973.

2 E. R. Hendrickson et al., *Control of Atmospheric Emissions in the Wood Pulping Industry*, Vol. 2, PB-190-352, March 15, 1970.

3 H. W. Parker, *Air Pollution*, Prentice-Hall, Englewood Cliffs, N.J., 1977.

4 *Atmospheric Emissions from Pulp and Paper Manufacturing Industry*, Atmospheric Quality Improvement Technical Bulletin No. 69, National Council of the Paper Industry for Air and Stream Improvement, New York, February 1974.

5 K. T. Semrau, "Correlation of dust scrubber efficiency," *Journal of the Air Pollution Control Association*, 10 (6):200–207 (1960).

6 T. T. Collins, *Paper Industry*, 28 (8):680–686 (1947).

7 C. V. Callaghan, *Pulp and Paper Magazine of Canada*, 54 (3):106–107 (April 1953).

8 A. B. Adams, *Journal of the Air Pollution Control Association*, 26 (4):303–307 (1976).

9 R. W. Boubel, *Control of particulate emissions from wood fired boilers*, EPA Report, Research Triangle Park, North Carolina (prepared by Pedco Environmental, Cincinnati), 1979.

10 D. S. Dillard, *Tappi*, 63 (10):107–111 (1980).

11 E. N. Parquet, *Tappi*, 63 (4):23–24 (1980).

12 M. F. Szabo and R. W. Gerstle, *Operation and maintenance of particulate control devices in Kraft pulp mill and crushed stone industries*, EPA 600/2-78-210, U.S. Government Printing Office, Washington, D. C., October 1978.

13 P. J. Balasic and M. M. Peocos, *Tappi*, 63 (6):41–43 (1980).

14 H. White, *Industrial Electrostatic Precipitation*, Addison-Wesley, New York, 1963.

15 S. Oglesby, Jr., and G. B. Nichols, *A Mannual of Electrostatic Precipitation Technology*, Parts I and II, PB 196–381, August 1970, EPA, Washington, D. C.

16 *Environmental pollution control pulp and paper industry, Part I: Air*, EPA Technology Transfer Series, EPA-625/7-76-001, October 1976, Washington, D. C.

17 J. S. Henderson, *Tappi*, 65 (3):91–94 (1982).

18 J. S. Henderson, *Tappi*, 64 (1):55–58 (1981).

19 T. T. Collins, *Tappi*, 38 (8):172A–175A (1955).

20 J. Landry, *Tappi*, 46 (12):766–772 (1963).

21 E. R. Hendrickson and C. J. Harding, *Air Pollution Control Association Journal*, 14 (12):487–490 (1964).

22 R. O. Blosser and H. B. H. Cooper, *Survey of black liquor oxidation practices in the Kraft industry*, Atmospheric Pollution Technical Bulletin No. 39, National Council of the Paper Industry for Air and Stream Improvement, New York, December 1968.

23 P. T. Strahbien and T. M. Grace, *Tappi*, 65 (10):125–126 (1982).

24 K. V. Sarkanen et al., *Tappi*, 53 (6):766–783 (1970).

25 R. C. Tobias et al., "A non-conventional strong black liquor secondary oxidation system," paper presented at the West Coast Regional Meeting, National Council of the Paper Industry for Air and Stream Improvement, Portland, Ore., November 4, 1970.

26 L. L. Canovali and S. Suda, *Tappi*, 53 (8):1488–1493 (1970).

27 K. C. Ayers, L. W. Clutter, and A. B. Adams, Jr., *Tappi*, 57 (9):125–128 (1974).

28 P. J. Connery, "Thermal oxidation of Kraft pulping emissions," *1978 TAPPI Environmental Conference*, Technical Association of the Pulp and Paper Industry, Atlanta, 1978, p. 31.

29 J. C. W. Evans, *Pulp and Paper*, 52 (10):70–74 (1978).

30 R. F. Winplepleck and S. Suda, *Tappi*, 56 (6):73–77 (1973).

31 Hoerner-Waldorf Corporation, *Improved air pollution control for Kraft recovery boiler: Modified recovery boiler no. 3,* EPA-650/2-74-041A, Contract No. 68-02-0247. (Available from NTIS, PB 237 62 F/AS.)

32 G. A. Hansen, *Journal of the Air Pollution Control Association*, 12 (9):409–413 (1962).

33 J. R. Smith and K. C. Ayres, *Tappi*, 63 (8):53–57 (1980).

34 T. J. Parrett, *Tappi*, 63 (10):113–116 (1980).

35 G. G. Copeland, "A 12 year review of fluidized bed technology as applied to pulp and paper," *1975 Alkaline Pulping Conference*, Technical Association of the Pulp and Paper Industry, Atlanta, 1975, p. 53.

36 A. K. Jain, *Investigation of TRS emission and raw waste load impact of the use of the Rade cold blow process*, Technical Bulletin No. 403, National Council of the Paper Industry for Air and Stream Improvement, New York, 1983.

37 I. P. Morgan and F. E. Murray, *Pulp and Paper Magazine of Canada*, 73 (5):62–66 (1972).

38 M. I. Matteson, L. N. Johanson, and J. L. McCarthy, *Tappi*, 50 (2):86–91 (1967).

39 R. W. Coughlin et al., *Air Pollution and Its Control*, AIChE Symposium Series 126, Vol. 68, 1972, New York.

40 A. G. Cooper, *Sulfur oxides and other sulfur compounds, a bibliography with abstracts*, U.S. Department of Health, Education, and Welfare, Washington, D. C., 1965.

41 "Flue gas desulfurization," *Chemical Engineering Progress*, 72 (2):41–80 (1978).

42 "Air pollution control," *Chemical Engineering Progress*, 73 (8):31–73 (1979).

43 A. M. Springer and R. C. Peterson, "A systems analysis approach to process water conservation in manufacture of paper and board," paper presented at the AIChE Annual Meeting, San Francisco, November 25–29, 1979.

44 *Multimedia pollution assessment in pulp, paper and other wood products industry*, EPA Draft Summary Report, Contract No. 68-03-2605, prepared by the E. C. Jordon Co., Inc., Portland, Maine, February 1979.

45 H. I. Dyer, *Paper Trade Journal*, 162 (15):27–29, 41 (1978).

46 K. L. Patrick, *Pulp and Paper*, 53 (5):68–74 (1979).

23
Ambient Air Monitoring of Paper Industry Emissions

BACKGROUND

The Clean Air Act of 1977 reinforced the goal of obtaining air clean enough to harm neither public health nor public welfare. Primary air standards were established to protect public health and secondary air standards to protect public welfare.

The federal government published a set of Air Quality Criteria Documents that describe in detail the effects of different concentrations of air pollutants. These documents summarize current knowledge obtained from both epidemiological (health effects observed in the field) and toxicological (laboratory data) studies on specific pollutants. States were to use these documents to set AAQS. The states were also required to submit implementation plans showing how they would achieve compliance. The federal government issued a set of primary and secondary standards, and the standards determined by each state cannot be more liberal than the federal ones. Table 23-1 gives the federal standards and, for comparison, the Ohio state standards. The pollutants regulated are suspended particulates, sulfur dioxide, carbon monoxide, photochemical oxidants, nonmethane hydrocarbons, and nitrogen dioxide. Depending upon the health consequences of a particular pollutant, the requirements vary from as strict as a 1-hour mean concentration to as liberal as an annual mean.

The federal regulations and procedures for testing for ambient pollutants are listed in the *Federal Register* and the *Code of Federal Regulations* (1, 2). They are also specified in *Methods of Air Sampling and Analysis*, published by the American Public Health Association (3), which is the *Standard Methods* equivalent for air sampling and analysis. Many other textbooks on the subject are also available (4, 5). A good reference on setting up an ambient air monitoring program, which obviously involves more than just the analytical procedures, is *Air Pollution Sampling and Analysis Deskbook* (5).

AMBIENT PARTICULATE SAMPLING

The most common device for sampling airborne particulate matter is the high-volume sampler and shelter illustrated in Figure 23-1; its internal parts are

Table 23-1 Air Quality Standards

Pollutant	Duration[a]	Restriction	Ohio EPA	Federal EPA[b] Primary (Health)	Federal EPA[b] Secondary (Welfare)
Suspended particulates	Annual mean (G)	Not to be exceeded	60 $\mu g/m^3$	75 $\mu g/m^3$	60 $\mu g/m^3$
Suspended particulates	24-hour concentration	Not to be exceeded more than once per year	150 $\mu g/m^3$	260 $\mu g/m^3$	150 $\mu g/m^3$
Sulfur Dioxide	Annual mean (A)	Not to be exceeded	60 $\mu g/m^3$ 0.02 PPM	80 $\mu g/m^3$ 0.03 PPM	60 $\mu g/m^3$ 0.02 PPM
Sulfur dioxide	24-hour concentration	Not to be exceeded more than once per year	260 $\mu g/m^3$ 0.10 PPM	365 $\mu g/m^3$ 0.14 PPM	260 $\mu g/m^3$ 0.07 PPM
Carbon monoxide	8-hour mean (A) concentration	Not to be exceeded by more than 1 8-hour period per year	9.0 PPM 10 $\mu g/m^3$	9.0 PPM 10 $\mu g/m^3$	9.0 PPM 10 $\mu g/m^3$
Photochemical oxidants	1-hour mean (A) concentration	Not to be exceeded more than once per year	0.06 PPM	0.08 PPM	0.08 PPM
Photochemical oxidants		Not to be exceeded by more than 1 consecutive 4-hr period per year	0.04 PPM	—	—
Photochemical oxidants	24-hour mean (A) concentration	Not to be exceeded by more than 1 day per year	40 $\mu g/m^3$ 0.02 PPM	— —	— —
Nonmethane hydrocarbons	3-hour mean (A) concentration	Not to be exceeded between 6 AM and 9 AM	0.19 PPM	0.24 PPM 100 $\mu g/m^3$	0.24 PPM 100 $\mu g/m^3$
Nonmethane hydrocarbons	24-hour mean (A) concentration	Not to be exceeded by more than 1 day per year	0.50 PPM	—	—
Nitrogen dioxide	Annual mean (A)	Not to be exceeded	0.05 PPM	0.05 PPM 100 $\mu g/m^3$	0.05 PPM 100 $\mu g/m^3$

Source: Reference 1.
[a](A), arithmetic; (G), geometric.
[b]Primary standard—for protection of public health; secondary standard—for protection of public welfare.

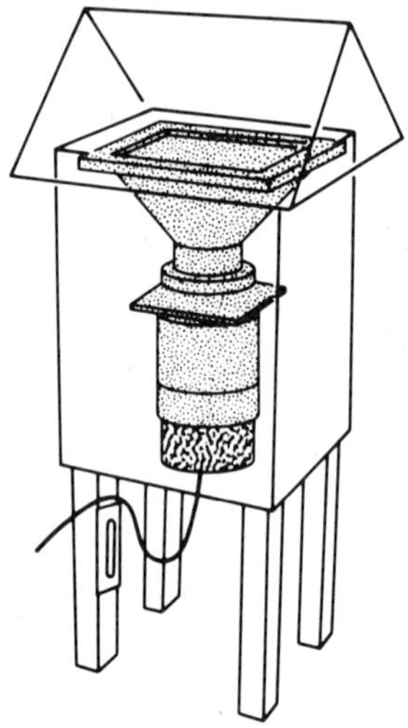

Figure 23-1 High-volume sampler

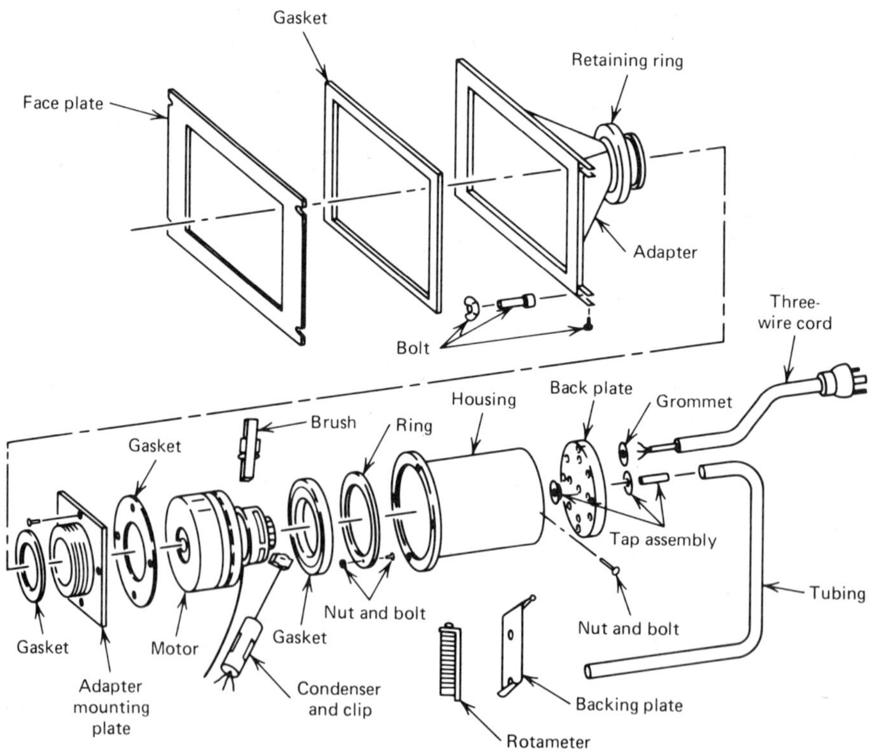

Figure 23-2 High-volume sampler—parts

shown in Figure 23-2 (1). In this device, air is drawn into a covered housing and through a filter by means of a high-rate suction blower. The flow rate through the filter is 1.13–1.7 m^3/min (40–60 ft^3/min), which allows particles less than 100 μm in size to pass into the device and be collected on a previously tared glass fiber filter surface. The filter traps particles down to 0.1 μm. The mass concentration is calculated after a 24-hour collection period and is reported as micrograms per cubic meter of air. The sampler must be calibrated, and detailed operating procedures have been published (1). Routine maintenance, quality control, troubleshooting, and other suggestions have also been reported (6). If proposed regulations specific to particle sizes less than 10 μm are promulgated, the high-volume sampler with the wedding head modification will be used. Other measurement techniques are available for dust fall, particle size classification, optical microscopic examination, chemical analysis, and radioactivity (3, 4).

The high-volume sampler is the device of choice for ambient particulate sampling, but it does not differentiate the sources of the particulates it collects. Recent work (7) indicates that problems are often caused not by point sources but by fugitive dust loadings. In sorting out fugitive dust problems, the sticky paper visual procedure for particulates may be useful. Particulates carried by wind are collected on adhesive paper, and the number of particles per square millimeter (P/mm^2) is estimated by comparison with visual standards (8). Particles may be examined to determine their origin. The procedure has a sensitivity of 1–15 P/mm^2 and a range of 2–100 P/mm^2. The samples can be preserved with clear enamel and are then stable for about 3 months.

The majority of particulates from the recovery furnace are Na_2SO_4, which can be extracted from the filter of the high-volume sampler with distilled water, in which the sulfate is soluble. The sulfite can then be determined in a variety of ways (9, 10), one of which is with a reagent that contains equal amounts of barium chloride and methyl thymol blue at a pH of 2.8. After mixing, the pH is raised to 12.4 with potassium hydroxide and the unreacted barium forms a chelate with the dye. The excess dye becomes yellow and is determined colorimetrically. The procedure is applicable over the range 1.0–150 μg SO_4^{-2}/ml with a sensitivity of $\pm$ 1.0 μg/ml. The sulfate solution must be free of particulates, heavy metals, and oxidizing substances, all of which interfere with the procedure.

SULFUR DIOXIDE ANALYSIS

Because of the great concern over SO_2 pollution, many methods have been devised for determining its ambient levels (11). Colorimetry, coulombic titration, electrochemistry, flame photometry, and emission and absorption spectroscopy are among the many techniques that have been employed. The one chosen depends on the amount of money and personnel available and on whether continuous or grab samples are needed. A number of commercial instruments based on one or another of these techniques have been developed. The most severe limitation of all commercial instruments and many manual methods is their lack of sensitivity. The realistic lower detection limit of most instrumental methods is about 0.01 ppm. Since a typical urban concentration of SO_2 is less than 0.02 ppm, a real need exists to improve the lower detection limit of these instruments (4).

The recommended manual procedures for SO_2 determination are the West–Goeke colorimetric technique (12) and the hydrogen peroxide method (13). However, if the sampler is located near a Kraft mill, it may be desirable to use the gas chromatographic method for determining SO_2, because it can be adapted to determine TRS compounds as well. With proper column selection, an air stream can be injected from which the compounds will separate before entering the flame detector. Hydrogen sulfide and carbonyl sulfide are eluted first, followed by sulfur dioxide, methyl mercaptan, ethyl mercaptan, dimethyl sulfide, and propyl mercaptan. After each analysis cycle, a stripper column is backflushed to ensure that heavier sulfur compounds do not reach the chromatographic column. The operating range of this instrument is 0.0–1.0 ppm (0.0–2,618 $\mu g/m^3$). Carbonyl sulfide causes interference but can be removed with silver wool and scrubbers. The minimum detectable limit for different compounds is given in Table 23-2 (10). This approach is very advantageous and has good sensitivity if the instrument can be made to operate in a reliable, trouble-free manner.

The EPA reference method for the determination of sulfur dioxide is the pararosaniline method. Air is bubbled through a solution of 0.04 *M* potassium tetrachloromercurate (TCM). The SO_2 reacts with the TCM to form a stable monochlorosulfonatomercurate complex, which is even stable to strong oxidants. This complex is then reacted with acid-bleached pararosaniline dye and formaldehyde to form an intensely colored pararosaniline methyl sulfonic acid. The quantity present is determined spectrophotometrically at 548 nm and is directly related to the amount of SO_2 collected. The lower limit of the procedure is 0.01 ppm (25 $\mu g/m^3$), and the upper limit is 0.43 ppm (1,130 $\mu g/m^3$) in a standard 30-l air sample. The precision is about 5%. Interference from NO_x can be eliminated by addition of sulfamic acid, and from heavy metals by the

Table 23-2 Sensitivity of Gas Chromatographic Determinations of SO_2 and TRS

Compound	Minimum Detectable Sensitivity ($\mu g/m^3$)
SO_2	13.09
H_2S	6.95
CH_3SH	9.81
$(CH_3)_2S$	38.01
$(CH_3)_2SH$	46.59

addition of ethylenediaminetetraacetic acid (EDTA) and phosphoric acid. Ozone will eliminate itself with a time delay (14).

CARBON MONOXIDE ANALYSIS

It is relatively easy to detect carbon monoxide because of its infrared (IR) absorption properties. The amount of absorbence is proportional to the concentration. With an IR spectrophotometer using a 10-mm path length IR cell, concentrations in excess of 10 ppm can be measured with an accuracy of 10%. The usual IR frequency used is 4.67 μm. The most common method is the nondispersive IR method, in which a two-sided cell is used. The mixture containing the carbon monoxide is placed in one side and an IR-transparent gas in the other. When IR radiation is beamed through the cell, the absorbed energy is converted to heat that produces a change in the pressure or volume of the gas in the detector cell. This change is converted into an electrical signal directly related to the CO concentration. The principal interferences are from water and carbon dioxide, but they can be greatly reduced by using filter cells. The Intersociety Committee has published a complete tentative method and calibration procedure (3).

PHOTOCHEMICAL OXIDANTS ANALYSIS

"Oxidants" are broadly defined as those compounds present in air that will oxidize a reference reagent that is not oxidized by atmospheric oxygen (15). The commonly used reference compound is potassium oxide (KI). The oxidant concentration is equivalent to the amount of iodine released after exposure to a fixed amount of oxidant-containing air. The main photochemical oxidants of concern are ozone (O_3), peroxides, peroxyacylnitrates (PAN), and NO_2, with ozone predominating. These photochemical oxidants are secondary pollutants, that is, they are formed by chemical reactions of primary gaseous pollutants within the atmosphere. They are usually thought of in terms of the reactions occurring in smog. Minor amounts of these compounds exist naturally or are emitted as primary pollutants, but the majority are produced as secondary pollutants. They can interact with one another to sustain their concentrations. For example, ozone is formed as a result of the photodissociation of NO_2 to give oxygen atoms that combine with molecular oxygen. NO is reconverted into NO_2 via free radical reactions, so that the NO_2 is continuously replenished and the ozone concentration increases. The reactions are:

$$NO_2 + \text{UV light} \rightarrow O + NO$$

$$O_2 + O \rightarrow O_3$$

The KI procedure for oxidants is a good one but is interfered with by competitive reducing compounds such as H_2S, SO_2, and aldehydes. Tests for total oxidants reflect the state of the air. Readings of 20–50 ppb are normal; lower readings signal the presence of reducing agents, while readings greater than 100 ppb may indicate the presence of photochemical pollution.

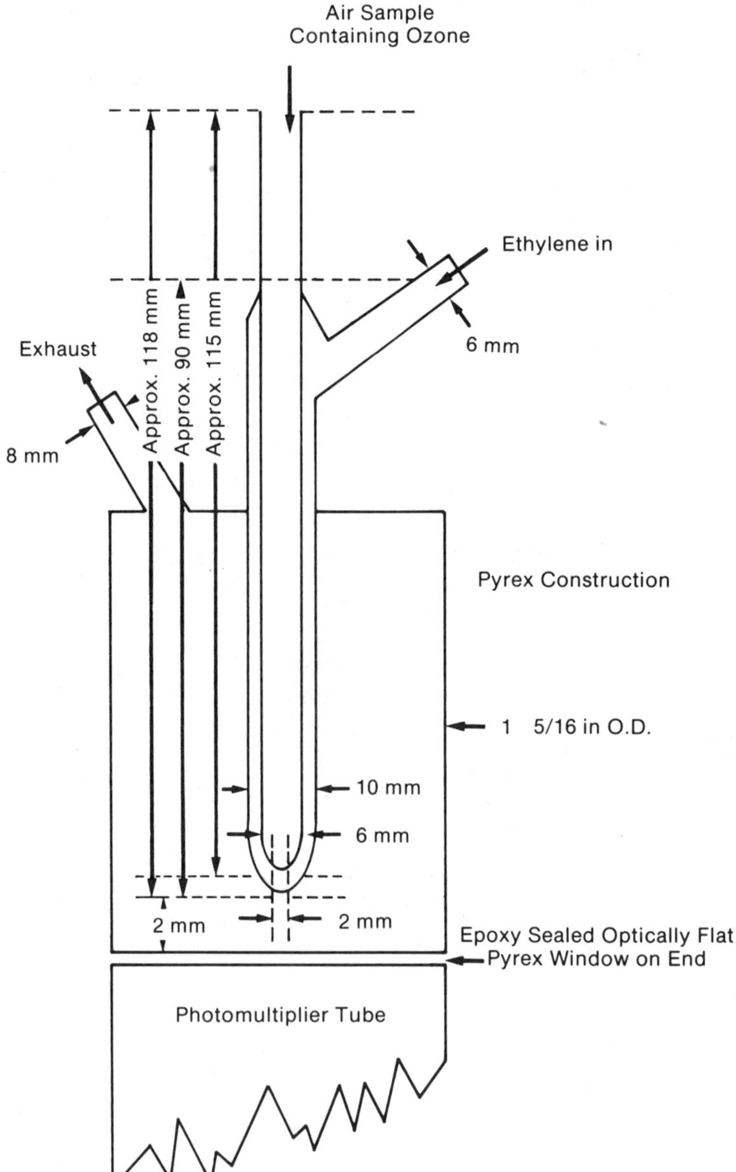

Figure 23-3 Detector cell for ozone. Copyright © 1981. TAPPI. Reprinted from *Introduction to Environmental Management for the Pulp and Paper Industry*, Study Guide, Volume 6, with permission

In taking samples to measure photochemical oxidants, it is best to use Teflon equipment. Metal, rubber, and polyvinyl chloride (PVC) must be avoided. The I_2 liberated in the KI procedure for total oxidants is commonly measured colorimetrically, but it is even better to measure it coulometrically.

If only ozone, the major component of photochemical pollution, is of interest, there is an excellent chemiluminescence method for its measurement (4). Ozone reacts rapidly in the gas phase with ethylene (C_2H_4), and the reaction produces an emission of light in the 350- to 600-nm wavelength region. Figure 23-3 is a schematic diagram, of the detector assembly. A mixture of ozone and ethylene flows to the cell, producing chemiluminescence that is proportional to the ozone concentration; this light is detected by a sensitive photomultiplier tube. The technique is good for ozone concentrations ranging from 0.005 to 100 ppm (9.8–1,960 $\mu g/m^3$) with a sensitivity of 0.005 ppm (9.8 $\mu g/m^3$). There are no known interferences with the method. The precision of the method depends on the maintenance of constant gas flow rates and on suppression of and compensation for the "dark" current of the photomultiplier tube. A precision of 2% at 50 ppb ozone can easily be achieved. This chemiluminescence method is slowly being replaced by a newly developed UV adsorption method.

HYDROCARBONS

Hydrocarbons present as atmospheric pollutants are difficult to measure as specific compounds because so many different ones exist. Most hydrocarbons themselves are not particularly hazardous environmentally. Of concern are their possible conversion products in photochemical smog. Two exceptions to this generalization are carbon monoxide, discussed previsously, and polycyclic aromatic hydrocarbons, which are suspected carcinogens (16, 17). Hydrocarbon pollutants can exist in two physical forms—either as volatile hydrocarbons or as the hydrocarbon content of particulate matter, that is, as gases or solids. This part of the discussion concerns only gaseous hydrocarbons.

There is a question about the source of atmospheric hydrocarbons, and some information is available. The only naturally occurring atmospheric hydrocarbon is methane, which is present worldwide at a level of about 1.3–1.4 ppm. For this reason, it is popular to report nonmethane hydrocarbons. A study in the Los Angeles area (18) provided some information about hydrocarbons from man-made sources. These sources and their relative importance were automobile exhausts, 47%; evaporative losses of gasoline, 31%; geogenic gas leaks, 14%; and leaks in the commercial natural gas supply, 8%. In the Los Angeles area, therefore, 78% of the hydrocarbon pollution appears to be transport related, and the dominant position of motor vehicles as the source of hydrocarbon emissions is also true nationwide. Hydrocarbon emissions consist of a wide variety of compounds with differing stabilities, but most are sufficiently inert to be dispersed through complex processes involving transport by air, water, particulate matter, and the food chain.

A common way to measure hydrocarbon content is by gas chromatography. The best detector is the flame ionization detector. In one popular procedure, one aliquot of a sample is measured for total hydrocarbon content, and another aliquot is stripped of its water vapor, carbon dioxide, and all hydrocarbons except methane and carbon monoxide, which can then be separated and measured by gas chromatography. This method allows a report of total hydrocarbon content corrected for methane. It is possible to use a single gas chromatograph capable of switching from total hydrocarbons to methane and carbon monoxide for this analysis. Many other techniques have also been developed for the measurement of individual hydrocarbons. Techniques for differentiating between methane and other hydrocarbons have often proven unreliable because the final result is the difference between two large numbers. Thus most of the analytical techniques have been abandoned in favor of emission factors based on analysis of the fuel to obtain more reliable information.

OXIDES OF NITROGEN

As was discussed previously in more depth, oxides of nitrogen are by-products formed during the burning of fossil fuels at high temperature in the presence of nitrogen and oxygen. The two oxides of concern are NO and NO_2. The methods of analysis for them are closely related, because all chemical methods convert the NO to NO_2 and then measure the NO_2 concentration. The colorimetric procedure for measuring nitrite ion (3) is a popular method for the determination of NO_2. Nitrogen dioxide is collected by bubbling air through a sodium hydroxide solution to form a stable solution of sodium nitrite. The nitrite is then determined with a procedure based on the Griess–Isolvay reaction, in which a pink-colored azo dye complex is formed between sulfanilic acid, nitrite ion, and x-naphthylamine in an acid medium. The main uncertainty in this procedure arises over the stoichiometry of conversion of NO_2 to nitrite. The value of the nitrite equivalence, or Saltzman factor, has been questioned, since values ranging from 0.72 to 0.764 have been reported. The particular factor to use depends on which modification of the procedure is employed. In general, this test can measure NO_2 concentrations, ranging from 0.1 to 2.0 μg/ml with a sensitivity of 0.02 μg/ml (9).

A new procedure for the measurement of nitrogen dioxide in the atmosphere, gas phase chemiluminescence, has been approved by the EPA. The light intensity at wavelengths greater than 600 nm that results from the chemiluminescent reaction of nitric oxide (NO) with ozone (O_3) is measured photometrically. NO_2 is quantitatively reduced to NO by a converter. Measuring a sample of unconverted air gives the ambient concentration of NO, and measuring a sample of converted air gives the NO_2 concentration by difference. The technique also measures other nitrogen containing compounds such as PAN, but they are usually in low concentration compared to NO and NO_2. The method should not be used in areas of high smog, however. A stable source of ozone is the key to the success of this measurement. Reference 19 provides further details on this procedure.

Coulometric analyses for NO_2 based on the oxidizing action of NO_2 on KI solutions have also been developed. However, it is necessary to remove other oxidizing and reducing pollutants efficiently in order to yield good results, since the coulometric procedure is subject to more interference.

A variety of systems have been devised for the oxidation of NO to NO_2. The recommended procedure involves passing the sample through a tube containing chromic oxide supported on an inert organic material (20).

AIR POLLUTION INDEX

Since quantitative numbers on the amounts of various pollutants present are not easily comprehended by the general public, a procedure has been developed that translates these data into a uniform air quality index. The index ranges in value from 0 to 500; 0–50 is termed good air quality, 51–100 is moderate, 101–199 is unhealthy, 200–299 is very unhealthy, and over 300 is hazardous. The index is based on the worst-case scenario of the five pollutants—total suspended particulates, sulfur dioxide, nitrogen dioxide, carbon monoxide, and ozone. In addition, a factor is computed for the synergistic cross-reaction of total suspended particulates and sulfur dioxide. For each of these six factors, a graph has been created relating the concentration to the index value (0–500). These data have been published in the *Federal Register* (21). For each pollutant, the index value 100 is the NAAQS, 200 is the alert status, 300 is the warning status, and 400 is the episode status. An index value is determined for each of the six factors, and the highest value becomes the uniform air quality index for the specified area. The values are not summed. This index makes air pollution levels somewhat more comprehensible to the layman.

REFERENCES

1 "National primary and secondary air quality standards," *Federal Register* 36 (84):8187–8201 (April 30, 1971).

2 "Ambient air quality monitoring, data reporting and surveillance provisions," 40 CFR Part 50, Appendices A–H, U.S. Environmental Protection Agency, Washington, D. C.

3 Intersociety Committee, *Methods of Air Sampling and Analysis*, American Public Health Association, Washington, D. C., 1972.

4 R. Perry and R. Young, *Handbook of Air Pollutant Analysis*, Halsted Press, New York, 1977.

5 P. H. Cheremisinoff and A. C. Morresi, *Air Pollution Sampling and Analysis Deskbook*, Ann Arbor Science, Ann Arbor, Mich., 1978.

6 L. Duncan, *Compendium of methods for measuring ambient air quality and process emissions, Section 4—Ambient air monitoring for particulate matter*, Atmospheric Pollution Technical Bulletin No. 37, National Council of the Paper Industry for Air and Stream Improvement, New York, December 1968.

7 F. E. Courtney, *Tappi*, 65 (4):84–85 (1982).

8 J. B. Clements, *Analytical methods*, U.S. Environmental Protection Agency, National Environmental Research Center, Research Triangle Park, N.C., 1972.

9 G. B. Morgan, E. C. Tabor, C. Golden, and J. B. Clements, "Automated laboratory procedures for the analysis of air pollutants," *Automation in Analytical Chemistry*, New York, 1966.

10 A. O'Keefe, *Sulfur dioxide, hydrogen sulfide, methyl mercaptan and high molecular weight gaseous sulfur compounds*, U.S. Environmental Protection Agency, National Environmental Research Center, Research Triangle Park, N.C., 1972.

11 J. Forrest and L. Newman, *Journal of the Air Pollution Control Association*, 23 (9):761–768 (1973).

12 P. W. West and G. C. Goeke, *Analytical Chemistry*, 28 (12):1816–1819 (1956).

13 M. B. Jacobs and L. Greenburg, *Industrial and Engineering Chemistry*, 48 (9):1517–1521 (1956).

14 F. P. Scaringelli, B. E. Saltzmann, and S. A. Frey, "Spectrophotometric determination of atmospheric sulfur dioxide," *Analytical Chemistry*, 39:1709 (1967).

15 *Air quality criteria for photochemical oxidants*, National Air Pollution Control Administration Publication No. AP-63, U.S. Dept. of Health, Education and Welfare, Washington, D. C., 1970.

16 A. Dipple, "Formation, metabolism, and mechanism of action of polycyclic aromatic hydrocarbons," *Cancer Research*, 43 (suppl 5):2422–2425 (1983).

17 J. Santodonato et al., "Health and ecological assessment of polynuclear aromatic hydrocarbons," *Journal of Environmental Pathology and Toxicology*, 5 (1):1–364 (1981).

18 H. Mayrsohn and J. M. Crabtree, *Atmospheric Environment* (Oxford), 10 (2):137–143 (1976).

19 E. C. Ellis, *Technical assistance document for the chemiluminescence measurement of nitrogen dioxide*, EPA-E600/4-75-003, EPA, Washington, D. C., 1975.

20 D. A. Levaggi, E. L. Kalhny, T. Belsky, E. R. DeVera, and R. K. Mueller, *Environmental Science and Technology*, 8 (6):348–351 (1974).

21 *Federal Register*, May 10, 1979.

24
Stationary Source Sampling

AAQS are met by controlling emissions from both stationary and mobile sources. Stationary sources dominate within the pulp and paper industry. A brief treatment of the basics of point source sampling is provided in this chapter; for more detailed information, chapters 14 and 15 in volume III of reference 1 should be consulted.

PLANNING THE SAMPLING PROGRAM

Emission sources are studied for several possible reasons:

1 Determination of the mass emission rates of particular pollutants from a particular source and of the effect of process variations on these rates
2 Evaluation of control devices for the reduction of pollutant emissions
3 Collection of data on emissions from several sources for input into air quality management models

It is of paramount importance to plan the sampling program carefully. Proper planning requires a thorough knowledge of the process to be sampled, particularly of process cycles. For a specific source, two requirements must be satisfied:

1 Accurate measurement at a given point in time in order to characterize the emission
2 Sufficient measurements at different positions and different times in order to characterize the total emission pattern

Sources can be classified into four categories based on whether the emissions are steady with respect to time and whether they are uniform in position (across the stack).

Class I: Steady and Uniform
Theoretically, for this class, only one measurement at a single point in time is needed to characterize the emissions. An example would be a very steady, turbulent gas stream such as from a gas well.

Class II: Steady and Nonuniform

For this class, measurements must sample the stack systematically across its section, but should only need to be performed at one time. An example would be the particulate emissions from a large, continuous-feed, coal-fired power station.

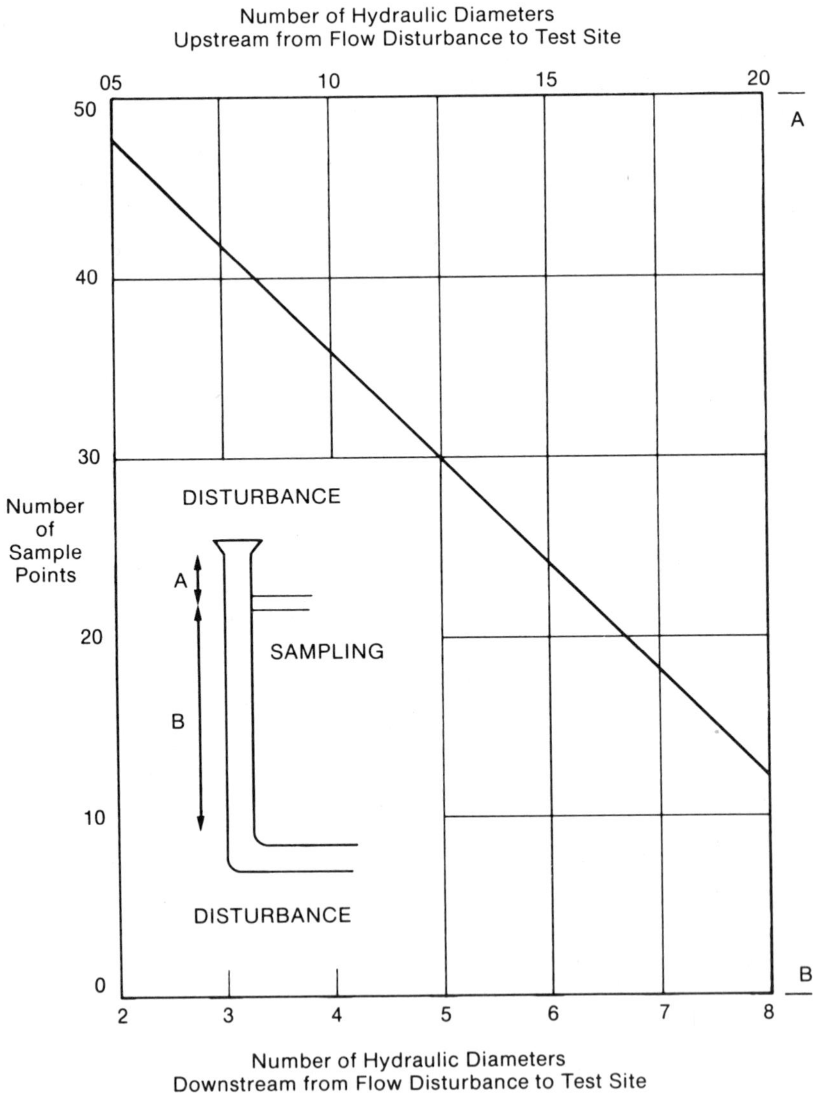

Figure 24-1 Number of sample points. (Copyright © 1981. TAPPI. Reprinted from *Introduction to Environmental Management for the Pulp and Paper Industry*, Study Guide, Volume 6, with permission.)

Class III: Unsteady and Uniform

Here only one point needs to be sampled, but it must be sampled over at least one full cycle of nonuniformity. The cyclic blowing of a digester with the stack gases in a turbulent flow would be an example.

Class IV: Unsteady and Nonuniform

This is a very difficult case because sampling is needed not only at several positions in the stack but also over the cycle of variability if one exists. If the

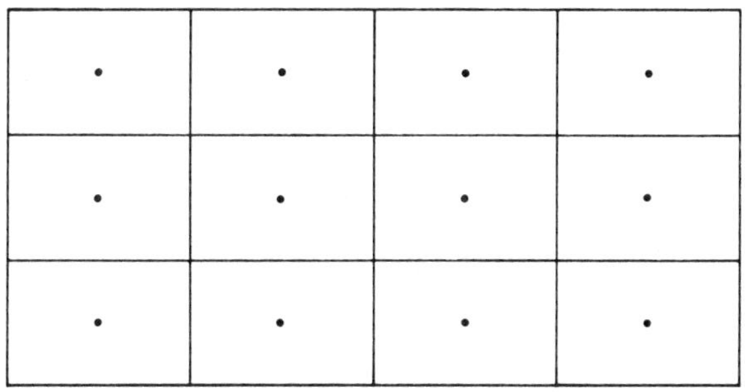

Figure 24-2 Sampling cross section for a rectangular stack. (Copyright © 1981. TAPPI. Reprinted from *Introduction to Environmental Management for the Pulp and Paper Industry*, Study Guide, Volume 6, with permission.)

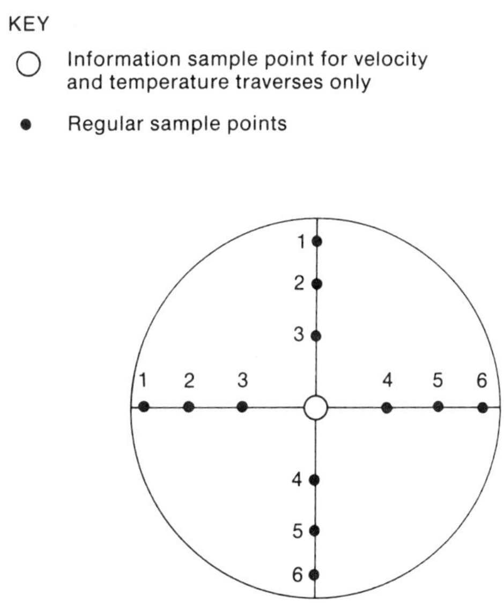

Figure 24-3 Sampling cross section for a circular stack. (Copyright © 1981. TAPPI. Reprinted from *Introduction to Environmental Management for the Pulp and Paper Industry*, Study Guide, Volume 6, with permission.)

Table 24-1 Percentage of the Circular Stack Diameter from the Inside Wall to the Traverse Point

Traverse Point Number Along Diameter[a]	Number of Traverse Points on a Single Diameter[b]											
	2	4	6	8	10	12	14	16	18	20	22	24
1	14.6	6.7	4.4	3.3	2.5	2.1	1.8	1.6	1.4	1.3	1.1	1.1
2	85.4	25.0	14.7	10.5	8.2	6.7	5.7	4.9	4.4	3.9	3.5	3.2
3	—	75.0	29.5	19.4	14.6	11.8	9.9	8.5	7.5	6.7	6.0	5.5
4	—	93.3	70.5	32.3	22.6	17.7	14.6	12.5	10.9	9.7	8.7	7.9
5	—	—	85.3	67.7	34.2	25.0	20.1	16.9	14.6	12.9	11.6	10.5
6	—	—	95.6	80.6	65.8	35.5	26.9	22.0	18.8	16.5	14.6	13.2
7	—	—	—	89.5	77.4	64.5	36.6	28.3	23.6	20.4	18.0	16.1
8	—	—	—	96.7	85.4	75.0	63.4	37.5	29.6	25.0	21.8	19.4
9	—	—	—	—	91.8	82.3	73.1	62.5	38.2	30.6	26.1	23.0
10	—	—	—	—	97.5	88.2	79.9	71.7	61.8	38.8	31.5	27.2
11	—	—	—	—	—	93.3	85.4	78.0	70.4	61.2	39.3	32.3
12	—	—	—	—	—	97.9	90.1	83.1	76.4	69.4	60.7	39.8
13	—	—	—	—	—	—	94.3	87.5	81.2	75.0	68.5	60.2
14	—	—	—	—	—	—	98.2	91.5	85.4	79.6	73.9	67.7
15	—	—	—	—	—	—	—	95.1	89.1	83.5	78.2	72.8
16	—	—	—	—	—	—	—	98.4	92.5	87.1	82.0	77.0
17	—	—	—	—	—	—	—	—	95.6	90.3	85.4	80.6
18	—	—	—	—	—	—	—	—	98.6	93.3	88.4	83.9
19	—	—	—	—	—	—	—	—	—	96.1	91.3	86.8
20	—	—	—	—	—	—	—	—	—	98.7	94.0	89.5
21	—	—	—	—	—	—	—	—	—	—	96.5	92.1
22	—	—	—	—	—	—	—	—	—	—	98.9	94.5
23	—	—	—	—	—	—	—	—	—	—	—	96.8
24	—	—	—	—	—	—	—	—	—	—	—	98.8

Source: Copyright © 1981. TAPPI Reprinted from *Introduction to Environmental Management for the Pulp and Paper Industry*, Study Guide, Volume 6, with permission.

[a] Points numbered from the inside wall to the opposite wall.

[b] The total number of points along two diameters would be twice the number of points along a single diameter.

flow pattern is random rather than cyclical, this situation can yield completely nonreproducible results. The best choice for this class is continuous monitoring.

Once the classification of a particular stack has been determined, the next step is to choose a sample site. Criteria used in site selection must include:

1 Safety of the location for the test personnel
2 Relationship of the sampling site to other points of particular interest, for instance, sampling downstream of a control device
3 Availability of a platform for personnel and equipment
4 Accessibility of the platform to the personnel and equipment
5 Access to the stack interior through a suitable port
6 Provision of a power supply for the sampling equipment
7 Satisfaction of flow disturbance criteria

Various criteria have been used for flow disturbance, but a general rule is that a sampling location in a vertical flue that is at least eight equivalent flue diameters downstream and/or two equivalent flue diameters upstream of a flow disturbance such as a bend, inlet, or outlet is considered satisfactory. If this condition cannot be met, the number of samples is increased to compensate. Figure 24-1 shows the increase in samples needed if the sampling point is not greater than eight diameters downstream of a disturbance.

It is recommended that sample points be located at the centroids of equal areas across the stack. For rectangular stacks, sample points are located at the centroids of smaller similar rectangles. Figure 24-2 illustrates the selection of proper sampling points for 12 points across a rectangular stack. For a stack with a circular cross section, the points are located at the centroids of equal annular segments, as illustrated in Figure 24-3. Table 24-1 is useful in selecting the proper positions for sampling points along the diameters of a circular stack.

VELOCITY AND MOISTURE MEASUREMENTS

The velocity of the gases passing through a stack is an important basic measurement that must be known, since much sampling must be done isokinetically, that is, at the same velocity as that occurring within the stack. Velocity is usually measured with a pitot tube; a Staubschube or type S pitot tube is preferred. Figure 24-4 shows a conventional pitot tube and Figure 24-5 illustrates the S type. The S type is easier to use, less susceptible to plugging, and less directional, but it must be calibrated against the conventional pitot tube. Computer programs have been developed that make it easier to calculate the isokinetic variation (3). The EPA-approved procedure is method 1 in reference 4.

Knowing the moisture content of the stack gases is also important because

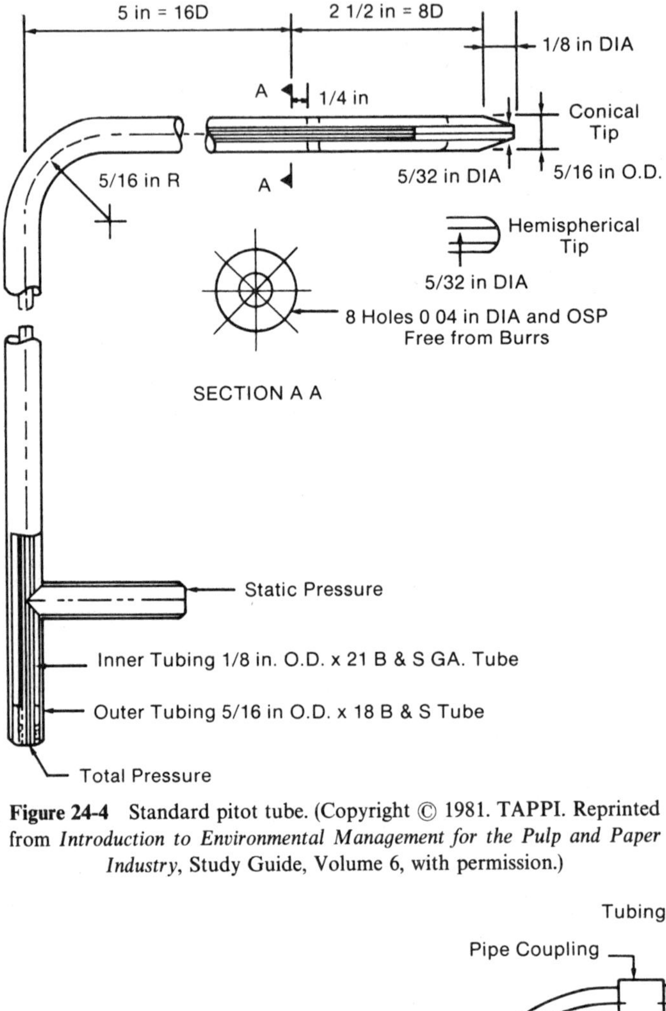

Figure 24-4 Standard pitot tube. (Copyright © 1981. TAPPI. Reprinted from *Introduction to Environmental Management for the Pulp and Paper Industry*, Study Guide, Volume 6, with permission.)

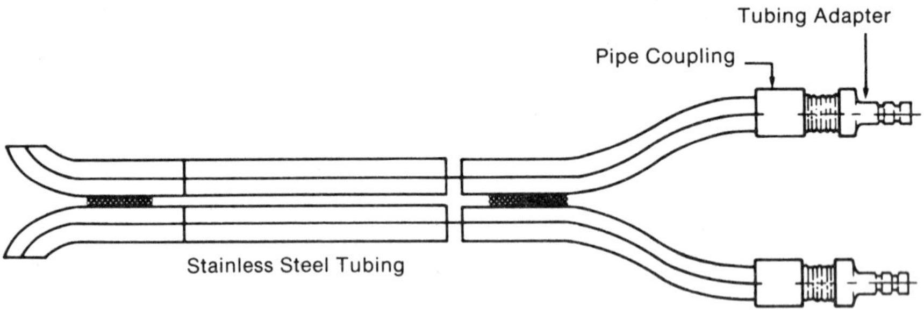

Figure 24-5 Type S pitot tube. (Copyright © 1981. TAPPI. Reprinted from *Introduction to Environmental Management for the Pulp and Paper Industry*, Study Guide, Volume 6, with permission.)

many pollutant concentrations are reported on a moisture-free gas basis. Moisture measurements are made by standard wet and dry bulb cycrometric measurements and by reference to a humidity chart. The EPA-approved method is method 4 in reference 4.

STACK SAMPLING SYSTEM

In the pulp and paper industry, the important pollutants that need to be monitored are particulates, oxides of sulfur, and TRS. Very sophisticated and expensive procedures have been developed to monitor these emissions on both a batch and a continuous basis. Continuous monitoring is generally much more costly than batch monitoring, so a combination of techniques has been developed in the industry.

We will first consider a generalized gas-handling and conditioning system for an externally located gaseous monitoring system. Figure 24-6 illustrates such a system, which consists of six major components:

1. Sample probe
2. Conditioning device
3. Flow control and flow measurement device
4. Prime mover
5. Sampling lines
6. Detection system

A common sample probe is a stainless steel tube with a 90° bend. If particulates are present in the sample, a glass wool filter is needed in the tube, and the filter must be cleaned periodically.

The handling and conditioning device is needed to remove the sample from the flue gas and transfer it to the detector for subsequent analysis without changing the concentration or character of the constituents to be measured.

Two approaches are used for conditioning the sample in order to handle the water problem. One is to maintain the sample at a temperature above the dew point by insulating and electrically heating the Teflon lines to 110°C (230°F) and the probes to 150°C (300°F). The other is to condense out the water by cooling the stream to 30°C (86°F) or lower. If SO_2 is being monitored, the condensed moisture must be acidified to a pH of less than 2 to minimize absorption.

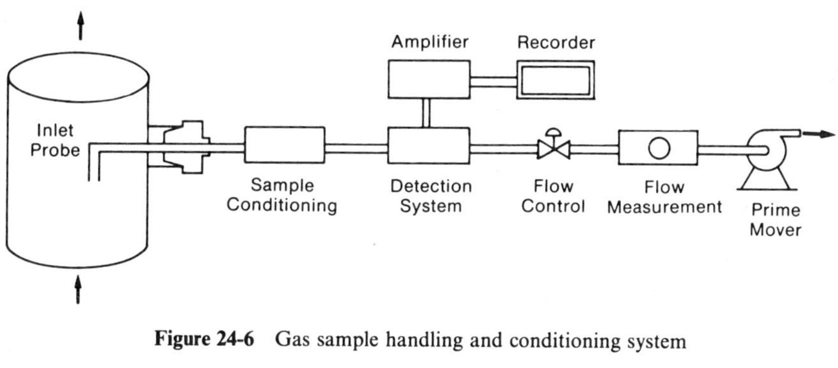

Figure 24-6 Gas sample handling and conditioning system

To monitor the flow, it is common to use a rotometer, or, less commonly, an orifice meter.

The device used to move the flow through the sampling train is the prime mover, usually a vacuum pump. It is commonly located downstream of the detector, so it does not need to be leakproof. However, if a larger sample is drawn than the detector can handle or if the detector has to be operated under positive pressure, then the prime mover must be located upstream of the detector and must be leakproof. Leakage in the sampling train is always a problem.

In the sampling lines, it is desirable to have both minimum pressure drop and minimum detention time. These are incompatible needs, so a compromise must be devised. Electrically heated Teflon tubing is available commercially and is commonly used for sampling lines.

DETECTION SYSTEMS FOR STACK SAMPLING

TRS Detection Systems

In a generalized form, the sulfur components of the gas stream to be measured are SO_2, H_2S, R-SH, and R-S-R. These components can be measured collectively or individually. It is common to scrub out the SO_2 and to detect the remainder as TRS. However Table 24-2 lists selective scrubbing systems that have been developed that allow the separation and measurement of each component.

The methods available for detection of the TRS components include:

1 Coulometric titration
2 Electrochemical membrane sensing
3 Flame photometry
4 Ultraviolet spectrophotometry

Coulometric titration is the most commonly used method, but electrochemical

Table 24-2 Selective Scrubbing Systems for TRS Determination

Gas Removed	Scrubbing Solution	Concentration (% by Weight)
SO_2	$KHC_8H_4O_4$ Potassium acid phthalate	3
$H_2S + SO_2$	$CdSO_4 - H_3BO_3$	1–2
$H_2S + SO_2 + RSH$	NaOH	10
$SO_2 + H_2S + RSH + RSR$	$AgNO_3$	0.5

Source: Copyright © 1981. TAPPI. Reprinted from *Introduction to Environmental Management for the Pulp and Paper Industry*, Study Guide, Volume 6, with permission.

membrane sensing is becoming more popular. The latter technique is good for measuring SO_2, and the flue gases can be sent through a combustion furnace at a temperature of about 815°C (1,500°F) to convert the reduced sulfur to SO_2.

The coulometric titration technique is often referred to as "Barton titration" after the person who developed it. This method operates on the principle that H_2S, SO_2, and organic sulfur compounds present are oxidized to sulfate ion by the action of a halogen electrolyte. An electrical current is required to maintain a constant concentration of halogen gas generated from the electrolyte. The current level is proportional to the overall concentration of reactive gases passing through the detection cell. Coulometric titration is sensitive to all compounds that can be oxidized by the halogen in varying degrees particular to each compound and is not specific to any one compound.

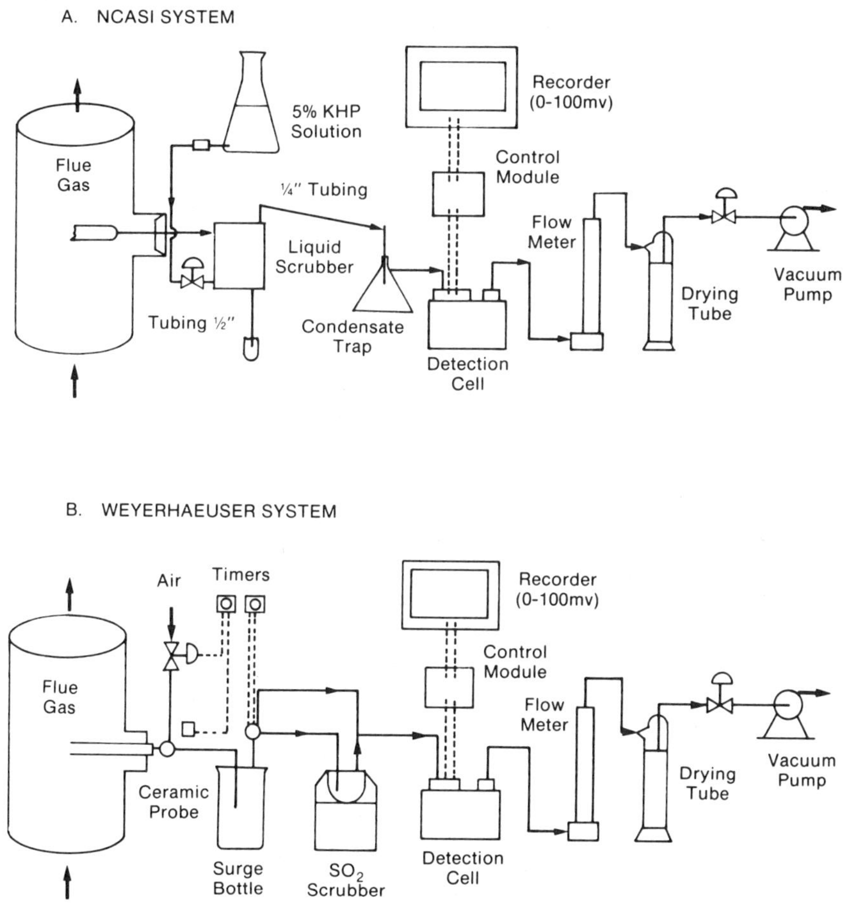

Figure 24-7 Continuous source monitoring system for TRS emissions using coulometric titration. (Copyright © 1981. TAPPI. Reprinted from *Introduction to Environmental Management for the Pulp and Paper Industry*, Study Guide, Volume 6, with permission.)

Commercially available instruments employ a 16% solution of hydrogen bromide (HBr). The detection cell can generate varying levels of bromine gas. The attenuation setting corresponds to concentrations ranging from 1 to 800 ppm H_2S as the full-scale measurement. Figure 24-7 illustrates two different continuous monitoring systems that employ the coulometric titration technique.

A recent study that compared TRS continuous monitoring systems showed the Barton titrator and the Sampling Technologies, Inc. (STI), systems to be the most reliable (5). The principle of the STI system is oxidation of all of the TRS to SO_2 in a thermal oxidizer followed by detection of the resulting SO_2 by UV fluorescence. This system appears to be simpler than Barton titration, and several of them have been installed on Kraft recovery boilers and lime kilns. There is a slight positive interference from NO of about 0.5 ppm per 100 ppm NO.

Gas chromatographic systems (EPA method 16) are also beginning to be employed to monitor TRS. They are much more complicated to operate, but they have evolved to a stage of development that allows them to be used continuously on line successfully. Gas chromatographic techniques have also been developed to monitor SO_x emission from stacks (6).

Some mills have experienced excessive variation in stack TRS values as measured by gas chromatography. One set of tests showed 67–165% variation from the average value (7). This problem has led to the development of an easier alternative method, the basic apparatus of which is shown in Figure 24-8. The gas sample collected by a heated probe is filtered to remove particulate matter.

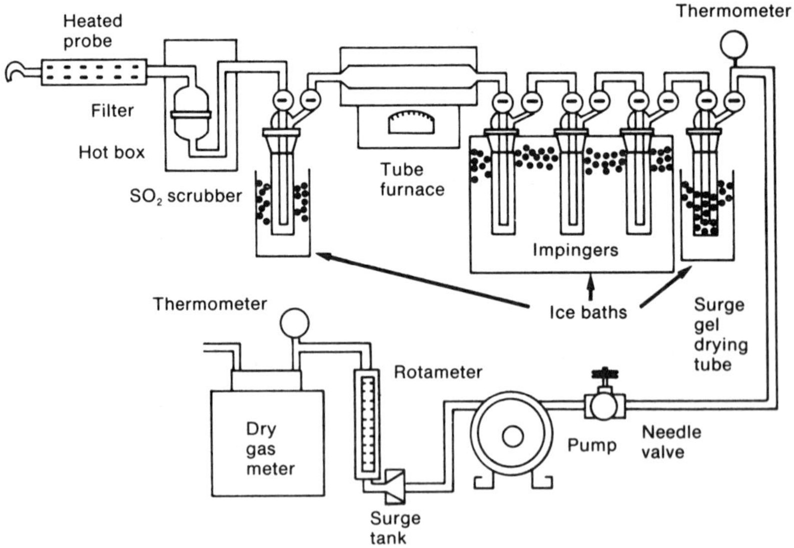

Figure 24-8 Simplified TRS stack sampling system. (Copyright © 1982, reprinted from *Tappi Journal*, 65(3): 95(1982) Technical Association of the pulp and Paper Industry, with permission.)

It then passes through a 500-cm^3 Greenburg impinger containing 200 cm^3 potassium citrate–citric acid solution to absorb any SO_2 present in the sample. Next, it passes through a combustion tube furnace where the TRS are oxidized to SO_2 at 815°C (1,500 °F). The gas stream then passes through three impingers in series containing hydrogen peroxide solution that oxidizes the SO_2 to sulfate, which can easily be measured. The peroxide trap, oxidation, and detection of sulfate are basically EPA method 6 for SO_2 determination. The average TRS concentration is determined for the stack by analyzing three different samples, each collected over a 2-hour period, a total testing period of 6 hours (7). Method 16 requires grab samples taken over a 3- to 6-hour period during which at least 16 determinations are made; this process is repeated three times to establish a compliance test. Thus the alternative method, which has been approved by the EPA, seems less time-consuming.

Particulate Monitoring

In particulate monitoring, both the particle size distribution and the absolute mass are important. Cascade impactors can be used to determine particle size classifications. The small particulates are the most dangerous because they penetrate farthest into the lungs.

"Particulate matter" is normally defined as any material emitted into the atmosphere in either a solid or liquid state, including dusts, fumes, fly ash, soot, tars, droplets, and mist. Changes in chemical form such as oxidation of SO_2 to H_2SO_4 in sampling train impingers after collection can cause especially serious difficulty in interpreting emission standards. The definition accepted by the EPA is that particulate matter is any material collected on a filter of 0.45 μm that has been heated to 121°C (250°F). The particulates from pulp and paper sources vary from less than 0.01 to more than 100 μm.

Sampling under isokinetic conditions is normally necessary because of the inertial properties of particles greater than 5 μm. A pitot tube is placed next to the sampling probe and is used to keep the sampling velocity at the stack velocity. Batch sampling methods usually include a combination of filtration and liquid impingement. Various opinions exist about how best to collect and measure particulates.

The simplest method is the one specified by the American Society of Mechanical Engineers (ASME), which relies on an alundum thimble as the collection element (8). The thimble collects particles larger than 1 μm but passes smaller material, particularly during the initial collection period. Figure 24-9 illustrates such a sampling train.

The particulate sampling train specified by the EPA (9) employs an externally located, two-stage collection system, as shown in Figure 24-10. The primary collection stage consists of a glass cyclone and a 5-cm (2-in.)-diameter glass fiber filter of 0.30–0.45 μm ($1.26–1.8 \times 10^{-6}$ in.) porosity that is heated to 121°C (250°F) to avoid condensation. The secondary collection stage consists of a series of Greenburg–Smith liquid impingers containing water. The impingers

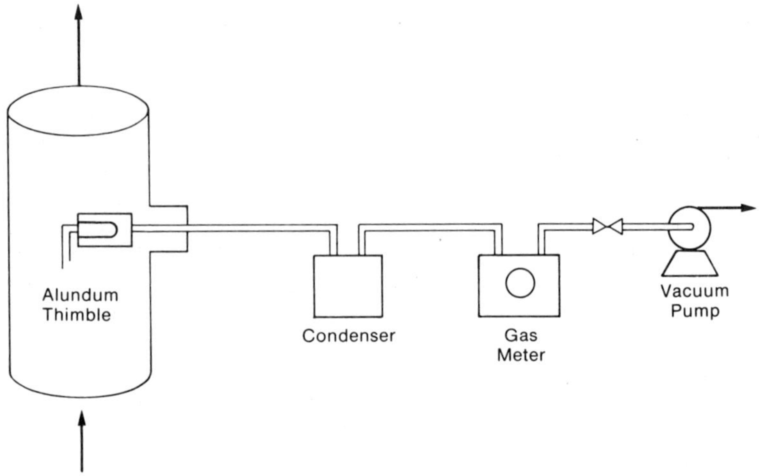

Figure 24-9 ASME batch particulate sampling train. (Copyright © 1981. TAPPI. Reprinted from *Introduction to Environmental Management for the Pulp and Paper Industry*, Study Guide, Volume 6, with permission.)

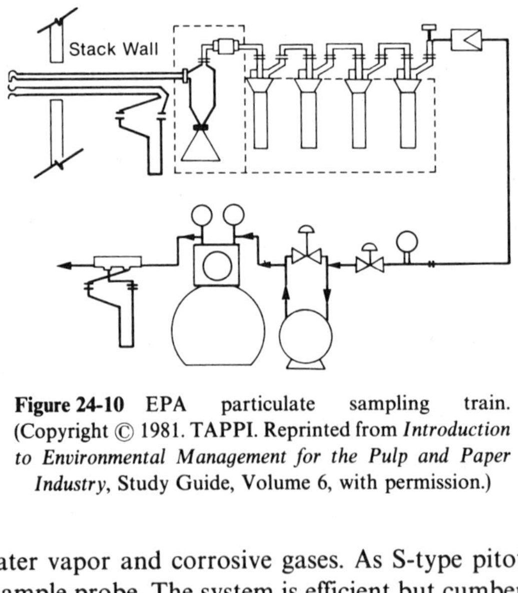

Figure 24-10 EPA particulate sampling train. (Copyright © 1981. TAPPI. Reprinted from *Introduction to Environmental Management for the Pulp and Paper Industry*, Study Guide, Volume 6, with permission.)

take out the water vapor and corrosive gases. As S-type pitot tube is located parallel to the sample probe. The system is efficient but cumbersome to handle. The small filters are subject to rapid plugging at high loadings, and large pressure drops develop after long periods of sampling. The details of EPA-approved particulate sampling are given in reference methods 1, 2, 3, 4, and 5 (4), which address, respectively, the sample collection and velocity measurement, the determination of stack gas velocity and flow rate, the gas analysis, the determination of moisture content, and the sample collection and analysis procedure.

An alternative to EPA method 5 for particulate analysis is EPA method 17, which employs an in-stack filter at an isokinetic sampling rate to monitor particulates. The method is similar to that using the alundum thimble, but the filter is a special glass fiber filter pad. The entire filter assembly fits inside the stack and the particulate sample is collected at stack conditions, avoiding the need for special temperature and condensation precautions. This method seems to be simpler than method 5 and would appear to be a viable substitute.

Continuous monitoring of particulates is gaining in importance. The methods now in use include:

1 Wet chemical techniques that use conductivity and specific ion determination
2 Optical bolometry and transmissivity

Additional techniques that may become commercially available in the future include β ray attenuation, piezoelectric crystallography, electronic sensing, optical mephelometry, holography, electric ion capture, optical lasers, and radar.

In the recovery furnace, the majority of the particulate matter is Na_2SO_4. It is therefore possible to draw off the Na_2SO_4, dissolve it in deionized water, and measure its conductivity (10)—in effect, a chemical procedure for particulate determination. Interference results from the dissolution of SO_2 and CO_2. If a specific ion electrode for sodium is used to measure the dissolved sodium, this potential interference is bypassed.

Particulates can be measured optically, in stack gases free of entrained moisture, by locating bolometers and transmissometers in the stack. The degree of light attenuation is a function of the particulate concentration in the duct. The system directs a light beam across the duct to a detector in which the resulting electrical signal is amplified, transmitted, and printed out on a recorder (11). The sensitivity and accuracy of the optical systems are affected by the path length, intensity and wavelength of the light beam, amount of moisture present, and the mass concentration and size distribution of the particulates being measured (12). Optical systems are particularly useful as warning devices to indicate possible malfunctions of particulate emission control equipment. Performance specifications and test procedures for continuous emissions monitoring systems are given in Appendix B of 40 *Code of Federal Regulations* (CFR) 60.

Optical techniques cannot be used on wet-stack gases. A recent survey (13) that examined nine measurement principles and looked at both in situ and extractive commercial systems found that most failed to perform on wet-stack gases. None of the in situ monitors were adequate. Of the extraction techniques, the method referred to previously for the recovery furnace proved the most useful. This technique can also be used on the lime kiln, smelt-dissolving tank vents, and other wet sources. It is not applicable, however, on a power boiler. References 14 and 15 summarize some industry experience.

The principle is to take a sample and dissolve all of the material, and then use a specific ion electrode to determine the concentration of a surrogate ion whose concentration is related in some known way to the particulate concentration. Sodium is often the ion of choice, although others are possible. For the technique to be applicable, the following six conditions must be satisfied (13):

1 The point of sampling must be representative of the particulate concentration in the entire stack
2 The relationship between the ion monitored and the mass of particulates must be known and must remain constant
3 Monitor flow rates must be maintained within critical limits
4 Moisture in the gas stream must remain within specified limits
5 The sensing probe must be operated at optimum conditions with respect to response range and chemistry
6 The particulate capture efficiency of the system must remain within known limits

This technique makes possible continuous particulate monitoring on wet as well as dry stacks.

CONTINUOUS MONITORING

A few remarks about the value and use of continuous monitoring are in order. Continuous monitoring is always more expensive than spot monitoring, but it does convey the following additional benefits (8):

1 It can be used to optimize capital investment for air pollution abatement equipment
2 It can be useful in detecting malfunctions of control equipment, leading to early correction of the problem
3 It is useful in evaluating new equipment
4 It provides data to establish compliance with air pollution regulations

In many instances, the benefits outweigh the additional costs, and thus more continuous monitoring equipment is being installed. It is part of an ongoing trend in the industry to upgrade its instrumentation in order to improve the operation and control of its processes.

There are a number of locations in a Kraft mill where continuous monitoring might be used to advantage. Opacity could be measured at the recovery furnace and the power boiler stacks. TRS could be monitored continously at the recovery furnace, lime kiln, digester, brown stock washers, evaporator, stripper system, and black liquor oxidation system. Continuous monitoring of the combustion temperature at the point of incineration of effluent gases emitted from any digester system, the brown stock washer, the multiple effect evapoator, or the condensate stripper could be useful.

MOBILE TESTING LABORATORIES

Another type of equipment that is useful in establishing compliance is that provided by the mobile testing laboratory. Several of these laboratories have been described in the literature (16, 17). A mobile laboratory is very useful in making comprehensive measurements of ambient and point source concentrations of both particulate and gaseous emissions. It is generally equipped with devices to monitor particulates, gas chromatographs, coulometric titrators, chemical and analytical hardware, and so on. Mobile units are very useful in assessing the relative magnitude of each source at an emissions site and in mapping the dispersion pattern at the site and in its vicinity. These units enable better assessment of the consequences of an emission on the surrounding area. They also enable observation of the effects of weather patterns on dispersion of emissions.

REFERENCES

1 A. C. Stern, *Air Pollution*, 3d ed., Vols. 1–5, Academic Press, New York, 1976.

2 *Source Testing Manual for the Determination of Discharges to the Atmosphere*, 3d ed., Pollution Control Branch, B.C. Water Resources Services, Victoria, British Columbia, 1974.

3 D. C. Smith, "Percent isokinetic variation," *Tappi*, 62(10):139–140 (1979).

4 40 CFR 60, Appendix A.

5 D. Rovell-Rixx et al., "Results of a laboratory and field evaluation of commercially available continuous monitoring systems for total reduced sulfur emissions," *Proceedings of the TAPPI Environmental Conference*, Savannah, Ga., April 9–11, 1984, pp. 239–42 (TAPPI Press, Atlanta, 1984).

6 A. Jain, *A study of SO_x measurement procedures and their use at Kraft recovery furnaces*, Atmospheric Quality Improvement Technical Bulletin No. 106, National Council of the Paper Industry for Air and Stream Improvement, New York, 1980.

7 A. S. Rosenfeld, J. M. Knowles, and J. E. Walther, "New sulfur testing meets federal standards," *TAPPI*, 65(3):95–97 (1982).

8 *Determining dust concentrations in a gas stream*, Performance Test Code 27-1957, American Society of Mechanical Engineers, New York, 1957.

9 "Standards of performance for new stationary sources, method 5," *Title 40 Code of Federal Regulations*, Part 60, Chapter I, July 1, 1979.

10 R. O. Blosser and H. B. H. Cooper, *Analytical equipment and monitoring devices for gases and particulates*, Atmospheric Pollution Technical Bulletin No. 35, National Council of the Paper Industry for Air and Stream Improvement, New York, March 1968.

11 H. P. Beutner, "Measurement of opacity and particulate emissions from stacks," paper presented at the Symposium on Instrumentation for Continuous Monitoring of Air and Water Quality, Department of Pulp and Paper Technology, Miami University, Oxford, Ohio, June 20, 1973.

12 S. Larssen, D. S. Ensor, and M. J. Pilat, *Tappi*, 55(1):88–92 (1972).

13 A. Caron, *An analysis of the options available for continuous particulate measurement in wet stack gases*, Technical Bulletin No. 377, National Council of the Paper Industry for Air and Stream Improvement, New York, August 1982.

14 V. J. Treller, "Use of continuous monitors for soda loss and malodorous sulfur loss in process control," *Tappi*, 62(12):324 (1979).

15 P. E. Brooks, "Use of specific ion electrodes in the Kraft pulp mill," paper presented at the ISA Pacific Northwest Instrumentation Symposium and Exhibition, Instrumentation Society of America, Portland, Oregon, October 1974.

16 J. E. Walther and H. R. Amberg, *Tappi*, 51(11):126A–129A (1968).

17 G. E. Waddington, *Pulp and Paper Magazine of Canada*, 74(6):58–61 (1973).

25
Basics of Air Quality Mathematical Models

MODEL UTILITY

In Chapter 4, it was mentioned that the purpose of modeling in SIPs was to provide a basis for allocating available emission margins equitably. Thus modeling plays a major role in determining the permit limits for a given emission source, which in turn has a direct impact on the control technology that will be needed.

The object of any air pollution control system is to improve ambient air quality. After a control system is chosen and installed, the air quality can be measured to determine the effect of the control equipment, but it would be desirable to be able to estimate the change in the air quality before the exact control system is decided. Such an estimate could influence the choice between alternative control systems. This is the role that air quality modeling can play. It is a means by which pollutant emissions can be related to atmospheric pollutant cocentrations. The model must involve considerations of emission characteristics (rate, stack height, size, specific pollutants present), meteorology (wind speed, mixing height, stability), chemical transformations, and removal processes in order to be an accurate reflection of reality. In addition to providing aid in deciding between alternative control systems and in setting permit limits, an air quality model may also be helpful in:

1 Assessing responsibility for existing levels of air pollution
2 Planning locations for future air pollution sources
3 Planning control systems to be used in case a pollution episode develops
4 Establishing reasonable, cost-effective air pollution control regulations

If the model is constructed so as to reflect reality, there will be little question of its usefulness. The problem is to develop an accurate model that reflects reality rather than just an empirical, data-fitting mathematical expression. In modeling there is always a trade-off between exactness and the difficulty of using the model. Air quality modeling is more an art than a science. There are so many parameters in the models and so much judgment must be exercised in their use that an experienced professional analyst should always be employed.

Much development is still needed before the modeling technique is capable of handling all situations. Some idea of the complexity of the problem can be seen in Figure 25-1, which illustrates the different plume patterns that a stack can exhibit. The dashed line in the temperature plot is the adiabatic lapse rate, and the solid line is the actual atmospheric temperature pattern that produces each type of plume behavior. A good model should be able to handle all of these situations accurately. The complexity of the alternatives available for

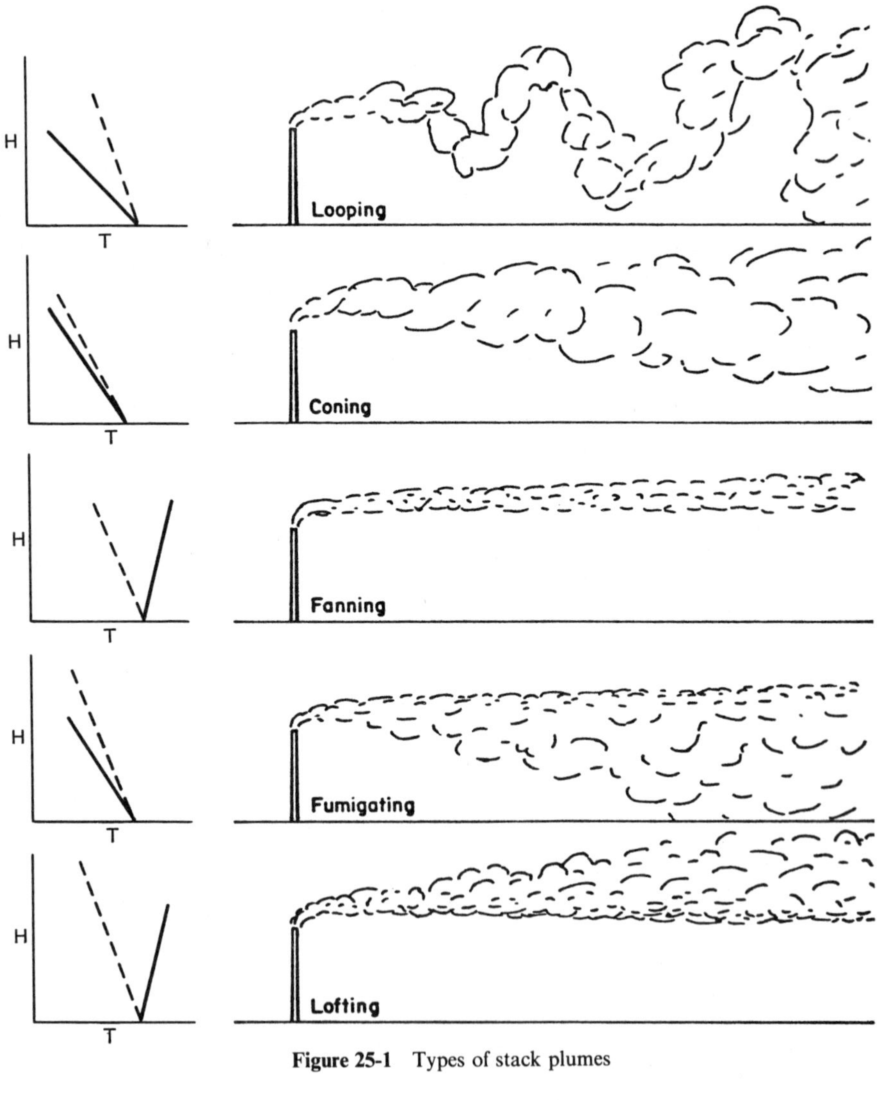

Figure 25-1 Types of stack plumes

modeling is also demonstrated in reference 1, in which 58 alternative models of plume transport in complex terrain are compared.

TYPES OF MODELS

A great deal of effort has gone into the development of models. Models can broadly be classified into two types, physical and mathematical.

Physical models try to simulate air pollution situations by a small-scale representation of the actual situation. These methods usually involve building a scale model of the physical terrain and placing it in a wind tunnel. Urban models of this type have been developed (2, 3). The main factor in this type of model are the criteria used to scale down the physical situation. The proper way to do this, using the principles of dimensional analysis and matching boundary conditions, is covered in reference 4. The French National Meteorological Institute (5) has developed a system using laminar water flow in a channel to simulate wind and an acid solution release system to simulate emission. The model is coated with an indicator that changes color based on the amount of acidity encountered. The color pattern thus indicates the pollutant loading experienced by different locations.

Even with elaborate approaches, it is difficult to duplicate the actual scale of turbulence found in the atmosphere, which limits the usefulness of physical models (6). Thus, in most instances, mathematical models must be used. The availability of inexpensive, high-speed computers has also aided in making mathematical modeling the preferred approach.

Mathematical models can be divided into two groups, statistical models based on a previously collected data base and phenomenological models based on the fluid mechanics of the atmosphere and the chemical reactions that occur. Statistical models generally are based on several years of monitoring data and often employ a linear rollback method of extrapolation. These models assume that a linear relationship exists between a maximum concentration found over a particular time period and the emission intensity. The level above background is directly proportional to the amount emitted. Statistical models are used less commonly than phenomenological models because the required data base does not usually exist for the area surrounding a mill site. Statistical models are also site specific in that they cannot be generalized from one site to another.

The scenario to which the model is to apply must also be chosen. A time-dependent, continuous picture of air quality would be ideal but would consume an inordinate amount of computer time for its operation. Usually either the most probable case or the worst-case scenario is chosen. The most probable case is that combination of emissions and atmospheric interactions that produces the most frequently encountered air quality impact. The worst case is the combination that produces the worst possible air quality. The usual determining factor is meteorology. Emission rates usually stay fairly constant, but daily variations in wind speed and direction, atmospheric stability, and mixing height occur. Meteorology also varies considerably with the season of the year. For daily variation, the worst times are usually early morning or late evening, when radiation inversions can be a factor. Good meteorological data are necessary for good air modeling. The performance of a model depends on the user, the input data, the model mathematics, and the application to which it is put.

PHENOMENOLOGICAL MODELS

A phenomenological model must incorporate four different types of information—emission sources and rates, meteorological dispersion considerations, chemical reactions, and removal processes.

Emission Sources and Rates

In order for a model to function properly, the emission rates from various sources in the area must be calculated. There are three different approaches possible for obtaining this information—a material balance on the process, source emission measurements, or calculation of emission factors. The best approach for determining emission rates is to use source emissions test measurements and to generalize the results.

The approach using emission factors is explained in references 7 and 8. The emission factor is a statistical average of the rate at which pollutants are emitted from the burning or processing of a given quantity of material. There are several approaches to emission factors, so it is important to know how the data were prepared and how they are to be used.

Meteorological Dispersion

The next step is to determine how the quantified emissions are dispersed in the atmosphere. The field of meteorological dispersion is very complex, with a number of differing opinions about the best methods to use. The subject will be discussed only briefly here; more detailed treatments of the alternatives and trade-offs can be found in references 6 and 9–15.

Box Model. The simplest phenomenological model possible is the box model. It is not acceptable for formal air quality analysis but is useful for a preliminary evaluation. It considers most of the variables essential to air quality modeling and, unlike other types of models, is easy to comprehend.

Consider a paper plant surrounded by a box-shaped shed of air. The height of the box is equal to the meteorological mixing height over the plant. The width might represent the plant boundary perpendicular to the wind direction. The wind blows fresh air through the box. Instantaneous mixing of wind and emissions in the box is assumed. Figure 25-2 illustrates the situation. Let $V=$ box volume, $C=$ pollutant concentration in the box and leaving the box weight/

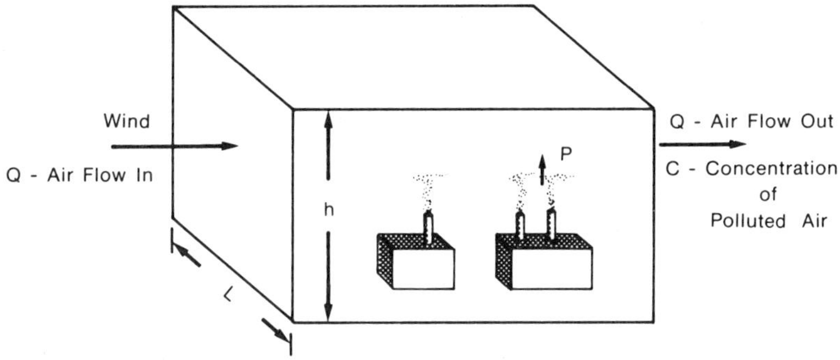

Figure 25-2 Box model

volume, P = pollutant emission rate (weight/time), and Q = air flow rate (volume/time). A simple mass balance over the box gives

Rate of change of pollution in box	=	rate of pollution entering box	−	rate of pollution leaving box

or, mathematically,

$$VdC/dt = P - QC \tag{25-1}$$

If the initial concentration of pollution in the box is zero, this simple first-order differential equation can be solved to give

$$C = (P/Q)(1 - e^{-(Q/V)t}) \tag{25-2}$$

which shows a simple exponential buildup of concentration in the box with time. The limiting concentration at infinite time is

$$C = P/Q \tag{25-3}$$

$$Q = vLh \tag{25-4}$$

where v = wind speed. Thus

$$C_{\max} = P/vLh \tag{25-5}$$

This represents a very reasonable relationship, for the pollutant concentration decreases as the wind speed increases or the mixing height increases. The relationship underscores the importance of meteorology in determining final pollutant concentrations. Emission rates (P), meteorological considerations (v, h), dispersions conditions (complete mix), and terrain effects (flat) have been included.

The box model is not used in practice except for quick or order-of-magnitude calculations, but a modification of it called "proportional modeling" is used.

Proportional Modeling. The concept behind proportional modeling is that air quality will deteriorate in proportion to an increase in emissions and improve in proportion to a decrease. Assumptions involved in this model include the following:

1 The meteorology will be similar to that of the base year in the future.
2 Changes in temporal distribution are insignificant.
3 Changes in the quantity of emissions occur proportionally throughout the area.
4 The maximum concentration measured at the monitoring station is representative of the air quality within the area.

The simple box model for the base year (BY) is

$$C_{\mathrm{BY}} = P_{\mathrm{BY}}/vLh \tag{25-6}$$

and for the future year (FY) is (note that wind velocity is assumed to remain the same)

$$C_{\mathrm{FY}} = P_{\mathrm{FY}}/vLh \tag{25-7}$$

Dividing Equations (25-7) by (25-6) gives an expression for future emission concentrations:

$$C_{\mathrm{FY}} = C_{\mathrm{BY}}(P_{\mathrm{FY}}/P_{\mathrm{BY}}) \tag{25-8}$$

This concept was used as a guideline for rolling back emission rates in order to meet NAAQS. For conservative species the model may have validity, but for nonconservative species the justification for such a simple approach is questionable. The EPA did develop an approach using the rollback model for photochemical oxidants, the "Appendix J" rollback model (16). Now more sophisticated approaches are usually used.

On a local scale, wind velocity is very important, and the proportional model is easily modified to account for changes in it. This modification is called the "Hanna–Gifford model," in which a calibration constant K from base year data is applied to wind velocity (17).

$$C_{\mathrm{FY}} = K(P_{\mathrm{FY}}/v_{\mathrm{FY}}) \tag{25-9}$$

$$K = (C_{\mathrm{BY}}v_{\mathrm{BY}})/P_{\mathrm{BY}} \tag{25-10}$$

The average wind speeds should be determined on a time scale consistent with the emission rate and concentration average. This model at least allows for wind speed variation and does not demand exact meteorological similarity from year to year.

These simple models have yielded to more complex approaches based either on Gaussian diffusion assumptions or on numerical methods for solving the mass conservation equations.

Gaussian Models. The idea behind certain models is to treat each emission source separately as a Gaussian plume. The major assumptions are a Gaussian plume distribution in the horizontal and vertical directions, constant mean wind speed, uniform emission rates, and no deposition or reaction at the surface. The coordinate system that shows the Gaussian plume distribution is given in Figure 25-3, and the equation that describes the dispersion is

$$X(x, y, z) = \frac{Q}{2\pi\sigma_y\sigma_z\mu}\{\exp[(-1/2)(y/\sigma_y)^2]\cdot[\exp(-1/2)(z - H/\sigma_z)^2] + \exp(-1/2)(z + H/\sigma_z)^2\} \tag{25.11}$$

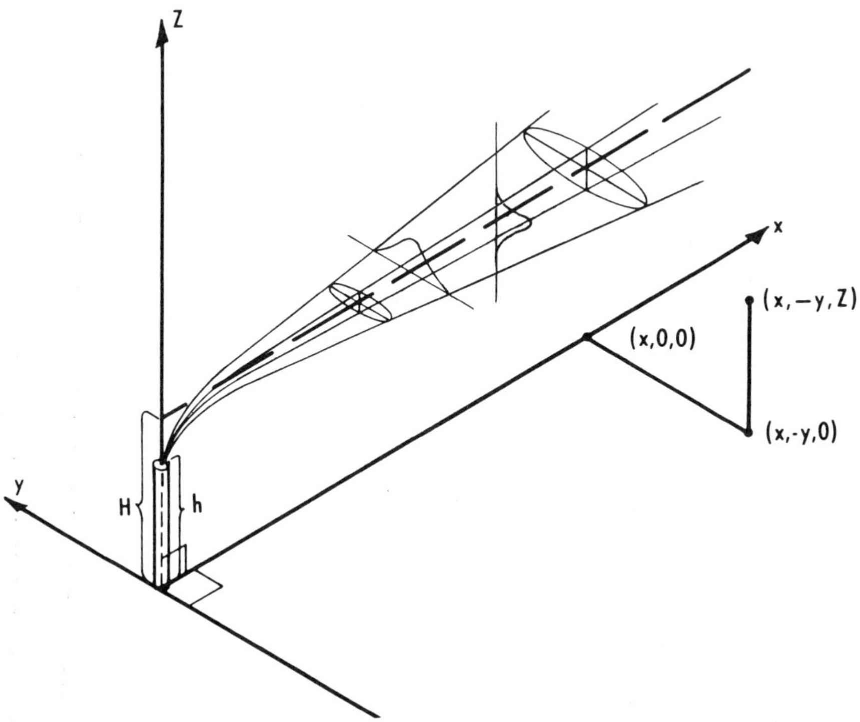

Figure 25-3 Coordinate system for Gaussian distribution. (From J. P. Casey (Ed.), *Pulp and Paper: Chemistry and Chemical Technology*, 3rd edition, volume 2. Copyright © 1980 by John Wiley & Sons, Inc. Reprinted by permission of John Wiley & Sons, Inc.)

where X = concentration at (x, y, z) in grams per cubic meter, Q = source in grams per second, H = effective height of the emission in meters, μ = wind speed in meters per second, and σ_y and σ_z are standard deviations of a plume concentration distribution in the two planes in meters. The concentration, X, of the gas or particulate (less than 20 μm in diameter) at x, y, z from a continuous source, Q, with an effective height of emission, H, is given by this equation.

Different types of Gaussian distributions are available, depending on the type of source. The distribution given in Equation (25-11) is for an elevated source, but there are equations available for ground and line sources as well. Reference 17 gives a more complete set of these equations. These equations can be solved without the aid of a computer and can be understood by novices as well as by the modeling professional. Programs for solving these equations on personal computers are available (18).

For the elevated-stack form of the Gaussian equation, H is an important parameter. Note that the effective height, H, is the sum of the actual height, h, and the plume rise. The later may be estimated using Holland's equation:

$$\Delta H = (V_s d/\mu)(1.5 + (2.68 \times 10^{-3} P((T_s - T_a)/T_s)d) \tag{25-12}$$

where ΔH = plume rise above the stack in meters, V_s = stack gas exit velocity in meters per second, d = inside diameter of the stack in meters, μ = wind speed in meters per second, p = atmospheric pressure in millibars, T_s = stack gas temperature (°K), and T_a = ambient air temperature (°K). It is recommended that ΔH be increased by 10–20% when conditions are unstable and decreased by the same amount when stable conditions occur.

At least three other ways to calculate plume rise exist: the Briggs equation (19), the ASME equation (20), and the TVA equation (21). The ASME and the TVA

Table 25-1 Key to Stability Categories[a]

Surface Wind Speed (at 10 m), m/sec	Day: Incoming Solar Radiation			Night	
	Strong	Moderate	Slight	Thinly Overcast or ≥ 4/8 Low Cloud	≤ 3/8 Cloud
< 2	A	A–B	B		
2–3	A–B	B	C	E	F
3–5	B	B–C	C	D	E
5–6	C	C–D	D	D	D
> 6	C	D	D	D	D

Source: Copyright © 1981. TAPPI. Reprinted from *Introduction to Environmental Management for the Pulp and Paper Industry*, Study Guide, Volume 6, with permission.
[a]The neutral class, D, should be assumed for overcast conditions during the day or night.

have developed extensive information on Gaussian dispersion models; for the best results, the appropriately corresponding equation for plume rise and the corresponding dispersion coefficient should be used.

The equations presented here could be used for rough predictions, but their main purpose is to illustrate principles and explain concepts rather than to provide a complete basis for modeling.

Stability is a function of wind speed. Solar radiation plays a significant role in determining the atmospheric dispersion. The values for the horizontal dispersion (σ_y) and vertical dispersion (σ_z) coefficients vary with the distance from

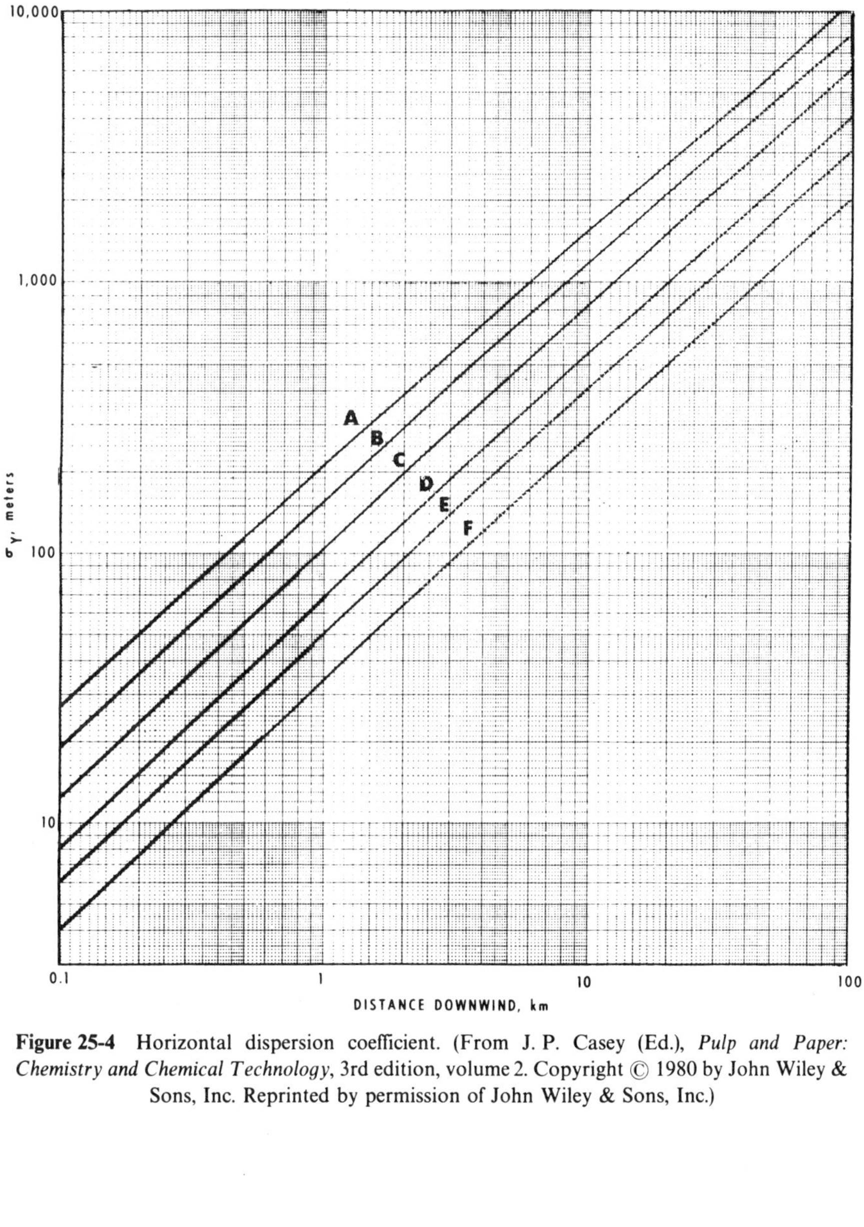

Figure 25-4 Horizontal dispersion coefficient. (From J. P. Casey (Ed.), *Pulp and Paper: Chemistry and Chemical Technology*, 3rd edition, volume 2. Copyright © 1980 by John Wiley & Sons, Inc. Reprinted by permission of John Wiley & Sons, Inc.)

source, the sampling time for the estimated concentration, the degree of surface roughness, the height above the surface, and the amount of atmospheric turbulence or stability (22). The classification of atmospheric stability is given in Table 25-1. Class A is the most stable and class F the least stable.

Figures 25-4 and 25-5 give values for the two dispersion coefficients as a function of downwind distance and stability classification. From these estimates, the concentration of a pollutant at the point of interest can be calculated.

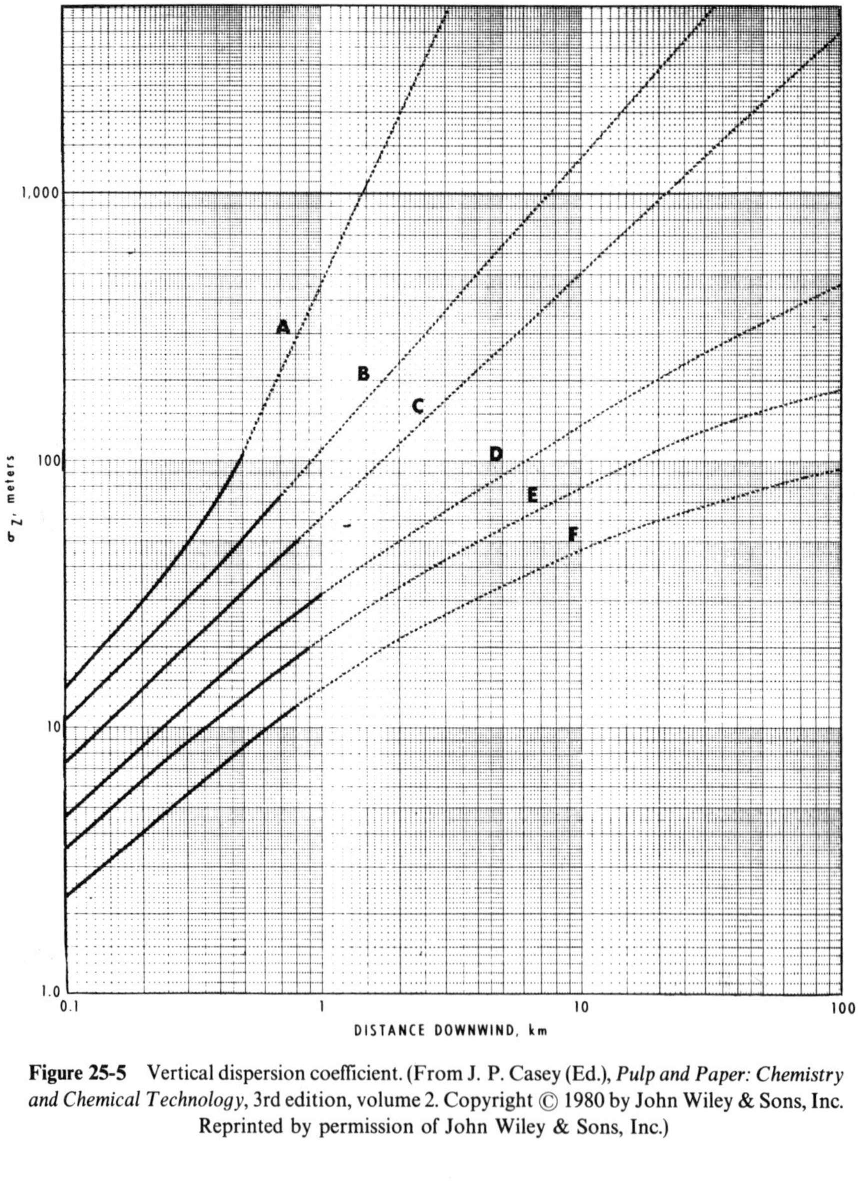

Figure 25-5 Vertical dispersion coefficient. (From J. P. Casey (Ed.), *Pulp and Paper: Chemistry and Chemical Technology*, 3rd edition, volume 2. Copyright © 1980 by John Wiley & Sons, Inc. Reprinted by permission of John Wiley & Sons, Inc.)

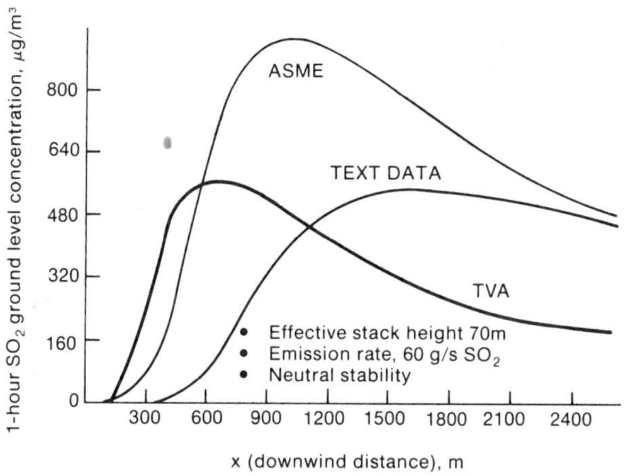

Figure 25-6 Comparison of Gaussian model predictions for different dispersion coefficients. (From *Environmental Impact Analysis Handbook*, by J. G. Rau and D. C. Wooten, 1979, McGraw-Hill. Reproduced with permission.)

In addition to the data in these two figures, the ASME (20) and the TVA (21) have developed sets of dispersion coefficients as a function of distance. Figure 25-6 shows the results of using three different sets of dispersion coefficients to calculate the SO_2 concentration as a function of distance. It is obvious that different coefficients give very different results, illustrating the importance of consistently choosing parameters compatible with the model development and of validating the model against field data to verify its applicability.

Figure 25-7 is most useful for estimating the maximum concentration to be expected and the distance downwind where it will be found. More details, as well as discussions of accuracy of the estimates, can be found in reference 23.

The Gaussian technique is simple to use the requires short computation times, but it has some basic disadvantages (9):

1 The concentrations are not time dependent because it is a quasi–steady state approach.
2 The spatial variability of meteorological parameters is difficult to incorporate.
3 Difficulties are encountered when the wind is light and ill defined in direction.
4 The approach cannot be used for reactive or secondary pollutants.

This approach, however, continues to be popular and is receiving further refinement. Table 25-2 gives a brief description of some of the Gaussian-based air quality models and their uses. The following need to be remembered if one of these models is used (17):

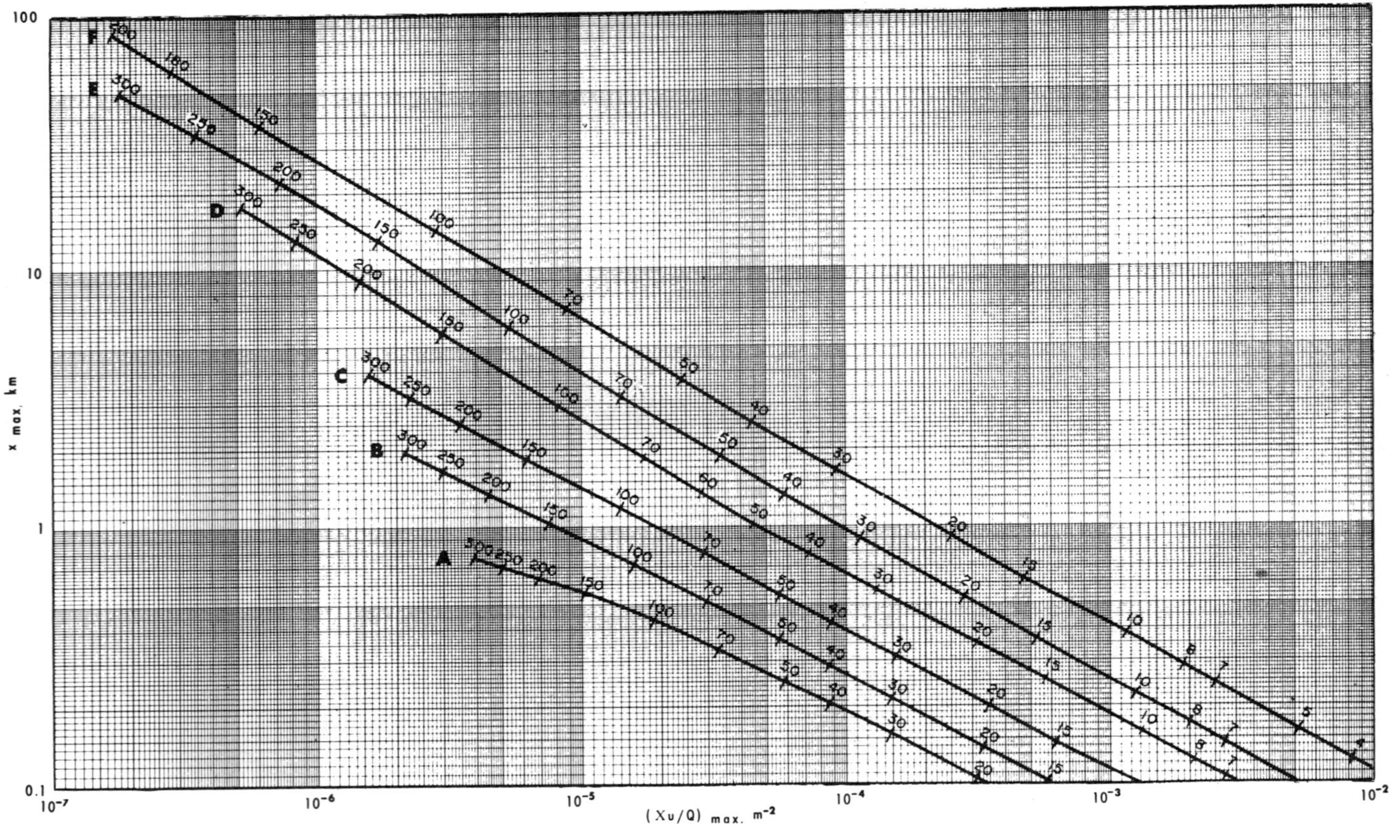

Figure 25-7 Gaussian prediction of maximum concentration. (From J. P. Casey (Ed.), *Pulp and Paper: Chemistry and Chemical Technology*, 3rd edition, volume 2. Copyright © 1980 by John Wiley & Sons, Inc. Reprinted by permission of John Wiley & Sons, Inc.)

Table 25-2 Air Quality Models Incorporating Gaussian Concepts

Model	Description
VALLEY	A program often used to model pulp and paper mills. It is designed for multiple point and area sources. It calculates pollutant concentrations for each frequency designated in an array defined by 6 stabilities, 16 wind directions, and 6 wind classes for 112 program-designated receptor sites on a radial grid of variable scale. The output concentrations are appropriate for either a 24-hour or an annual period, as designated by the user. The model contains the concentration equations, the Pasquill–Gifford dispersion coefficients, and the Pasquill stability classes. Plume rise is calculated according to Briggs. Plume height is adjusted according to terrain elevation for stable cases. An EPA UNAMAP program.
CRSTER	A steady-state, Gaussian plume dispersion model designed for point source applications. It calculates pollutant concentrations for each hour of the year at 180 receptor sites on a radial grid. The hourly concentrations are averaged to obtain concentration estimates for time increments of specified length, such as 3-hour, 8-hour, 24-hour, and annual periods. The model contains the concentration equations, the Pasquill–Gifford dispersion coefficients, and the Pasquill stability classes. Plume rise is calculated according to Briggs.
PTMTP	An interactive program that estimates short-term concentrations for a number of arbitrarily located receptor points at or above ground level due to a number of point sources. Plume rise is determined for each source. Downwind and crosswind distances are determined for each source–receptor pair. Concentrations at a receptor from various sources are assumed to be additive. Hourly meteorological data are used; both hourly concentrations and averages over any averaging time from 1 to 24 hours can be obtained. Uses Briggs plume rise methods and Pasquill–Gifford dispersion methods. An EPA UNAMAP program.
PTDIS	An interactive program that estimates short-term concentrations directly downwind of a point source at distances specified by the user. The effect of limiting vertical dispersion by a mixing height can be included, and gradual plume rise to the point of final plume rise is also considered. An option allows the calculation of isopleth half-widths for specific concentrations at each downwind distance. Uses Briggs plume rise methods. An EPA UNAMAP program.
PTMAX	An interactive program that analyzes the maximum short-term concentrations from a single point source as a function of stability and wind speed. Uses Briggs plume rise methods and Pasquill–Gifford dispersion methods. An EPA UNAMAP program.

Source: From *Environmental Impact Analysis Handbook* by J. G. Rau and D. C. Wooten, 1979, McGraw-Hill. Reproduced by permission. Other data from references 24–27.

1. Review publications in the literature.
2. Review reports of agency-supported studies.
3. Contact groups in responsible local, state, and federal agencies for advice and guidance.
4. Contact groups responsible for code development and validation.

These models also do not account for downwash caused by winds swirling around buildings or rough terrain, which can cause a problem if they are applied under such conditions. When these conditions occur or when there are reactions in the gas stream, it is best to apply the mass conservation approach and develop numerical models.

Numerical Models. The mass conservation approach is a more fundamental way to proceed, but it is necessary to make several assumptions in order to solve the partial differential equation. Only the basic concepts will be introduced; the detailed, in-depth approach will be left to other treatments (6, 9).

The equation of continuity for one dimension only is as follows:

$$\underset{\text{(A)}}{(\partial C/\partial t)} = \underset{\text{(B)}}{-(\partial/\partial x)(uC)} + \underset{\text{(C)}}{D(\partial^2 C/\partial x^2)} \tag{25-13}$$

This equation states that the change in concentration with time (term A) is due to two factors: material flowing in or out (term B) and material diffusing in or out (term C). If this equation could be solved with appropriate boundary conditions, the result would be the pollutant concentration as a function of position and time. Solving it is not a simple matter, however. Since the flow in the atmosphere is turbulent, the equation must be modified by splitting the velocity and concentration terms into two parts, an average term and a perturbation portion, as shown by the following two equations:

$$C - \bar{C} + C' \tag{25-14}$$

$$u = \bar{u} + u' \tag{25-15}$$

Inserting these into Equation 25-13 yields Equation 25-16:

$$\underset{\text{(D)}}{(\partial\bar{C}/\partial t)} = \underset{\text{(B)}}{(-\partial/\partial x)(\bar{u}\bar{c})} - \underset{\text{(F)}}{(\partial/\partial x)(\overline{u'C'})} + \underset{\text{(G)}}{D(\partial^2/\partial x^2)(\bar{C})} \tag{25-16}$$

This equation states that pollutant concentration changes with time (term D) due to transport of material by fluid flow of wind (term E), transport by turbulence (term F), and transport by molecular diffusion (term G). All of these terms can be handled easily except term F, turbulent diffusion, for which an assumption is made, as expressed the following equation:

$$\overline{u'C'} = -K(\partial\bar{C}/\partial x) \tag{25-17}$$

This assumption results from an analogy to molecular diffusion. The term K is the eddy diffusivity, which in three dimensions would be a tensor quantity. The turbulent diffusion so dominates the molecular diffusion term that the latter is often neglected; the resulting equation to be solved is therefore:

$$(\partial\bar{C}/\partial t) = -(\partial/\partial x)(\bar{u}\bar{C}) + (\partial/\partial x)K(\partial\bar{C}/\partial x) \tag{25-18}$$

In one-dimensional form, this is the basic "K theory" equation that is the starting point for most models using the mass conservation approach.

This K theory equation is usually solved by finite difference methods, and it can be solved either from a fixed-coordinate system that is stationary with respect to the fluid or from one that moves at the average velocity of the fluid. So-called Eulerian models result if the fixed-coordinate system is chosen. These models divide space into a series of boxes and analyze what occurs in each box. Thus the Eulerian approach is often called the "multibox model," whereas the other approach that gives solutions from a coordinate system moving with the fluid is often called the "moving-cell model" or, more formally, the "Lagrangian model." In this approach, attention is focused on the polluted air itself in a Lagrangian or moving frame, and horizontal dispersion is neglected. Both approaches contain many assumptions whose validity can be questioned, so it is not clear which method is superior.

Numerical models require a large data base. Both emissions data and meteorological variables must be geographically coded. These models are valuable in

Table 25-3 Numerical Air Quality Models

Model	Description
LIRAQ-2 (*LI*ve*R*more Air *Q*uality model)	A grid (Eulerian) model for regional photochemical oxidant. The current computer code is specific for the San Francisco Bay Area. There is only one layer between the surface and the inversion base; a mean transport wind is calculated by integrating the wind profile function. Consequently, the model is in essence a two-dimensional model. The required meteorological input includes hourly values for regional windfield (wind speed and direction), inversion height, and turbulent-eddy mixing coefficients. Hourly values for the regional emissions distribution are also required.
REM (*R*eactive *E*nvironmental simulation *M*odel)	A two-dimensional trajectory (Lagrangian) model for photochemical oxidant with a fairly complex air chemistry submodel. One layer between the surface and the inversion height. Meteorological input includes hourly values for regional windfield, inversion height, and stability class. Hourly values of regional emissions distributions are also required. Expensive to run many trajectories to provide regional description of air quality.

(*Contd.*)

Table 25-3 (*Contd.*)

Model	Description
DIFKIN (*DIF*usion/ *KIN*etics)	A three-dimensional trajectory (Lagrangian) model for photochemical pollutants. Uses a multiple-layered column of air transported by a mesoscale windfield. A gridded emissions inventory with hourly variation is required. Horizontal dispersion is neglected. Expensive to run many trajectories to provide a regional description of air quality. Meteorological input includes hourly values for regional windfield (wind speed and direction), inversion height, and stability class. Principle areas of application have been the Los Angeles and San Francisco metropolitan areas.
IMPACT (*I*ntegrated *M*odel for *P*lumes and *A*tmospherics in *C*omplex *T*errain)	A three-dimensional grid (Eulerian) model for photochemical oxidant. The air chemistry submodel can also simulate sulfate/SO_2 processes. IMPACT utilizes the DIFKIN photochemical air chemistry. Options are available for the detailed treatment of the air chemistry of major point source plumes. Meteorological input includes hourly values for regional windfield (wind speed and direction) and atmospheric stability. The inversion height is input implicitly via the stability parameter; up to 20 layers between the surface and the inversion height can be specified by the user. Hourly values for the regional emissions distribution are also required. Model output includes hourly average values for each grid cell and layer in the region.
SAI (Urban Airshed Model)	A three-dimensional grid (Eulerian) model for regional photochemical oxidant; also applicable to regional simulation of conservative pollutants. The primary area of application has been the Los Angeles area, with some applications in the Denver, Colorado, area. This model has been modified frequently, and there are several configurations of the meteorology, air chemistry, and emissions submodels. The required meteorological input includes hourly values for regional windfield (wind speed and direction), inversion height, and stability class. Hourly values for the regional emissions distribution are also required. Several options are available to treat major point and line sources on a subgrid cell basis. The species resolution of the hydrocarbon emissions depends on which configuration of the model is used. The model output includes estimates of hourly averages for each grid cell for carbon monoxide, oxides of nitrogen, hydrocarbons, and photochemical oxidant.

Source: From *Environmental Impact Analysis Handbook* by J. G. Rau and D. C. Wooten, 1979, McGraw-Hill. Reproduced by permission. Other data from references 28–32.

that they can incorporate chemical reaction processes and unusual wind flow patterns. Because of their complexity, they require above-average technical expertise for their operation and consume large amounts of computer storage space; they can be costly to operate. Table 25-3 lists some of the numerical air quality models available.

Attempts have been made to devise simpler models to achieve the same results. A model using the box modeling approach but incorporating the various chemical reactions has been developed that gave a reasonable fit to urban ozone data around Denver, Colorado, and Houston, Texas (33). The normal complex model used to analyze these data would have been the empirical kinetic modeling approach (EKMA) (33). Agreement between the simple model and the EKMA was quite good. There is always some trade-off between model sophistication and accuracy.

The EPA has actively encouraged the development and use of models. At its research center in Research Triangle Park, North Carolina, it maintains a system called the "Users Network for Applied Modeling of Air Pollution (UNAMAP)." It contains such models as PTMAX, PTDIS, PTMTP, CRSTER, VALLEY, and HIWAY (34). Plans have been made to maintain, update, and add to these models in order to provide good modeling capabilities for anyone desiring to use the system.

Chemical Reactions

Once the emission rates and fluid mechanics have been determined, they must be coupled with the chemical reactions that can occur in the atmosphere to produce an accurate picture. The important processes by which secondary pollutants are produced must be included in the model. Gas to solid transformations, such as SO_2 to sulfate, must be included, and, if of concern, the photochemical reactions that generate aerosols (35, 36).

Processes That Remove Pollutants

The final element that should be included in an air quality model is the processes that remove pollutants from the atmosphere. However, because little is known about these processes and because eliminating them can only lead to a model that overstates rather than understates the pollution problem, they are often omitted from the model. It is unfortunate that factors such as dry surface deposition of particulates, and precipitation scouring are not included, because in certain instances they are important mechanisms for removal of pollutants.

MODEL CALIBRATION AND VALIDATION

Because the physical situations are so complex and because modeling is still in the process of development, it is extremely important to calibrate any model used and then to validate it independently. Calibration consists of adjusting the model parameters to better fit a particular set of empirical data. Models are then validated by comparing their predictions with another set of actual data collected under the same conditions as those input to the model. Reference 37 should be a useful aid in verifying that the theory being applied is correct.

Calibration

In any complex model, there are enough empirically derived values so that the model can be manipulated to achieve any desired result. One modeler has claimed that he could model the path of a crazed elephant wandering through the jungle using six parameters. Thus it is important to use judgment in calibrating a model. In the calibration process, variables such as wind speed, wind direction, dispersion coefficients, reaction rates, and even source emission rates may be adjusted, but they should be adjusted to correspond to reality rather than to fiction or best fit (17). Fit is important, but only to the extent that it corresponds to a physically real situation. From the calibration procedure, some knowledge of the confidence limits of the model should be gained.

Validation

Once the model has been calibrated, it must be validated. It is desirable to perform validation comparisons over as wide a range of meteorological conditions and emission rates as possible. The NCASI has made efforts to validate some of the EPA models (34, 38). Validation of the main portion of a model is best accomplished through special experiments designed to isolate the specific phenomenon being simulated and to obtain data on it. It is often a good idea to validate independently certain subportions of the model, such as those covering the chemical reactions or the removal processes.

Dispersion models currently being used by regulatory agencies have not been well documented due to the lack of sufficient field data to verify the models and the lack of well-defined performance measures for evaluation purposes (39).

In performing the validation, it is useful to apply both paired and unpaired comparison techniques in order to evaluate the model's performance critically. Extreme value theory is a powerful technique used to assess the performance of models (39).

A number of studies have been reported that validate existing models (40–42). There is considerable contention over the issues of calibration and validation, and Dr. F. Pasquel has very aptly put this disagreement into verse (6):

> Calibration may be consoling
> But is a practice we shouldn't condone,
> And for validation to have virtue
> The physics and chemistry must be known.

MODEL USE

Mathematical modeling has been developed to the point where it is integrated into SIPs and is used to apportion the allowable levels of emission permitted from various sources. Anyone contemplating the expansion of an existing plant or the construction of a new plant that will emit more than 50 tons/year of any of the criteria pollutants must carry out an air quality modeling program. Thus, if a new or expanded pulp or paper mill can be expected to emit more than 50 ton/year of either particulates or sulfur dioxide, a modeling exercise is required using EPA-prescribed models to determine if more than an acceptable quantity of the PSD increment is consumed. The NCASI has conducted some modeling of this type for the industry (34, 38) and has concluded that the single most important factor in determining the amount of the increment consumed is the terrain surrounding the mill. In the future, modeling is likely to see increasing use as a decision-making aid and a regulatory tool.

Most of the early modeling effort was devoted to regional models used to assess the impact of local sources on the local environment. Recently, however, with the advent of issues such as global CO_2 balance and the impact of acid deposition on the environment, long-range or interregional transport models have assumed more importance. These models are of the numerical type and can be classified into three broad categories: Eulerian grid, statistical trajectory, or Lagrangian trajectory. Several review articles have been written on this subject (43–46), and many problems with the accuracy of these long-range models still exist.

REFERENCES

1 H. M. Ellis, P. C., Liu, and C. Runyan, *Journal of the American Pollution Control Association*, 30 (6):670–675 (1980).

2 F. H. Chandry and J. E. Cermak, *Proceedings of the Conference on Air Pollution Meteorology*, American Meteorological Society, Boston, 1971, p. 126.

3 J. E. Cermak and S. P. S. Arya, *Meteorology*, 1:40–60 (1970).

4 W. H. Snyder, *Meteorology*, 3:113–134 (1972).

5 L. Facy, P. Brechambout, A. Daury, and R. Quinino, *Review of Meteorology*, (71):69 (1963).

6 J. H. Sunfeld, *Air Pollution; Physical and Chemical Fundamentals*, McGraw-Hill, New York, 1975.

7 *A guide to estimating heat input for combination boiler emission rate calculations*, Atmospheric Quality Improvement Technical Bulletin No. 78, National Council of the Paper Industry for Air and Stream Improvement, New York, March 1974.

8 "Standards of performance for new stationary sources," 40 CFR Part 60, July 1, 1979.

9 A. C. Stern, *Air Pollution*, 3d ed., Vol. 1, Academic Press, New York, 1976.

10 H. Moses, *Mathematical urban air pollution models*, ANL/ES-RPY-001, Argonne National Laboratory, Argonne, Ill. 1969.

11 J. H. Sunfeld, P. M. Rath, and S. D. Reynolds, *Advance in Chemistry Series*, 113:58–100 (1972).

12 M. I. Hoffert, *AIAAJ*, 10:377–387 (1972).

13 L. T. Fan and Y. Horie, *Critical Review of Environmental Control*, 431–457 (1971).

14 R. J. Bibbero and I. C. Young, *Systems Approach to Air Pollution Control*, Wiley, New York, 1974.

15 J. Mathis and W. Grose, *A review of methods for predicting air pollution dispersion*, National Aeronautics and Space Administration Bulletin NASA SP 322, Washington, D. C., 1973.

16 *Federal Register*, 36 (158):15486–15506 (August 14, 1971).

17 J. G. Rau and D. C. Wooten, *Environmental Impact Analysis Handbook*, McGraw-Hill, New York, 1979.

18 H. M. Cota, "A basic computer program for the Gaussian equation for a point source," *Journal of the American Pollution Control Association*, 34 (3):253 (1984).

19 G. A. Briggs, "A plume rise model compared with observations," *Journal of the American Pollution Control Association*, 15 (9): (September 1965).

20 *Recommended Guide for Prediction of the Dispersion of Airborne Effluents*, American Society of Mechanical Engineers, New York, May 1968.

21 S. B. Carpenter, T. L. Montgomery, J. M. Leavitt, W. C. Colbaugh, and F. W. Thomas, "Principal plume dispersion models: TVA power plants," *Journal of the American Pollution Control Association* 21 (8):491 (1971).

22 E. V. Somers, "Dispersion of pollutants emitted into the atmosphere," in W. Strauss, ed., *Air Pollution Control*, Part 1, Wiley, New York, 1971, pp. 1–30.

23 D. B. Turner, *Workbook of atmospheric dispersion estimates*, U.S. HEW, Public Health Service Publication No. 999-AP-26, National Air Pollution Control Association, Cincinnati, 1970.

24 E. Burt and J. Mersch, *Valley model computer program*, EPA source magnetic tape, EPA/DF-78/002 (NTIS PB 275-700), U.S. Environmental Protection Agency, Washington, D. C., September 1977.

25 R. Lee and J. Mersch, *Single source (CRSTER) model computer programs*, EPA source magnetic type EPA/DF-78/004 (NTIS PB 275-701), U.S. Environmental Protection Agency, Washington, D.C., 1977.

26 D. B. Turner and A. D. Busse, *Users guide to the interactive versions of three point source dispersion programs: PTMAX, PTDIS and PTMTP* (NTIS PB 229-771), U.S. Environmental Protection Agency, Washington, D. C., June 1973.

27 P. Galomb, S. Ballerman, and J. Gruhl, "Air quality screening model with long term averaging," *Journal of the American Pollution Control Association* 33 (3):215–219 (1983).

28 M. C. MacCracken and A. Stauter, *Development of an air pollution model for the San Francisco Bay area*, Final Report, NSF, Lawrence Livermore Laboratory Report UCRL-51920, Vols. 1 and 2, 1975, Los Angeles.

29 L. G. Wayne, A. Kokin, and M. I. Weirburd, *Controlled evaluation of reactive environmental simulation model (REM)*, Vol. III, Final Report, EPA Contract 68-02-0345, Pacific Environmental Services, Inc., Santa Monica, Calif., 1973.

30 J. R. Martinez, R. A. Nordsick, and M. A. Hershberg, *User guide to diffusion/kinetics (DIFKIN) code*, Final Report, EPA Contract 68-02-0336, General Research Corp., Santa Barbara, Calif., 1973.

31 A. Fobrick, R. Sklrew, and J. Wilson, *Point source model evaluation and development study*, Report to the California Air Resources Board and the California Energy Resources Conservation and Development Commission, Contract No. A5-058-87, March 1971, Los Angeles.

32 S. D. Reynolds, M. K. Liu, T. A. Hecht, P. M. Roth, and J. H. Seinfeld, "Mathematical modeling of photochemical air pollution. III. Evaluation of model," *Atmospheric Environment*, 8:563 (1974).

33 A. R. Laird, R. W. Meksad, and P. Middleton, "A simple model for urban ozone impact predictions," *Journal of the American Pollution Control Association*, 32 (12):1221–1225 (1982).

34 *Estimated ground level concentration of sulfur dioxide and particulate matter for selected Kraft mill sites using EPA dispersion models*, Atmospheric Quality Technical Bulletin No. 100, National Council of the Paper Industry for Air and Stream Improvement, New York, February 1979.

35 T. A. Hecht, J. H. Sunfeld, and M. C. Dodge, *Environmental Science and Technology*, 8:327–339 (1974).

36 P. M. Roth, T. A. Hecht, and J. H. Sunfeld, *Existing needs in the experimental and observational study of atmospheric chemical reactions*, EPA Report R4-73-031, U.S. Environmental Protection Agency, Research Triangle Park, N.C., 1973.

37 *Workbook for comparison of air quality models*, EPA 450/2-79-028a,b (NTIS PB 292 249 and PB 80-120108), U.S. Environmental Protection Agency, Washington, D.C., 1979.

38 *A review of terrain and other physiographic features at U.S. pulp mill sites and their implications in defining atmospheric dispersion modeling needs*, Atmospheric Quality Improvement Technical Bulletin No. 104, National Council of the Paper Industry for Air and Stream Improvement, New York, December 1979.

39 S. Trivikrama Rao and J. R. Visalli, "On the comparative assessment of the performance of air quality models," *Journal of the American Pollution Control Association*, 31 (8):851–860 (1981).

40 H. M. Ellis and P. C. Liu, "Review of the performance of the RAM model in predicting highest measured concentration," *Journal of the American Pollution Control Association*, 31 (2):148–152 (1981).

41 T. M. Heron, J. F. Kelly, and P. B. Hootaya, "Validation of the industrial source complex (ISC) dispersion model in a rural setting," *Journal of the American Pollution Control Association*, 34 (4):365–369 (1984).

42 R. F. Hodanbasi and R. Brodkey, "Evaluation of RAM model for Cleveland, Ohio, Part II," *Journal of the American Pollution Control Association*, 34 (4):371–374 (1984).

43 A. Eliassen, "A review of long range transport modeling," *J. Appl. Meteorol.*, 19:231 (1980).

44 F. B. Smith and R. P. Hunt, "The dispersion of sulfur pollutants over Western Europe," *Philosophical Transactions of the Royal society* (*London*), A290:523 (1979).

45 "Atmospheric transport, transformation and deposition," *Air quality criteria for sulfur oxides and particulate matter*, Chapter 6, U.S. Environmental Protection Agency, Research Triangle Park, N.C., December 1980.

46 W. B. Johnson, "Interregional exchanges of air pollution: Model types and applications," *Journal of the American Pollution Control Association*, 33 (6):563–573 (1983).

26

Acid Deposition and the Forest Products Industry

Acid deposition, often dubbed "acid rain," has received a great deal of attention in the popular press. Acid materials can be deposited from the air either in wet or in dry form. The major cations and anions that need to be considered in acid deposition are H^+, NH_4^+, K^+, Na^+, Ca^{2+}, Mg^{2+}, HCO_3^-, SO_4^-, NO_3^-, Cl^-, and PO_4^{2-} (1). These ions determine the ionic balance in a natural system and the resultant pH. The ions can be dissolved in rain (acid rain or acid fog) or deposited in the form of dry particulate matter. The limited data available suggest that wet and dry deposition account for roughly equal amounts of acid deposition nationwide (2). There are actually three processes of deposition—adsorption or absorption, settling of dry particulate matter, and wet deposition in fog, rain, dew, or snow. The latter two are of most importance. The science of dry deposition monitoring is in its infancy, so most of the information available applies to wet deposition. Near emission sources, dry deposition is more important. Dry deposition usually decreases with the distance from the emission source.

Precipitation formed in the atmosphere ranges in pH from 4.8 to 6.5, depending on the amount of CO_2 and other substances dissolved (3). Water saturated with CO_2 would have a pH of 5.6. CO_2 saturation occurs naturally as a rain drop falls through the atmosphere. Therefore a possible arbitrary definition of acid rain would be any precipitation with a pH of less than 5.6. Figure 26-1 shows the annual mean pH values for precipitation across the United States and Canada. Observe that in the midwestern and eastern United States, the pH of the precipitation is well below this limit, being in the range 4.1–4.6 for most of the eastern part of the country. In specific storms, very low pH values have been observed—for example, a pH of 2.7 in Kane, Pennsylvania, and 1.5 in Wheeling, West Virginia (4). It has also been reported that the pH of some fogs is lower than that usually measured for rain (5).

There is no question that this pattern of acidic deposition exists. Its cause, however, is only speculative at present. There are no known natural causes to account for the pattern illustrated in Figure 26-1. The pattern does correspond, however, to the area of heavy industrialization and urbanization along the Ohio river valley and the eastern seaboard. Figure 26-2 gives representative values of SO_2 emission, and Figure 26-3 gives the spatial distribution of wet sulfate

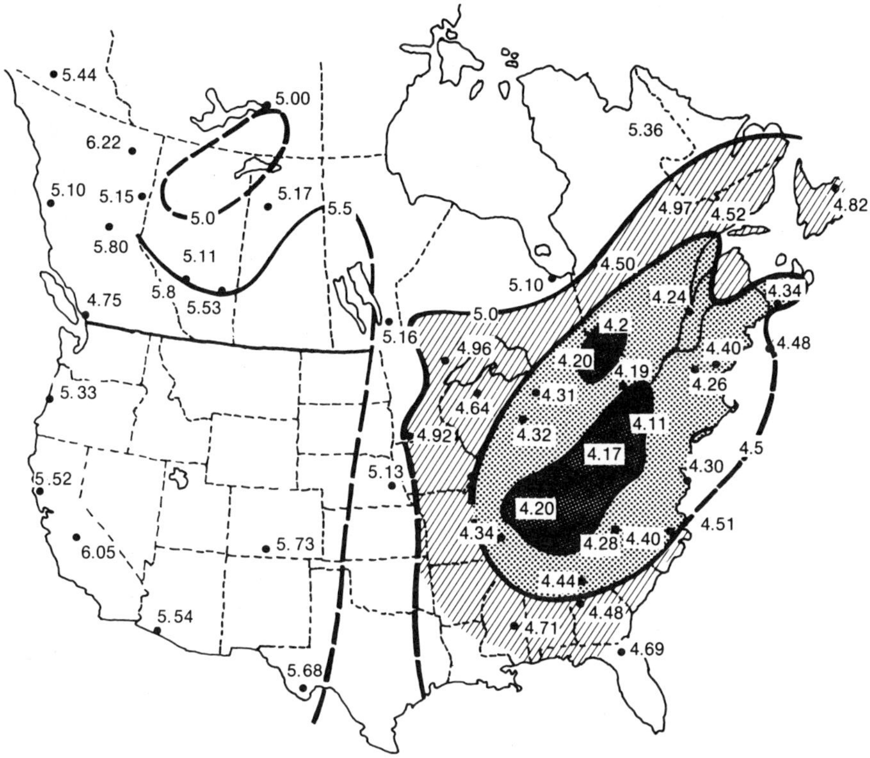

Figure 26-1 Annual mean values for the pH of precipitation

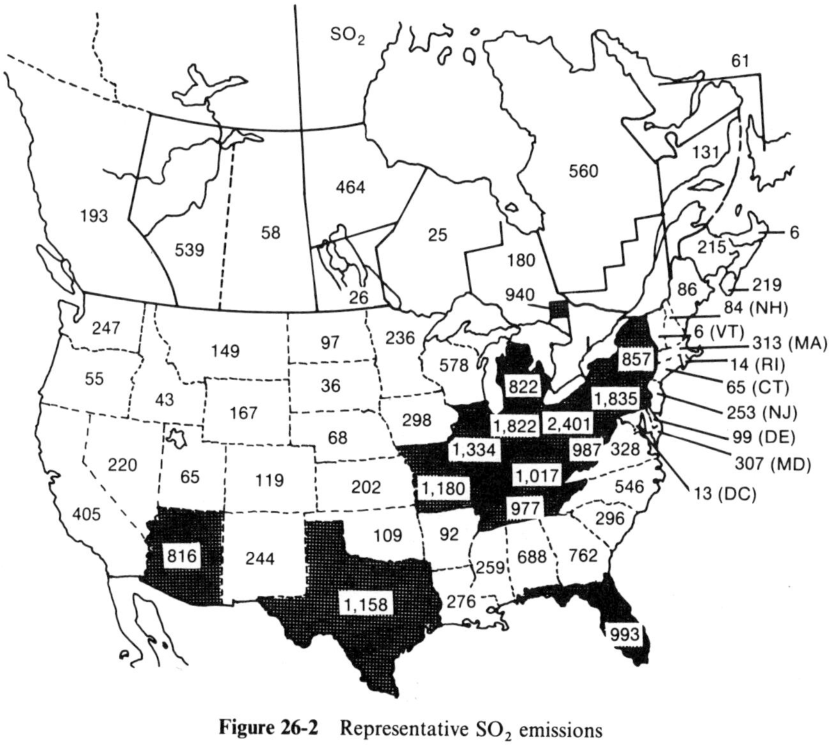

Figure 26-2 Representative SO_2 emissions

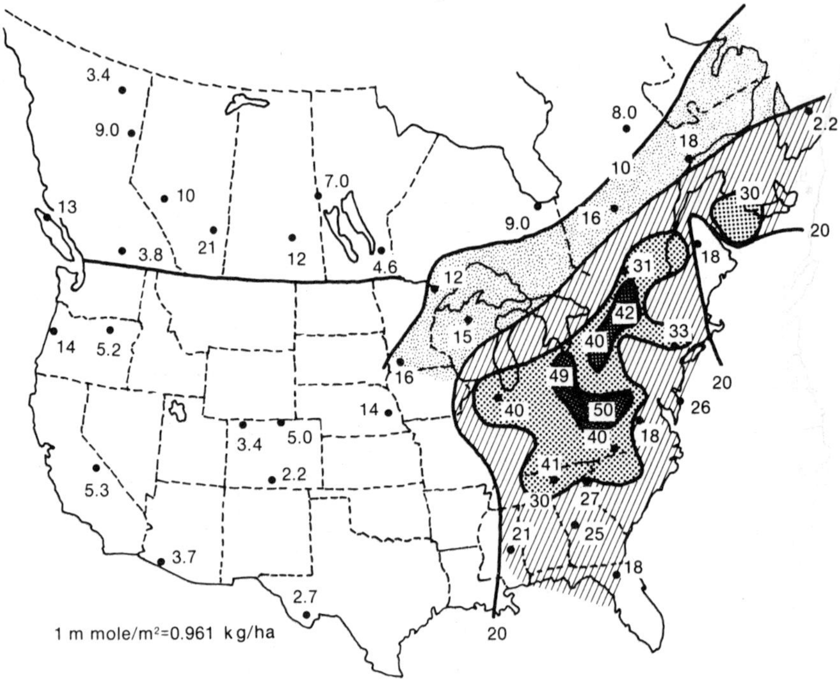

Figure 26-3 Spatial distribution of annual mean wet sulfate deposition

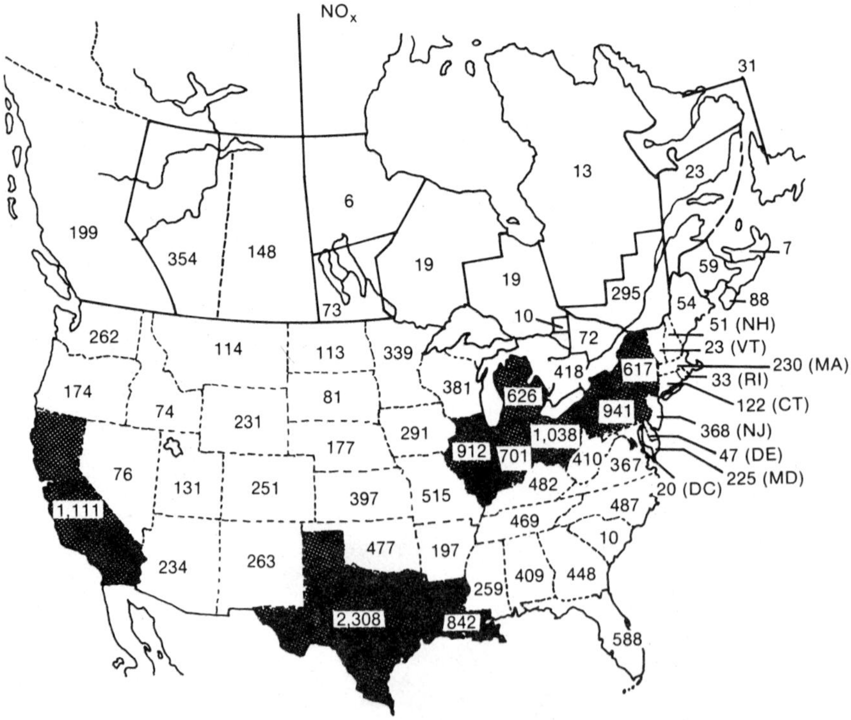

Figure 26-4 Representative NO_x emissions

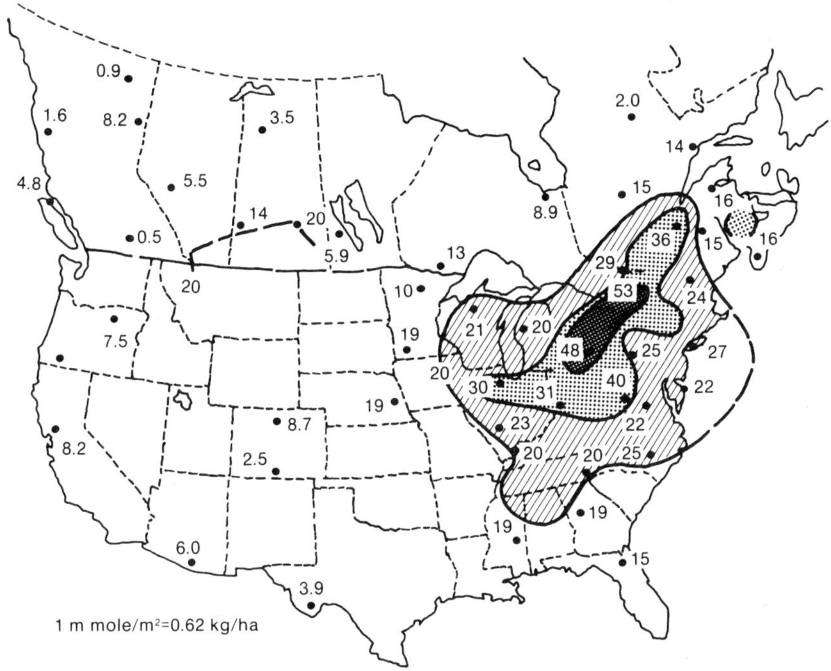

Figure 26-5 Spatial distribution of annual mean wet nitrate deposition.

deposition in nanomoles per square meter. The patterns are similar to that of Figure 26-1, but this fact does not prove that SO_2 and sulfate are completely responsible for the problem. Figure 26-4 gives NO_x emission, and Figure 26-5 gives the geographic distribution of nitrate. These patterns are also similar to that of Figure 26-1. It is this kind of evidence that led the National Academy of Sciences to consider that there was overwhelming circumstantial evidence linking power plant emissions to acid deposition (5). The exact amount of the problem that is attributable to SO_2 or NO_x is not yet clear, nor is the amount of the problem that is related to wet deposition (SO_2, NO_x) or dry deposition (SO_4^{2-}, NO_3^-). What is clear is that these materials are those primarily involved in acid deposition and that some means to minimize these emissions must be implemented in order to improve the situation.

The National Academy of Sciences report (6) recommended a 50% reduction in the amount of emissions, concluding that this decrease would result in a 50% reduction in acid deposition. This conclusion was questioned by the scientific community, but a later report (7) reaffirmed it. The present consensus is that emissions and acid depositions are related linearly.

HISTORICAL PERSPECTIVE

Experience with acid deposition is not new, nor is it restricted to the northeastern United States. In 1852 an English chemist, Robert Angus Smith, discovered the relationship between the air pollution over Manchester and the pH of precipitation in the region. He later wrote a 600-page book on the subject. The work was rediscovered in the 1950s by Dr. Eville Gorham, a Canadian ecologist, who elaborated on Smith's theories but was virtually ignored. It was not until the late 1960s, when Swedish soil scientist Svante Oden reported on the pattern of increasingly acid precipitation over a definite geographic area, that acid rain became a major concern. He used colorful language and characterized acid rain as "chemical war." Oden is thought of as the father of acid rain research, but it can be seen to have a much older heritage.

One of the early North American investigators was Dr. Gene Likens, an ecology professor at Cornell. He teamed up with Dr. F. Herbert Borrmann of Yale to conduct an interdisciplinary study of forest productivity in the Hubbard Brook Experimental Forest in New Hampshire. They have developed records of the pH of precipitation there since the early 1960s. Their record shows annual fluctuations in the acidity, but no clear worsening or improving trend through the early 1980s. The evidence of damage to aquatic resources is probably more pronounced than that to forests, but both should be of concern.

PAPER INDUSTRY PERSPECTIVE

The pulp and paper industry should be concerned about acid rain because it is a net receiver rather than a primary emitter of acid deposition. The paper industry owns a significant part of the forest lands and surface waters on which acid deposition falls. Maintaining productive forests is essential to the long-term well-being of the industry.

Even though the paper industry is a small emitter of SO_2 and NO_x in terms of total emissions of these substances, it is still an important local source of gaseous pollutants that may have a significant local impact. The issue of local contribution versus that from long-range transport remains an area of contention.

If the eventual cleanup effort centers on the power industry, the pulp and paper industry will feel the impact of higher power costs to the extent that it depends on purchased power.

SOURCES OF SO_2 AND NO_x

Natural sources of SO_2 and NO_x exist. In fact, natural sources contribute 35–85% of the total global sulfur. The source of approximately 80% of the natural SO_2 is the nutrient-rich ocean regions (2). Biogenic sources of SO_2 include both vegetation and microbial activities in upland terrestrial sites, tidal/intertidal areas, and the nutrient-rich ocean regions. There are also nonbiogenic natural sources such as geothermal activities, natural combustion, lightning, airborne soils, and water aerosols. These natural contributions provide only a small per-

Table 26-1 Gaseous SO_2 and NO_x Emissions (MillionTons/Year, 1982 Estimates)

	United States	Canada	Total
SO_2	24.1	4.8	28.9
NO_x	19.3	1.8	21.1

centage of the total deposition in the northeastern United States, however. In North America, 3–13 times more NO_x comes from artificial sources than from natural sources.

Table 26-1 gives the 1982 estimate of the amounts of artificial SO_2 and NO_x emissions (million tons per year) produced by the United States and Canada. The natural emissions in these areas would be less than about 5% of these amounts. It is apparent from this table that the United States is the major producer of acid pollutants and that it also dominates the production of NO_x. The electrical utility industry is the industry most responsible for SO_2 and NO_x emissions, producing over 65% of the SO_2 and 31% of the NO_x (8). Coal burning by electrical utilities accounts for 90% of the SO_2 and 71% of the NO_x they produce.

The utility industry has not ignored this problem. In the period 1940–1970, SO_2 emissions grew by 60% and NO_x emissions tripled. Total utility emissions grew from 2 million to 15 million tons/year or 700%. Industrial emissions over that same period grew by 50%. In the period 1970–1980, sulfur oxide emissions decreased by 20% in spite of a 50% increase in power generation from utilities. By 1983, approximately 100 scrubbers had been installed on utility boilers and 50 more are being built (9). Also, some utilities are using coal washing or are switching to low-sulfur coal. The question is whether these efforts are enough.

Motor vehicles are another source of NO_x emissions, accounting for 40% of artificial NO_x. Significant quantities of SO_2 and NO_x also come from the metal-smelting and cement industries. The largest single source of SO_2 in the free world is the nickel smelter in Sudbury, Ontario, which emits 2,500 tons of SO_2 per day (4)—an improvement over the 7,000 tons/day it used to emit. This smelter's stack is 1,250 feet high.

There is no guarantee that elimination of SO_2 emissions would solve the acid rain problem, for the NO_x emissions still present would also contribute to an unknown extent. Thus there is a real reluctance to spend the billions of dollars required to retrofit all utility boilers with scrubbers without a better understanding of the effects of this aspect of source control on the overall problem.

Figure 26-6 shows the wind patterns in summer and winter that transmit air emissions to different areas of the country. These wind patterns are the means of long-range transport. Long-range transport is also related to stack height. In the United States as of 1984, 179 stacks were over 500 feet high and, of these, 20 were over 1,000 feet high.

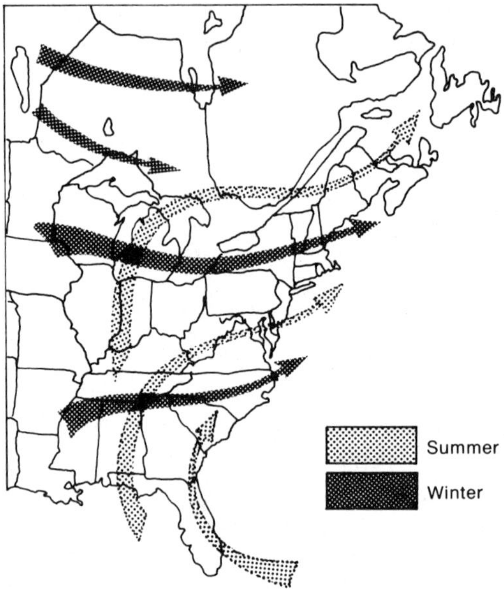

Figure 26-6 Wind patterns

An auxiliary issue involved in the burning of fossil fuels is the associated emission of heavy metal particulates (mercury, lead, zinc, cadmium) (10). Rain has been reported to contain lead in excess of safe drinking water standards in certain parts of the country (1). Stacks have the potential to emit fine particles of sulfate, nitrate, heavy metals, and toxic organics. Tracer experiments have followed particulates from the Midwest to the Northeast (11). The probability of dying from an air pollution–related disease is twice as high in the Northeast as in other parts of the country (11).

Intellectually, the acid deposition issue has similarities to earlier controversies over cigarette smoking, asbestos, and toxic substances. In each instance, it was difficult to establish definite cause-and-effect relationships, and certain groups of individuals or industries were affected more than others. One issue that makes acid deposition difficult to manage is that because of high stacks and long-range transport mechanisms, the effects of the emissions are felt at areas far distant from their sources. Similarly, the benefits of cleanup will not always be seen by those doing the cleaning up, but rather may appear several hundred miles away. Acid deposition is the first air pollution problem for which dilution does not appear to be an adequate solution.

EFFECTS OF ACID DEPOSITION

Acid deposition has effects on both the aquatic and terrestrial ecosystems. We will examine each of these systems briefly.

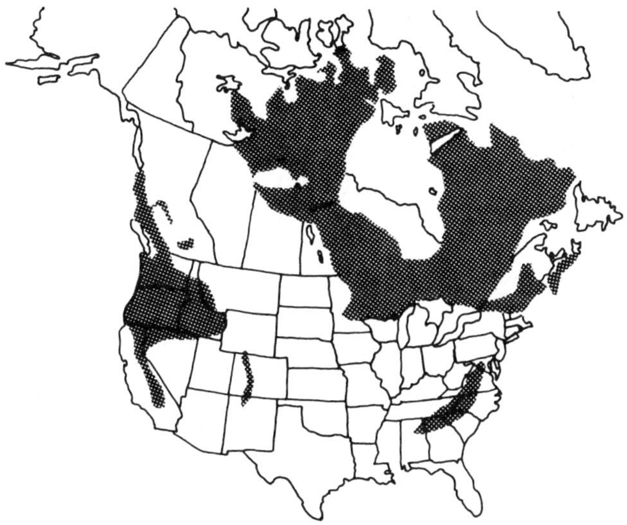

Figure 26-7 Areas containing lakes sensitive to acid precipitation

Aquatic Ecosystem

Severe degradation of the aquatic ecosystem has been recorded in Norway, Sweden, the United States, Scotland, and Canada (12). The number of lakes affected is expected to more than double by 1990. In Norway, 9,000 square miles of lakes are empty of fish and 12,000 square miles have reduced productivity. In Canada, 140 lakes are threatened with the extinction of fish life, and 9 rivers in Nova Scotia have lost their Atlantic salmon populations (2).

Lakes in the Adirondack Mountains of New York State in the 1930s had an average pH of 6.5. When revisited in 1975, 40% of the lakes had a pH of 4.5 or less, and more than 90 lakes no longer supported fish. The State of New York classified 3.6% of its lake area as in critical condition and 14% as endangered.

There is little question that the ecology of some lakes has been affected. It is also important to remember that each lake is a unique entity whose response to a given assault is difficult to predict. One very important factor is the amount of buffering capacity the lake possesses. Figure 26-7 shows lake areas in North America that are sensitive to acid precipitation. These areas indicate where the problems have occurred. The darkened areas of Figure 26-7 have igneous or metamorphic bedrock geology. The clear areas have calcareous or sedimentary bedrock geology. Because of the slower weathering rates in areas of igneous or metamorphic geology, lakes in these areas have low alkalinities (< 0.5 mEq HCO_3/l) and are thus more susceptible to acid rain.

Most aquatic life forms function best at a pH of around 7.0 neutral, and life forms disappear as the acidity increases. At a pH of between 5.5 and 7.0 small-mouth bass, brown trout, mussels, and mayflies disappear. When pH 4.5 is reached, lake trout, yellow perch, and salamanders, more tolerant species, are gone. Thus, at pH 4.5, most normal aquatic life has disappeared and the lake has become a biological desert.

Another factor contributing to biological desertification is the presence of aluminum ion. More acidic water leaches more aluminum ion from the soil. This aluminum is also more soluble because more of it is in the + 3 form rather than tied up as a hydroxide. In general, ionic forms of aluminum are more toxic than aluminum organic complexes. Aluminum collects on the gills of fry and causes the exudation of a mucus that strangles the fish. Other heavy metals also collect in the water due to acid leaching of soil, but aluminum seems to be the most important detrimental cation.

Spring is an especially difficult season for lakes because of the occurrence of pH shock. During the winter, acid snow accumulates that, at spring melting, provides a large input of low-pH water to lakes. The aquatic life adjusts to the lowest pH experienced during the year, and the spring pH shock often provides this minimum. Fish kills often occur in the spring because of pH shock. Fish are also spawning at this time, exposing the vulnerable eggs and fry to these acid conditions.

Water of low pH and containing heavy metals is less desirable for potable water supplies. When low-pH lake water was used in copper plumbing in Sweden, the water became blue tinged and, when used for washing, turned blond hair green (4).

Terrestrial Ecosystem

Soil is the backbone of the terrestrial ecosystem. The response of soil to acid deposition varies greatly depending on the soil type. Most forest soils are naturally mildly acidic from the organic acids produced during the rotting of vegetation on the forest floor. The increased acidity increases the leaching of plant nutrients such as aluminum, calcium, magnesium, potassium, iron, and manganese into the groundwater, decreasing soil fertility. Also, phosphorus in the soil is less available to plants under conditions of high acidity. In addition, acidification decreases the rates of many bacterial processes such as nitrogen fixation and the breakdown of organic matter (13).

Acid precipitation also has direct effects on plants. Laboratory, greenhouse, and field experiments have shown the following negative biological effects (1):

1. Certain crops are more susceptible to ozone damage when exposed to acid rain
2. Acid rain induces necrotic lesions on leaves
3. Nutrients are lost by leaching from foliar organs
4. Plants are more susceptible to infection by bacterial and fungal pathogens
5. Waxes on leaf surfaces are eroded at an accelerated rate

6 Nodulation of legumes is inhibited, leading to decreased nitrogen fixation by symbiotic bacteria
7 Formation of terminal buds is inhibited
8 Mortality of pine seedlings is increased

In addition, some data have shown adverse effects on cash crop productivity with corn and soybeans.

Not all the effects that have been seen are negative, however. Artificial acid rain environments have had either no effect or even beneficial effects on a large number of plant species (2).

Trees are, of course, the plant of most interest to the pulp and paper industry. So far, there is no direct evidence (field data) linking acid deposition to alterations in forest productivity (14). There is, however, enough circumstantial evidence in the form of documentation of forest die-off, both in Germany and in the U.S. New England states, to create the suspicion of detrimental acid rain effects. Such die-off may be just the beginning of a major future problem. Once the buffering capacity of the soil is exhausted, the soil could rapidly become acidic and affect forest productivity adversely.

It has been reported that 34% of West German forest land is damaged, but due to the complexity of the natural environment the damage could not be attributed to a single specific cause. Ulrich (15, 16) postulated that continuous deposition of acidic materials in the German forest ecosystem for many decades has resulted in progressive soil acidification. He speculated that the leaching of basic cations, calcium and magnesium, from the tree root zones has been damaging. The increased concentration of aluminum may have damaged the fine root structure. There appears to be a consensus in the German scientific community that the forest damage is related to air pollution, but other stress factors are also of importance.

In the United States, there has been both die-off of older trees and increased mortality of seedlings in the red spruce population and in other species in several high-elevation areas of New York, Vermont, and New Hampshire. Vermont reports 50% mortality in red spruce since 1965 (8). Stresses resulting from drought and from high acid deposition rates have been postulated as the likely causes (17–19). As in Germany, increased mobilization of aluminum combined with accelerated leaching of calcium in acid-saturated soils of the Adirondack, Green, and White Mountains are thought to have caused die-off in red spruce. The shallow-rooted red spruce failed to develop the fine root structure required to survive drought conditions. Nutrient leaching from needles by acid fog is also thought to be a contributing factor (20, 21).

Although direct field data proving a cause-and-effect relationship do not yet exist, many foresters are quite concerned about the potential effects of acid deposition and its long-term consequences for forest productivity.

Another issue related to acid deposition that might be broadly termed a terrestrial effect is the damage caused by airborne acid to artificial structures. This serious problem is beyond the scope of this book.

SUMMARY OF ACID DEPOSITION ISSUES

The principal factor is that the air pollutants produced from the combustion of fossil fuels, including fuels burned in motor vehicles, are precursors of acids. The principal pollutants involved are SO_2, SO_4^{-2}, NO_x, and NO_3^-, all of which can interact with water in the atmosphere to form strong acids. Acid deposition from the air can occur in either wet or dry form. Public utilities, particularly those burning coal, are the main source of these emissions. The high stacks that many utilities have built result in the longer-range transport of the pollutants emitted. Meteorological conditions, such as wind patterns and the quantity of rain (which scavenges pollutants out of the air), are important in determining the amount of pollution received by a certain area. Regions of the country that have low buffering capacity in their lakes and soils are most affected by acid deposition.

Definite cause-and-effect relationships are difficult to establish from available field data, but there is considerable circumstantial evidence to link acid deposition with the damage to lakes, forests, and crops that has been observed in recent years. Acid deposition is an issue that must be taken seriously.

SCIENTIFIC ISSUES

Several scientific issues about the acid deposition situation remain unresolved. The most basic question is: Does acid deposition significantly affect aquatic resources, crops, soils, forests, structural materials, and human health (2). It would be desirable to have irrefutable scientific evidence of these assumed effects rather than only circumstantial evidence. The relative contributions in different geographic areas of natural sources of airborne acids as opposed to those that are the result of human activities need further investigation. The relative contributions to acidification of wet and dry deposition is still uncertain. In any given geographic area, both local and distant sources contribute to the problem, but assessing their relative importance is difficult.

Local area mathematical models seem to be reasonably accurate, but those that involve longer distances need further development. Modeling is complicated by our incomplete understanding of the transformation reactions that take place in the plume and the atmosphere over the time involved in long-range transport (9).

Much of the data that exist on acid rain may have problems with the measurement techniques employed (22). Measurement problems should be resolved in time because a major monitoring network has been established that will employ adequate procedures. In the United States, the National Atmospheric Deposition Program (NADP) has been established; its Canadian counterpart is the Canadian Network of Sampling Precipitation (CANSAP). As part of the U.S. effort, the Cross Appalachia Tracer Experiments (CAPTEX) are being con-

Table 26-2 Retrofit Control Technologies for Coal-Fired Utility Boilers

Process	SO_2 and NO_x Removal Efficiencies	Status	Estimated Capital Costs (\$/kW)[a]	Applicability	Problems
Flue gas desulfurization (lime or limestone scrubbing)	90% SO_2 reduction	Commercial	200–300	Most large utility boilers; costs could sharply increase over estimated values for difficult retrofits	High costs, especially for difficult retrofits; Large quantities of wet waste to be disposed of
Physical coal cleaning	20–30% SO_2 reduction	Commercial	40–50	Centralized cleaning facilities can allow broad applicability	Limited SO_2 removal, especially for difficult to clean coals Large quantities of pyrite solid residue for disposal
Low-sulfur fuel switching	60–80% SO_2 reduction	Commercial	26–61	Most utility boilers	Boiler derating sometimes required Potential displacement of locally mined coal
Low-NO_x burners (second generation)	60–70% reduction	Early commercialization	2–7	Most modern utility boilers	NO_x removal only; no SO_2 removal
Limestone injection multistage (low-NO_x) burners	50–60%, SO_2 reduction; 50–60% NO_x reduction	Developmental	27–37	Most utility boilers	Technology will be commercial no earlier than 1985 Questions must be resolved regarding performance and boiler reliability

[a]Based on 300 MW, 3% sulfur coal, and December 1982 dollars.

Table 26-3 Control Technologies for Artificial Sources of Acid Deposition

Major Acid Precipitation Pollutants	Major Sources	Applicable Control Approaches	Principles of Operation
Precursor pollutants Sulfur dioxide (SO_2)	Electric utilities: primarily high-sulfur coal and oil power plants	Low-sulfur fuel switching (commercial)	Substitute naturally occurring low-sulfur coal or oil
		Physical coal cleaning (PCC) (commercial)	Remove physically bound sulfur (pyrite) from coal via separation processes before combustion
	Industrial fuel combustion: primarily high-sulfur coal and oil combustion	Flue gas desulfurization (FGD) (commercial)	React acidic flue gas with alkaline liquids yielding either a throwaway or a salable sulfur compound product
		Limestone injection multistage burners (LIMB) (developmental)	Inject limestone in boiler through low-NO_x burners to react with sulfur gases to produce a throwaway product
Nitrogen oxides (NO_x)	Electric utilities: coal, oil and gas fired	Low excess air (commercial)	Minimize air used for Combustion to reduce NO_x production from air N_2
		Staged combustion (commercial)	Add overfire air ports to allow fuel-rich early combustion conditions to reduce NO_x to elemental N_2 and O_2
	Industrial fuel combustion: coal, oil and gas boilers	Low-NO_x burners (first-generation commercial; second-generation burners in demonstration phase)	Replace old burners with specially designed burners to maximize fuel-rich early flame conditions for NO_x reduction

		Flue gas cleaning (FGC) (commercial in Japan)	Treat flue gas with NH_3 in selective catalysts to reduce NO_x to N_2 and O_2
	Stationary engines: large diesels, gas turbines, and internal combustion engines	Catalytic reduction (demonstration phase)	Use combustion gas catalytic systems to reduce NO_x to N_2 and O_2 either with NH_3 (fuel-lean engines) or without NH_3 (fuel-rich engines)
		Combustion modification (demonstration phase)	Modify engine to encourage staged combustion
	Mobile sources: autos and trucks	Exhaust gas recirculation (commercial since 1972)	Recirculate exhaust gas back to cylinders to lower temperature and lower oxidation of air N^2 and NO_x
		Three-way catalyst (commercial since 1982)	Special tailpipe catalytic system to reduce NO_x to N_2 and O_2
Directly emitted acidic pollutants			
Acid sulfates; predominatly sulfuric acid mist	Electric utilities: coal- and oil-fired power plants	Limited information suggests that FGD and low-sulfur fuel switching may provide some level of control for coal	See discussion of SO_2 above
	Nonutility fuel combustion: coal and oil combustors	Low excess air operation	Lowering combustion air tends to decrease oxidation of SO_2 to SO_3, with subsequent sulfuric acid formation
Hydrogen chloride (HCl)	Electric utilities: high-chlorine, coal-fired power plants	FGD	See discussion of SO_2 above
	Industrial boilers: high-chlorine coal	LIMB	

cluded, and their results will be very useful in model development and verification.

In 1983 the U.S. federal government was to have spent \$22.3 million on research related to acid deposition in a major effort to resolve more fully some of these scientific questions. It is estimated that 1,600 scientists in 32 countries were working on problems related to acid deposition during 1983 (2).

CONTROL TECHNOLOGIES

There is little question that at least part of the acid deposition problem is related to the burning of fossil fuels in power plants and internal combustion engines. Thus it would seem reasonable to employ technology at these sources to help alleviate the problem. This valid approach will be employed on new sources of SO_2 and NO_x emissions as they are built, but there is much concern over the billions of dollars it would cost to retrofit all existing coal-fired power boilers with appropriate control technology. The technologies required to control SO_2 and NO_x emissions exist and improved technologies are being developed; the controls must be made cost effective.

Table 26-2 lists alternative control technologies that can be retrofit on coal-fired utility boilers. The technology with the highest SO_2 reduction efficiency is flue gas desulfurization (FDG) by scrubbing with a lime solution. It is also the most costly technology. Operation of the scrubber requires a fuel penalty of 4–8% (8). Despite this fact, there are approximately 100 scrubbers in operation, and 50 more were under construction at utility plants as of 1983. The capital cost of the scrubber is about 7% of the cost of the entire power plant (4).

Dry scrubbing processes, which are expected to be less costly, are under development. Dry scrubbing would work by spraying lime, limestone, nahcalite, and trona into the gas stream. These particulates would adsorb the SO_2 and then be collected.

Coal washing is another technology considered in Table 26-2. It is practiced by some utilities to reduce emissions and to upgrade the fuel value of the coal used. It accomplishes an approximately 30% reduction in SO_2 and creates an additional solid residue that must be dealt with.

There has been considerable discussion about switching from high-sulfur to low-sulfur coal, which does reduce SO_2 markedly. A major disadvantage is that low-sulfur coal has a lower BTU content per pound, so that the power-generating capacity of the utility would be reduced. Another problem is the local economic disruptions that would be caused by the resulting shift of the coal mining base from the eastern to the western United States.

The most promising new technology is the limestone injection multistage burner, which is also a low-NO_x burner. This process is under development and is expected to be available for commercial application no earlier than 1985. It provides reasonable reduction in both SO_2 and NO_x emissions at an estimated installed cost of \$27–37/kW. This technology would likely be employed on most new coal-fired utility boilers.

NO_x emissions can best be controlled by close control of the air/fuel ratio and by reduction of the combustion temperature.

Tables 26-3 provides a general overview of control technologies for artificial sources of acid deposition. It covers the technologies just discussed for utilities and possible modifications of internal combustion engines to minimize NO_x emissions.

LEGISLATION

As of 1984, Canada, West Germany, and Japan had enacted legislation specifically concerned with acid deposition. In the United States, specific legislation has not yet been passed. It is acknowledged that the Clean Air Act is not adequate to deal with acid deposition, but exactly what should be done about the problem has not been resolved. Several bills are under consideration in the Congress, but which if any will finally be enacted is unknown. Part of the uncertainty lies in deciding whether it would be best for the legislation to be technology specific, emission specific, or area specific. There is also disagreement over whether interim mitigation measures should be enacted. It is to be hoped that whatever solution is finally chosen will have a significant impact on the problem, and not one that mandates major expenditures for no gain. It would also be desirable for the solution to provide some equitable means to finance the required investment in control technology that minimizes the short-term shock while still effecting a significant reduction in emissions. We must, however, recognize our responsibility for stewardship of the environment and not allow this problem, which could eventually affect the earth's natural carrying capacity, to continue unresolved.

REFERENCES

1 E. B. Cowling, "Discovering causes, consequences and implications of acid rain and atmospheric deposition," *Tappi*, 66 (9):43–46 (1983).

2 Interagency Task Force on Acid Precipitation, *Annual Report to the President and the Congress 1982*, Washington, D. C.

3 R. J. Charlson and H. Radhe, *Nature*, 295:683–685 (1982).

4 A. LaBastelle and T. Spiegel, "Acid rain—how great a menace?," *National Geographic*, 160 (5):652–681 (1981).

5 J. M. Waldman, J. W. Munger, D. J. Jacob, R. C. Flagan, J. J. Morgan, and H. P. Hoffman, "Chemical composition of acid fog," *Science*, 218:677–679 (November 1982).

6 Committee on the Atmosphere and the Biosphere Board on Agriculture and Renewable Resources Commission on Natural Resources, NAS, NAE, IOM, *Atmosphere-biosphere interactions: Toward a better understanding of the ecological consequences of fossil fuel combustion*, National Academy Press, Washington, D. C., 1981.

7 Committee on Atmospheric Transport and Chemical Transformation in Acid Precipitation, NAS, NAE, IOM, *Acid deposition atmospheric processes in eastern North America*, National Academy Press, Washington, D. C., 1983.

8 *State of the Environment 1982*, Report of the Conservation Foundation, Washington, D. C., 1982.

9 P. Stalkman, "Acid rain control strategies being assessed by EPA," *Proceedings of the TAPPI Environmental Conference*, Savannah, Ga., April 9–11, 1984, pp. 25–28.

10 K. K. Bertine and E. D. Goldberg, "Fossil fuel combustion and the major sedimentary cycle," *Science*, 173:233–235 (1971).

11 J. E. Bridge and F. P. Fairchild, *Northeast Damage Report Long Range Transport and Deposition of Air Pollutants*, NESCAUM, NEIWPCC, April 1981, Stanford, Massachusetts.

12 P. Drablös and A. Tallan, eds., *Ecological Impact of Acid Precipitation, Proceedings of an International Conference*, Sendeford, Norway, March 11–14, 1980. SNSF Project, Oslo-Aas, Norway.

13 P. W. Einspahr and M. L. Harder, "Acid rain: Its effects on the paper industry," *Tappi*, 64 (4):61–64 (1981).

14 J. E. Pinkerton, "Acidic deposition and its relationship to forest productivity—an update on the current state of knowledge," *Proceedings of the TAPPI Environmental Conference*, Savannah, Ga., April 9–11, 1984, pp. 29–34.

15 B. Ulrich, "Effects of acid deposition," in I. S. Beilke and A. S. Elshout, eds., *Acid Deposition*, D. Reidel, Boston, 1982.

16 B. Ulrich and J. Pankroth, eds., *Effects of Accumulation of Air Pollutants in Forest Ecosystems*, Reidel, Boston, 1983.

17 A. H. Johnson, *Journal of the American Pollution Control Association*, 33 (11):1049 (1983).

18 A. H. Johnson and T. G. Siccama, *Environmental Science and Technology*, 17 (7):294A (1983).

19 A. H. Johnson and T. G. Siccama, *Proceedings of the TAPPI Annual Meeting*, TAPPI Press, Atlanta, 1983, p. 301.

20 H. W. Vogelmann, *Natural History*, 91 (11):8 (1982).

21 G. H. Tomlinson, *Environmental Science and Technology*, 17 (6):246A (1982).

22 G. Lauer, "Acid rain, the pollutant of the year," *Proceedings of the TAPPI Environmental Conference*, Savannah, Ga., April 9–11, 1984, pp. 19–24.

Index